Reliability Theory and Technology in Manufacturing Process

Yihai He · Wei Dai · Jun Ai · Weifang Zhang

Reliability Theory and Technology in Manufacturing Process

国防工業出版社
National Defense Industry Press

Springer

Yihai He
School of Reliability and Systems Engineering
Beihang University
Beijing, China

Wei Dai
School of Reliability and Systems Engineering
Beihang University
Beijing, China

Jun Ai
School of Reliability and Systems Engineering
Beihang University
Beijing, China

Weifang Zhang
School of Reliability and Systems Engineering
Beihang University
Beijing, China

Translated by
Wei Dai
School of Reliability and Systems Engineering
Beihang University
Beijing, China

Yihai He
School of Reliability and Systems Engineering
Beihang University
Beijing, China

Li Yang
School of Reliability and Systems Engineering
Beihang University
Beijing, China

ISBN 978-981-96-2734-9 ISBN 978-981-96-2735-6 (eBook)
https://doi.org/10.1007/978-981-96-2735-6

Jointly published with National Defense Industry Press
The print edition is not for sale in China (Mainland). Customers from China (Mainland) please order the print book from: National Defense Industry Press.
ISBN of the Co-Publisher's edition: 978-711-81-2217-6

Translation from the Chinese language edition: "制造过程可靠性理论与技术" by Yihai He et al., © National Defense Industry Press 2021. Published by National Defense Industry Press. All Rights Reserved.

This Springer imprint is published by the registered company Springer Nature Singapore Pte Ltd.
The registered company address is: 152 Beach Road, #21-01/04 Gateway East, Singapore 189721, Singapore

Preface

On May 8, 2015, the State Council of China issued the "Made in China 2025" clearly put forward the "quality first" guiding ideology of manufacturing in China, requiring the majority of manufacturing enterprises to continue to improve the "quality and efficiency" and continue to reduce the "defective product rate". In order to achieve these quality objectives, strengthening the research and application of manufacturing process reliability technology is an inevitable step and basic requirement for strengthening the industrial basic capability.

On February 6, 2012, the State Council issued the "Quality Development Outline (2011–2020)", the eighth part of the "Reliability Enhancement Project" is clearly put forward, requiring the strengthening of product reliability design, testing and quality control of the production process relying on technological advances, management innovations and standardization, to enhance the level of reliability, and to promote the quality of China's products from conformity to applicability and high reliability. The research and application of manufacturing process reliability technology is a useful supplement for the majority of manufacturing enterprises to enhance the manufacturing process reliability assurance capability and product reliability level.

Article eight of the "Regulations on Quality Management of Weapons and Equipment" promulgated in November 2010 clearly stipulates that the State encourages the adoption of advanced science and technology and management methods to improve the quality of weapons and equipment; and article 11 clearly states that the performance indicators of equipment, i.e., the quality characteristics, include the functional characteristics, reliability, maintainability, supportability, testability, and safety. In the equipment of the whole characteristics, the whole process, the whole system of the new quality concept requirements, and what reliability technology in the manufacturing process can be used by the majority of contractors, which is an issue that must be resolved in order to implement the provisions of the "Regulations on Quality Management of Weapons and Equipment".

For the first time, this book systematically establishes a more complete cognitive theory of manufacturing process reliability and a technical system for analyzing, improving, and controlling the reliability of manufacturing processes, which is oriented to the reliability assurance of products. This book provides general theoretical and methodological support for guiding the analysis and optimization of the reliability of specific component-level mechanical and electronic manufacturing processes and host assembly processes and fills in the gaps in reliability system engineering technology. The book is divided into chapters to introduce the basic concepts and development of manufacturing process reliability, the basic theory of manufacturing process reliability, manufacturing process reliability analysis, improvement, control technology, and other content and look forward to the development and application of reliability technology in manufacturing process.

The preparation of this book has been supported by the National Natural Science Foundation of China's top-level project "Research on product reliability degradation mechanism and inhibition technology of manufacturing process based on QR extended chain (RQR)" (Project No. 61473017, 2015–2018), the Equipment Pre-research Fund project "Assembly process reliability modeling theory and analysis technology based on RQR chain" (Project No. 6140002050116HK01001, 2017–2018), the Equipment Pre-research Fund project "Research on Reliability Modeling and Verification Technology of Complex Equipment Manufacturing under Coupling Factors" (Project No. JZX7Y20190242012401, 2019–2020), and the National Defense Technology Foundation Project "Research on Process Reliability Optimization, Control and Evaluation Technology Based on Manufacturing Process Failure Mechanisms (2015–2017)".

This book was written by Dr. Yihai He, Dr. Wei Dai, Dr. Jun Ai, and Dr. Weifang Zhang. Dr. Yihai He was responsible for the preparation of Sects. 1.1, 1.4, 1.5, 2.2, 3.2.2, 3.2.3, 4.2, 4.3, 4.4, 5.1, 5.3 and 6.2. Dr. Wei Dai was responsible for the preparation of Sects. 1.2, 1.3, 2.1, 3.1, 3.2.1, 3.3, 4.1 and 5.2. Dr. Jun Ai was responsible for Sect. 6.1. Dr. Weifang Zhang was responsible for the preparation of Sects. 2.3 and 2.4. The overall manuscript of the book was conducted by Dr. Yihai He. In the process of writing this book, the authors also received guidance and assistance from Dr. Rui Kang and Dr. Guicui Fu, and we would like to express our sincere appreciation to them.

During the preparation of this book, a large amount of relevant literature at both national and international has been referred to, and we would like to express our sincere thanks to all the authors.

The connotation of manufacturing process reliability technology is very rich, due to the limited level of the author, there are inevitable omissions in the framework of reliability theory and technology in manufacturing process as well as the content thereof, and we sincerely invite all scholars both national and international to criticize and correct us.

Beijing, China

Yihai He
Wei Dai
Jun Ai
Weifang Zhang

Introduction

This book is divided into six chapters. Firstly, the development, connotation, and technical framework of manufacturing process reliability is presented; and the basic theory of manufacturing process reliability is introduced; then, the reliability analysis technology of the manufacturing process is expounded; the reliability optimization technology of manufacturing process is discussed; the reliability control technology of the manufacturing process is briefly described; the development and application trend of manufacturing process reliability technology are prospected.

This book could be used by researchers, engineers, and managers specializing in quality and reliability engineering, and all kinds of design and production technicians can also refer to it. This book can also be used as a textbook for graduates and undergraduates courses of related majors in colleges and universities.

This book was translated by Dr. Wei Dai, Dr. Yihai He, and Dr. Li Yang. Dr. Wei Dai was responsible for the translation of Chaps. 1, 5 and 6. Dr. Yihai He was responsible for the translation of Chaps. 2 and 3. Dr. Li Yang was responsible for translation of Chap. 4. The authors would like to express our sincere appreciation to those who provided suggestions and advice.

Contents

Chapter 1
Introduction

1.1 Life Cycle Evolution of Product Reliability

1.1.1 Bathtub Curve and Life Cycle Failure Rate

Xuesen Qian, founder of system engineering, clearly pointed out for the first time that "product reliability is designed, produced and managed" when summarizing the experience of reliability work at a reliability work conference of national defense science and technology industry. Through a large number of long-term engineering practice and exploration, people have gradually realized that product reliability comes from design, forming in production, showing in use. The failure rate varies dynamically with the change of life stages. The reliability of products with the same design specifications that are manufactured on different production lines is different, and always lower than expectation.

In order to clearly express the change rule of failure rate in different life stages of products, people present the classical bathtub curve as shown in Fig. 1.1 by counting the failure data of different life stages of products [1].

As shown in Fig. 1.1, the bathtub curve model is an intuitive reflection of the fault occurrence characteristics in the whole life cycle of products in a statistical sense. The early fault period describes the factors such as the defect of the material, the improper design and the deviation of the production in the early use of the product, which are closely related to the quality control effect of the manufacturing process. The worse the quality control effect, the higher the early failure rate of products. Therefore, the early failure is also called the quality failure. The early failure rate shows a rule from high to low. Burn-in test is an effective technical means to eliminate the early failure of electronic products, which can accelerate the decline of early failure rate and make it close to the design reliability index. After running-in in the early use stage, the high early failure rate gradually decreases and approaches to the reliability design index, and the failure rate curve is also stable, which is called product quality assurance stage or effective life period. The failure rate of this stage is mainly related

Y. He et al., *Reliability Theory and Technology in Manufacturing Process*,
https://doi.org/10.1007/978-981-96-2735-6_1

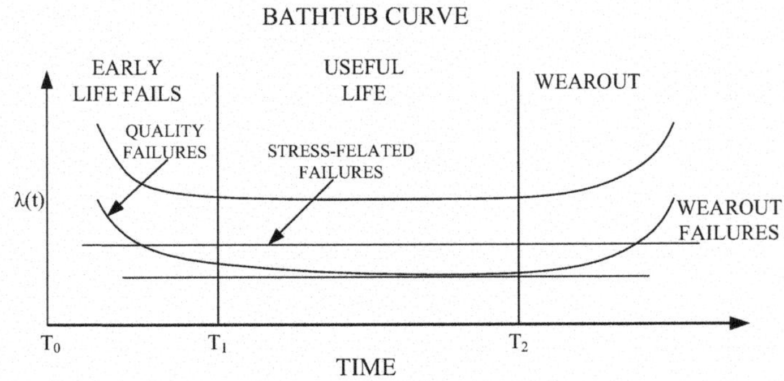

Fig. 1.1 Classical bathtub curve

to the strain change of the use environment. A stable use environment will effectively prolong the effective life period. The failure rate level of the effective life period is directly determined by the factors in the design stage. At the end of the product life cycle, a series of creep, fatigue or wear are produced with the degradation of the material or structure that constitutes the product. The failure rate curve increases gradually over time, and the cost of maintenance and use increases sharply until the product exits the life cycle.

Traditionally, reliability engineering technology mainly focuses on reducing the failure rate in the effective life period through the optimization of design scheme and reduce the increase rate of failure rate in the wear period by strengthening maintenance support technology. The research on manufacturing process, which is the key stage to form product reliability in physics, is ignored. Only technical means such as burn-in test are used to passively eliminate the formed manufacturing defects in the early fault period. The research on the technical means of prevention, analysis, improvement and control of manufacturing defect is lacked. Systematic manufacturing process product reliability assurance theory and technology have not been formed as well.

1.1.2 Classification and Evolution Process of Product Reliability in Life Cycle

In order to highlight the irreplaceable role of different stages of product life cycle in the formation and guarantee of product reliability, Murthy [2], a senior expert in the field of reliability, first presented the evolution and transmission rule of product reliability in the whole life cycle that includes design stage, manufacturing stage, sales stage and use stage from the perspective of system engineering in 2010, which are shown in Fig. 1.2.

Figure 1.2 shows that design reliability is the desired value of product reliability, and the level of it in the life cycle is the highest; The level of design reliability is

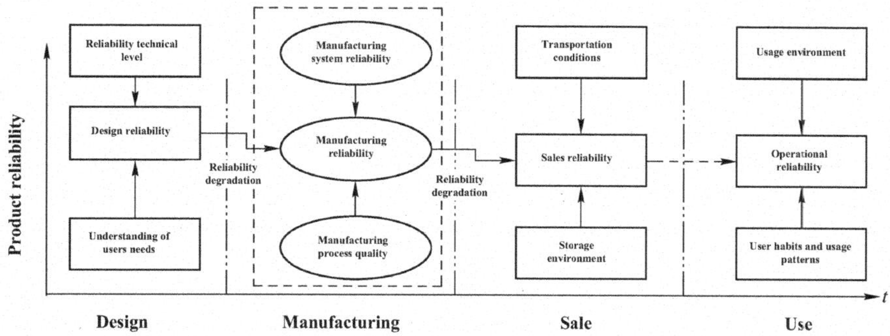

Fig. 1.2 The evolution model of product reliability in the life cycle

mainly affected by user requirement and the level of reliability technology. The design stage is critical to product reliability; With the same design scheme, manufacturing reliability is not more than design reliability due to component defects and assembly errors in the manufacturing stage, etc. Due to uncontrollable factors such as man, machine, material, method, measurement, environment, etc., reliability will decline to varying degrees according to the effect of quality and reliability control in the manufacturing process. The controllable factors in the manufacturing process can be classified into manufacturing process quality and manufacturing system reliability, which is also the main basis for expanding the reliability theory and technology in manufacturing process. After manufacturing, qualified products will enter the sales stage after quality and reliability test. Sales reliability is not more than manufacturing reliability due to transport and storage, etc. Strengthening the management of transportation and storage is the key to prevent reliability decline; Products are gotten by users through sales channels, and the product reliability begins to be intuitively displayed through the number of faults and their severity. Operational reliability is not more than sales reliability due to use habits and environment. Strengthening user training and customer communication is the key for product reliability to be precepted by users.

Unfortunately, due to the division of major and the limitation of understanding, the research of traditional reliability technology mostly focuses on the analysis and optimization of reliability in the design stage, the maintenance of reliability in the sales stage and the maintenance and guarantee of reliability in the use stage. However, reliability theory and technology in manufacturing process have been ignored for a long time. The transmission evolution chain of reliability design intention is interrupted in the manufacturing stage. As a result, even higher design reliability indexes and intentions cannot be effectively transmitted to the sales and use stage. In engineering, the product reliability will deteriorate to varying degrees in the manufacturing stage. The non-destructive transmission of design reliability can not be realized. Huge waste is also caused.

1.1.3 Connotation and Indication Function of Product Early Failure Rate

As shown in Fig. 1.2, the manufacturing stage is the core section in the formation of product reliability, which connects the design stage, the sales stage and the use stage. Manufacturing process capability directly affects whether the intention and requirements of reliability design can be completely transmitted to the sales and use stage, and the initial value of reliability in the sales and use stage is also determined. The performance of low manufacturing process capability is high early failure rate of products. Enterprises and customers are troubled by the high early failure rate, which is a new technical problem that must be solved in the field of reliability and systems engineering.

The fluctuation of human, machine, material, method, environment, measurement and other factors is transmitted and accumulated among various stations, resulting in deviation of manufacturing quality and explicit and implicit defects of products. Generally, the explicit defects of products will appear through the quality inspection of the manufacturing process, which will affect the qualified product rate which is called "product quality at zero time"; The implicit defects will not be found in routine quality inspection, and flow into the market in the form of pseudo "qualified products". But it will turn into explicit defects under actual stress conditions over time. Early failure also occurs.

In the book "design theory" published by Professor Suh of MIT in 1990, the theory and method for product model design called axiomatic design (AD) were presented in a paradigm mode, and the mapping models of user attribute (CA), function requirement (FR), design parameter (DP) and process variable (PV) are also built. In order to reflect the complexity of early fault formation mechanism, based on and expand axiomatic inter domain mapping theory such as functional domain, structural domain and process domain, a cumulative propagation framework of product life cycle quality fluctuation is presented based on axiomatic inter domain mapping theory such as functional domain, structural domain and process domain, which is shown in Fig. 1.3 [3].

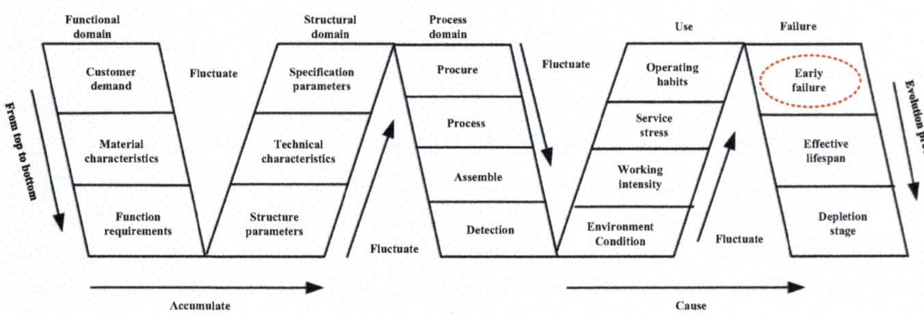

Fig. 1.3 A framework for quality fluctuation and deviation propagation based on interdomain mapping

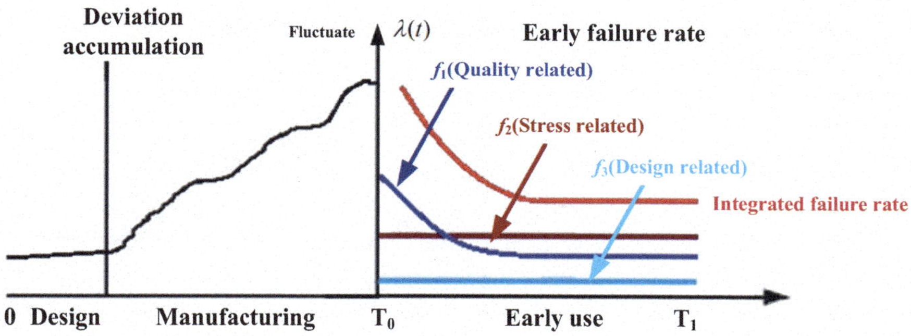

Fig. 1.4 Decomposition diagram of early failure rate curve under cumulative deviation effect

Figure 1.3 shows that the frequency of early failures is related to many stages of the product life cycle, including the technical level of design, manufacturing capacity and actual stress in use. The fundamental problem of early failure is determined by the reliability conceptual model in the design stage; The weak links of early failure are verified in the production process in the manufacturing stage; The manifestations of early failure are induced by the environmental stress in the use stage. Explicit faults are always used as the extraction object of reliability detection, which measured in time. Thus the product reliability eigenvalue is always defined as the failure rate function with time variable, and the index of early failure rate has become the core of reliability analysis and research in the early use stage, which is a key index to characterize whether the reliability of manufacturing process is controlled. Engineering practice shows that defects from design, materials, manufacturing and assembly can be mapped into different deviation state spaces. The transmission, accumulation and interaction of deviations result in many early failure problems of products under the excitation of certain use environment, stress, strength and operation habits. Figure 1.4 shows the variation curve of typical early failure rate under the cumulative effect of deviation in the design stage and the manufacturing stage.

As shown in Fig. 1.4, T_0 is the time when the product is delivered to the customer, and the early failure rate is the highest. On the left side of T_0, fluctuations from the design stage and the manufacturing stage accumulate and propagate over time, affecting the early use process of the product. The deviation effect in the manufacturing stage is the most obvious, which should be the focus on. From the right side of T_0 to the starting point of the effective life (T_1), the cumulative effect of deviation in different stages on the left side of T_0 is showed and quantified through the level of early failure rate. Compared to the level of design failure rate, high-level decreasing quality failure (f_1), constant stress failure (f_3) and design related loss failure (f_2) are integrated in the early stage of use. The early failure rate is much higher than the acceptable level set in the design stage. In addition, quality failure makes great contribution to the early failure rate, which should be the focus of analysis and control.

1.2 Influence of Manufacturing Process on Product Reliability

There is a huge gap in mechanism manufacturing technics between China and developed countries, especially high-end precision military equipment. According to the statistics of the U.S. Navy electronic laboratory, manufacturing defects account for only 10% of the causes of U.S. military equipment failures; In contrast, China's mechanism manufacturing technics has become an important factor that restricts the performance of equipment. A quality rectification work has been carried out in recent years. More than 3000 reliability problems exposed in the tests of more than 20 models of products have been statistically analyzed. About 50% of them are caused by processing and manufacturing; In the statistical analysis of 55 failures of certain mechanical parts in an aviation accessory factory, 49.1% of them are caused by processing and manufacturing; In the "tank biathlon-2014" international competition held in Russia in August 2014, the China 96A tank participated in the competition for the first time. With the advanced fire control system, all 19 shells hit the target and won the first place in the shooting project. However, there are more than 20 failures of the 96A tank in the 12 days high-intensity competition, and the reliability level of it was far lower than that of the Russian T-72B tank. The surface quality and surface integrity of parts are required more and more strictly to improve the performance of modern electromechanical products. The limitation of the requirements for the accuracy of manufacturing process and equipment have been reached or approached; At the same time, a large number of new technologies, new processes and new materials are widely used. Quality and safety accidents caused by the insufficient understanding of the physical mechanism of materials and processes occur from time to time. In order to narrow the gap between manufacturing and design and improve the inherent reliability of products, reliability engineering technology in the manufacturing stage has become a basic scientific problem to be solved urgently.

The failure rate change curve of product life cycle (also known as "the bathtub curve") is mainly composed of the effects of design, manufacturing and material burn-in on failure rate. The random failure rate is mainly determined by the design quality. Design reliability improvement will occur through reducing random failures; The early failure rate and inherent failure rate of products are mainly affected by manufacturing defects. Inherent reliability can be improved through optimizing process design and controlling manufacturing defects according to the generation mechanism and law of them; The failure rate during the wear out period of products is mainly affected by the degradation of materials in the service environment. Figure 1.5 shows that inherent reliability of products is always lower than design reliability. The reasons are as follow: First, the mechanism and law of manufacturing defects are not clearly understood, and there is no basis for setting the manufacturing process parameters and technical requirements of new products; Second, the process and technology of manufacturing is not capable at meeting the target of reliability, and

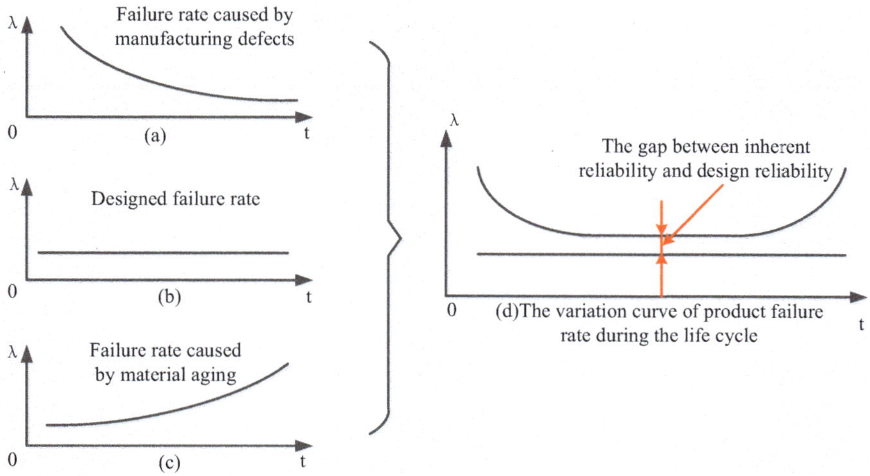

Fig. 1.5 Change curve of product failure rate and its components

design reliability requirements cannot be completely transformed into process characteristics requirements in the manufacturing stage [4]; Third, the process documents are not perfect, the methods of reliability test have shortcomings, and the implicit defects caused by residual and incidental phenomena in manufacturing process can not be effectively eliminated.

The inherent reliability of products is formed through design and manufacturing, which is presented under ideal use and guarantee conditions. Products with the same design specifications have different reliability due to different manufacturing processes and material properties. Xuesen Qian, the founder of system engineering, once said that "the reliability of products is designed, produced and managed" when summing up reliability work experience. Through a lot of engineering practice and exploration, people gradually realize that product reliability comes from design, forming in production, showing in use. For a long time, the research and application of reliability technology focus on design but ignore manufacturing. There is a complete set of design, control and verification methods and for product design reliability. However, it is difficult to study in manufacturing reliability due to those factors such as operators, materials, equipment, process, environment, measurement and detection that be involved. Therefore, there is little systematic further research on it. Manufacturing reliability is defined as the ability to perform a specified function under specified conditions and within the specified time. Based on the definition of manufacturing reliability, related research can be categorized into three levels.

In order for the manufacturing process to realize the specified functions, machining accuracy and surface quality need to be ensured, and early defects due to the process need to be reduced, so a research level aimed at quality control of the process system has been formed. According to the form in which manufacturing defects are manifested, this research level can be further divided into research on explicit and implicit defects. Explicit defects are mainly geometric defects. Geometric shape and position out of tolerance, geometric deformation, surface defects, etc. are also included. Explicit defects can be directly detected in the manufacturing quality detection. Implicit defects are mainly physical micro-defects. Precision shape and position defects, surface micro defects, stress defects, material variation, etc. are also included. Implicit defects detection is difficult or costly. Implicit defects will gradually evolve into explicit defects due to the stress in the use environment. Thus, implicit defects are the root cause of product failure and damage. The physical and mechanical properties such as wear resistance, corrosion resistance and fatigue strength of products are limited by implicit manufacturing defects, resulting in many reliability problems within the design life. The life cycle costs, the performance and the strength of equipment system are affected. So do personnel safety.

In order for the manufacturing process to be completed within the specified time, the machining allowances allocated in the process plan based on the process capability and process demand need to be balanced, and therefore a research level has been formed for the purpose of optimizing the process efficiency. The machining allowance is the thickness of the layer of material removed from the surface being machined during the machining process. Considering the process capacity of the current process and the need to remove the surface defect layer generated by the previous process, the reasonable allocation of machining allowance can improve the efficiency of the process and product quality, so as to achieve the specified functions within the specified time.

In order to complete the manufacturing process under the specified conditions, it is necessary to manage the operation and maintenance of the manufacturing system using various types of information in the manufacturing process, and therefore the research level aimed at realizing the intelligent scheduling of manufacturing missions has been formed. Process capabilities and process requirements for each station are different in common multi-station manufacturing processes. How to rationally plan the limited resources according to the manufacturing missions, and rationally allocate and schedule the various sub-missions are the core issues of this research level. The in-depth studies and effective solutions of these issues can improve the utilization of resources in the manufacturing process, reduce the cost of the manufacturing process, and help manufacturing enterprises to complete the manufacturing tasks under the specified conditions.

At present, China's economic development has accelerated. The structure of it has been optimized and the power has been transformed. The new situation should be recognized, adapted and led for manufacturing industry to change the dilemma of large but not strong. Quality development should be taken as a method to promote the development of manufacturing industry and even the country. The efficiency of manufacturing industry should be improved through improving quality and innovation capability. As a programmatic document for the medium and long-term strategy of quality power, "the quality development outline (2011–2020)" clearly proposes that quality technology innovation should be accelerated; reliability improvement project and quality comparison improvement project shall be developed; The goal that the technical quality indicators of key industrial products such as key parts, basic components and basic materials of major equipment and important consumer products meet the international advanced level should be achieved; Improving the manufacturing quality and reliability of products has become a top priority. The first China Quality (Beijing) conference was held in Beijing on September 15, 2014. Premier Li Keqiang attended the meeting and stressed "Quality is the concentrated reflection of the country's comprehensive strength". "Quality problems should be focused on; Technology innovation should be accelerated; Backward products should be eliminated; The world reputation of made in China should be formed". In order to achieve the above objectives, the bottleneck of manufacturing reliability technology must be broken through, and the international competitiveness of China's manufacturing industry must be improved as well.

1.3 Connotation of Manufacturing Process Reliability

Reliability theories and methods always focus on the life characteristics of products. The variation characteristics of failure mode and failure rate functions of products are analyzed as well, which are used to guide the enterprise design and manufacturing. Pertinence repair and maintenance can also be better carried out according to them. The traditional reliability technology is limited to the product design and use stage. Less attention is paid to the manufacturing stage of forming product reliability. The reliability degradation phenomenon that the reliability of quality-produced products can not meet the design requirements is caused, which is reflected in the high early failure rate of products that can indicate the reliability level of manufacturing process. However, there are few technical methods to prevent reliability degradation in the manufacturing stage.

Therefore, supported by the general project of NSFC "Research on product reliability degradation mechanism and suppression technology in manufacturing process based on QR extended chain RQR", the connotation of manufacturing process reliability is expressed combined with the QR extended chain proposed by the project team, as shown in Fig. 1.6. QR extended chain, namely RQR chain (reliability of manufacturing system (R), quality of manufacturing process (Q), reliability of manufactured product (R)), is considered as the basic cognitive framework to study the reliability of the manufacturing stage from the perspective of system engineering.

Figure 1.6 shows that manufacturing process reliability mainly includes the reliability of manufacturing system, the quality of manufacturing process, the reliability of manufactured products and the interaction among them. The three core elements of manufacturing process reliability derive three levels of research as follows:

- The research aiming to improve the inherent reliability of manufactured products, ensure the machining accuracy and surface quality of products, and reduce the early defects due to process reasons.

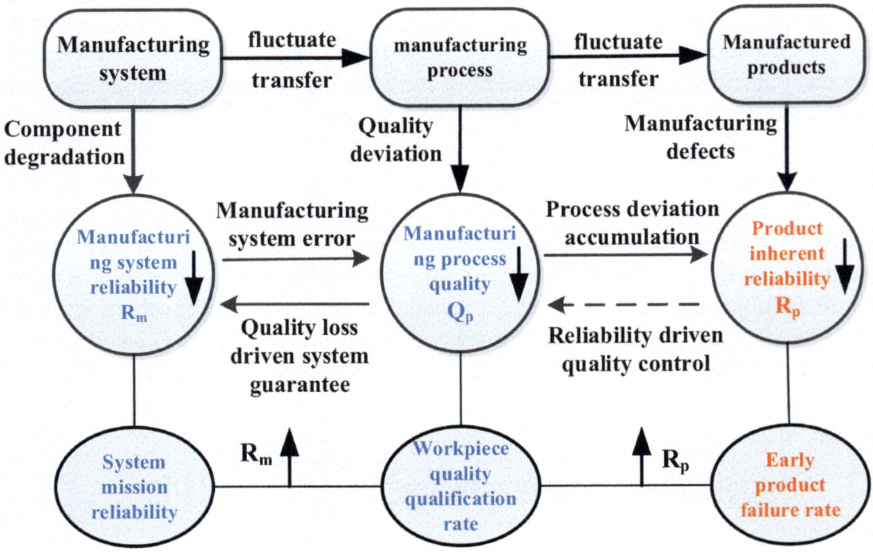

Fig. 1.6 The connotation of manufacturing process reliability

- The research aiming to improve the quality and efficiency of the manufacturing process, which need to balance the allocation of machining allowances in the process plan according to process capability.
- The research aiming to improve the mission reliability of manufacturing systems, which need to manage the health status of the manufacturing system by utilizing various types of information from the manufacturing process.

The manufacturing stage is the core link in the formation of product reliability, which connects the design stage and the sales and use stage. The level of manufacturing process reliability will directly affect whether the reliability intention in the design stage can be completely transmitted to the sales and use stage, and the initial value of reliability in the sales and use stage is also determined by it.

Manufacturing system refers to the system in which manufacturing equipment completes specified operations to produce products in the manufacturing process. Manufacturing equipment mainly refers to manufacturing system component (MSC), including tool, fixture, etc. Manufacturing system reliability is generally defined as the ability of a manufacturing system to complete the specified functions under the specified conditions and time. Manufacturing system reliability will fluctuate due to the influence of MSC and other system deviations. The manufacturing process is the executive process of manufacturing system. The decline of product reliability in the manufacturing process is mainly caused by the quality deviation of the manufacturing process. For example, early faults and the decline of product reliability are caused by the out of tolerance of dimensions and inaccurate assembly positioning in the processing and assembly process. As the main executive unit of the manufacturing system in the manufacturing process, the degradation and failure of MSC such as tools and fixtures will lead to deviations of quality in the manufacturing process. For example, the wear of tools will lead to lower machining accuracy and out of tolerance of dimensions. As shown in Fig. 1.6, The main reason for the decline of product manufacturing reliability is as follows: Lower manufacturing system reliability is caused by MSC degradation and other system anomalies; Deviations of manufacturing process quality are caused by lower manufacturing system reliability; Lower product reliability in manufacturing process is caused by those deviations of manufacturing process. Manufacturing process quality connects manufacturing system reliability and product reliability in manufacturing process.

Assuming that the distribution of faults considered in the design process is $F_D(t)$, then the design reliability is $R_D(t) = 1 - F_D(t)$. The fault density function $f_D(t) = dF_D(t)/dt$. The failure rate $r_D(t) = f_D(t)/R_D(t)$.

The failure rate $r_D(t)$ will inevitably increase over time due to equipment degradation, and the influence of manufacturing process quality deviation is not considered in design reliability.

The product reliability in manufacturing process $R_I(t)$ will be affected by the manufacturing system reliability and deviations of manufacturing process quality. Relative to the design reliability $R_D(t)$, the reduction of $R_I(t)$ is ΔR, as shown in the following formula:

$$\Delta R = R_D(t) - R_I(t) \tag{1.1}$$

When the design scheme is unchanged, ΔR will always be more than 0, which shows the decline of product reliability in manufacturing process relative to the design reliability. The analysis of the reliability decline mechanism of products in manufacturing process is to explore the quality deviation in the manufacturing process affecting $R_I(t)$ and the reliability decline ΔR caused by them. The key control nodes and control methods to prevent reliability decline can be provided through it, so as to make ΔR tend to 0 as much as possible.

1.4 The Development of Reliability Technology in Manufacturing Process

1.4.1 Review of Reliability Guarantee Technology in Manufacturing Process

In the 1960s and 1970s, reliability technology was introduced into the manufacturing process of mechanical products such as automobiles in the United States and Russia. With further research, a large number of product failures are found to be caused by process defects and manufacturing defects. Since then, the manufacturing process has been regarded as an important stage to form product reliability. Therefore, the research on reliability technology in manufacturing process has been paid more attention.

In 1972, in order to improve the reliability of electronic products, the application research based on statistical technology was carried out by Schroen et al. [5] in the manufacturing process, and remarkable results were achieved. The concept of manufacturing reliability was presented by Schroen; The definition of "process reliability" of equipment was proposed by Pronikov et al. [6] of the Soviet Union in the early 1980s: within the specified range and time, the capability of the equipment to maintain the value of process quality index which is determined by it. The reliability of equipment should be defined according to the purpose of analysis, which was also emphasized by Pronikov; The practical data were used by Soviet scholars to calculate the reliability index and utilization rate of specific production lines. At the same time, the topological connection diagram model is used to evaluate the reliability of the system, and the reliability standard document of the automatic production line is formulated as well; In 1991, a manufacturing reliability method was proposed by Crook [4] based on design reliability and test reliability, which is named building in reliability (BIR). Barringer et al. [7] believes that manufacturing reliability can be obtained from production data and process reliability is a method to identify the process with lower cost. Then Weibull analysis technology was used to analyze the process. Potential problems in the process were identified. Solutions were given as well; Chen et al. [8] defined "reliability of manufacturing process" as the ability to make processed parts qualified; Kong et al. [9] studied the process reliability evaluation means of microelectronic production process, and the reliability evaluation of process is realized through reliability evaluation technology, process control and statistical process control technology; The process reliability estimation model is established by Lin [10] using Bayesian estimation; The basic influence relationship reliability model of process system based on product reliability was proposed by Fu et al. [11] through establishing the relationship between process system, process product output parameters, product performance and product reliability; Aiming at the reliability of electronic assembly process, Wang [12] expounds the main failure forms and failure mechanism of electronic assembly process reliability, reliability design of solder joint and PCB, simulation analysis and life prediction of solder joint, process reliability experiment, process failure analysis, etc. which basically covers the reliability analysis technology of electronic assembly process.

Taking the modern electronic assembly process as research object, Fan [13] analyzed the specific electronic process reliability problems, such as the factors affecting the reliability of modern electronic assembly process, the formation of welding interface alloy layer and its influence on solder joint reliability, the influence of environmental factors on electronic equipment reliability and process reliability reinforcement, the process reliability problems affecting electronic products during service, the quality model of ideal solder joint and its influencing factors, process reliability of lead and lead-free hybrid assembly, the process reliability of Pb free process of electronic products. The process reliability design of wave soldering solder joint, the process reliability design of SMT reflow soldering solder joint, the common fault phenomena endangering reliability of PCBA and their analysis, engineering methods of PCBA solder joint failure analysis and process reliability test were also presented; Patrick and Andre [14], a senior expert in reliability, systematically described "reliability in manufacturing" in a separate chapter in his reliability masterpiece (world best seller) "practical reliability engineering" (5th Edition) revised and published in 2012. Manufacturing reliability technology is emphasized in this book, and the direction for strengthening the research and application of reliability technology in manufacturing stage is also pointed out. The inevitable relationship between manufacturing process and system and manufacturing product defects was proved by Inman et al. [15] through case study. The necessity and importance of process and manufacturing reliability were also confirmed once again, and the research of manufacturing process reliability theory and technology was promoted. For the reliability analysis of electronic process, starting with the physical principle of electronic manufacturing process, Gan et al. [16] analyzed the wafer processing process, wafer needle testing process and other reliability analysis methods of Front-End-of-Line process. Then the reliability test technology of Back-End-of-Line processes such as assembly and testing was given. In 2014, "process reliability of mechanical manufacturing" written by Jiang et al. was published, which takes the mechanical manufacturing process as the object, redefines "process reliability" as "the ability of the process quality index of the processing system to continuously meet the technical requirements under the specified time and conditions". Then the concept and index of process reliability were given by applying the classical reliability theory and method. Furthermore, combined with the requirements of reliability analysis of mechanical manufacturing process, process reliability modeling, Process FMEA analysis method, process reliability evaluation method and process reliability management are given. In this book, the process reliability index system and evaluation methods is systematically studied, and the relationship between equipment reliability and process reliability is deeply studied as well.

1.4.2 The Modeling Technology of Manufacturing Process and System Reliability

The output of the manufacturing process is product. The material basis for making the manufacturing process stable, controlled and reliable is the manufacturing system. The reliability of product is directly determined by the reliability of manufacturing system. The research of manufacturing system reliability includes reliability design, reliability test and evaluation, operational reliability improvement, etc. Modeling allocation and FMEA in reliability design are the basic links, which are placed in the initial position. Accurate reliability evaluation results can only be obtained through correct and reasonable models and analysis methods. Then, the weak links of the system can be found out according to the evaluation results, which is the objective basis for system reliability design, distribution and improvement.

The existing research categories of manufacturing system reliability modeling and analysis methods are as follows:

(1) Method based on classical reliability diagram. Following the complex system modeling theory, classical reliability diagram method mainly includes reliability diagram [17], fault tree analysis (FTA) [18], Markov model [19], Petri net [20, 21], GO method [22], etc. Su et al. [23] model and analyze the dynamic reliability of excavator manufacturing system through Petri net; Abhulimen [24] proposed the reliability modeling and analysis method of complex manufacturing system through Markov model method; Lin and Chang [25] established the reliability analysis and evaluation model of production line through Stochastic-flow manufacturing network.
(2) Method based on manufacturing system operation and maintenance data. Li and Ni [26] proposed the manufacturing system reliability estimation method through the maximum likelihood estimation based on the operation and maintenance data of manufacturing system, which can be the basis for preventive maintenance of manufacturing system; Li et al. [27] established reliability prediction model of manufacturing system through grey model; Lin et al. [28, 29] proposed the Capacitated Manufacturing Network which is a reliability analysis model of manufacturing system established using operation data such as failure and rework. Mokhtari et al. [30] established a manufacturing system operation and maintenance scheduling algorithm considering manufacturing system reliability.
(3) Method based on manufacturing process quality data. Chen et al. [31, 32] first proposed the reliability analysis model and maintenance strategy of manufacturing system considering process quality by using the interaction effect of process quality and system reliability (QR-co-effect); Sun et al. [33] proposed the reliability modeling and analysis method of manufacturing system for product dimensional quality; Zhang et al. [34] established the reliability modeling method of complex manufacturing process based on state space model; In this book, the modeling and analysis method of manufacturing system reliability based on process quality data and the correlation model between product quality and manufacturing system reliability in mechanical manufacturing process are studied.

1.4.3 Product Reliability Analysis Technology in Manufacturing Process

Product quality can always be described by multiple dimensions. As one of the dimensions of quality, reliability involves the dynamic characteristics of product performance changing over time [35]. Relative to the reliability requirements defined in the design stage, the reliability index of the manufactured product will be outputted in the end of the manufacturing process, which is manufacturing reliability. The manufacturing reliability is expressed as the early failure rate of products in use. The early failure rate of products is an indicator of the level of product reliability in the manufacturing process. For enterprises, on the one hand, the early failure rate can be used as an extension index of manufacturing process quality. On the other hand, it can be used as an advance and symptom index of product reliability. Its internal formation law and quantitative research play important role in eliminating the high early failure rate and ensure the quality and reliability of products. Therefore, the modeling and optimization of product early failure rate is the core of product reliability analysis technology in manufacturing process.

Product manufacturing includes the processes of raw material mobilization, component processing, component assembly and factory test. Through the operation and treatment of purchased and self-made parts, WIP and final products are formed. Therefore, the reliability of the final product can not only be analyzed according to the product maintenance data, but also be traced back to the whole supply chain system upstream [36]. In a word, the manufacturing process includes the processes of processing, assembly, testing etc. The fluctuation of human, machine, material, method, environment and testing is transmitted and accumulated in all links of manufacturing process. The deviation of manufacturing quality is leaded by it. Manufacturing process is an important stage to determine and affect the reliability of products. A series of quality data of product formation process can be used as extended data for product reliability analysis and improvement in manufacturing process.

In terms of data, the quality inspection data and the reliability test data in manufacturing process, the failure data used in the field, which are the most relevant data for evaluating the effect of manufacturing quality deviation [37]. Due to the testability and cost of data, the maintenance or failure data in the use have greater advantages in reflecting the actual reliability of products.

Jiang and Murthy [38] discussed the influence of manufacturing process fluctuation on the inherent reliability of products based on the maintenance or failure data of products in use. The conceptual model of product inherent reliability considering manufacturing fluctuation was constructed for the unqualified fluctuation of components and assembly deviation fluctuation in the process. The classification discussion of parametric estimation and nonparametric estimation were also carried out.

In terms of models, the distribution used to characterize product reliability indicators often has different structures and forms, such as piecewise Weibull distribution, generalized Weibull distribution, Mixture distribution, Competing Risk distribution, Polynomial distribution, etc.

Taking polynomial reliability model called Polynomial model as example, this kind of model is affected by different distributions, which can be used for fitting N-type or W-type failure rate curves under different failure mechanisms [39].

As for the problems related to traditional quality indicators and reliability indicators, the concept of "outbound point" can be used to analyze the risk factors leading to the decline of product reliability in the manufacturing process [40]. Then, the "outbound point" of quality was transmitted to "failure rate" in reliability through a new bathtub curve model with the number of out of bounds as the ordinate and the cumulative time as the abscissa. The guidance for establishing the relationship between "outbound point" and "failure rate" was also provided by this model [41].

Systematically speaking, the manufacturing reliability of products can be affected by front-end product design, raw material supply and the capability in machining assembly process. Therefore, several feasible and necessary measures to improve product reliability are to integrate the reliability analysis of products in the design stage, improve the quality of raw materials, strengthen supply chain management, and stabilize the manufacturing capability in whole process [42].

Generally speaking, the product reliability index can be statistically inferred based on the failure data collected in the whole life cycle. However, for a new product, the failure data is insufficient, and the prototype test is not only expensive, but also not completely accurate. Therefore, it is difficult to predict the reliability of new products. To solve this problem, Sanchez and Pan [43] put forwarded the idea of data source expansion combined with expert opinions and failure or maintenance information of similar products in the market, so as to realized the prediction and evaluation of the reliability of new products.

Kim [44] focused on the performance of reliability problems in the early use of products, analyzed the growth model of defects in the manufacturing stage and the influence relationship of aggregation effect on the decreasing failure efficiency curve of semiconductor devices, and established a product reliability function model considering manufacturing defects, which provides a reference for clarifying the relationship between deviation in the manufacturing stage and product reliability.

Shirley [45] divided defects into manufacturing quality defects and reliability defects, and proposed the proportion coefficient K between them. Each failure mechanism corresponds to a specific K. The quantitative relationship between quality and reliability is also established.

To characterize the correlation between yield and reliability of IC circuit, Chen et al. [46] have proposed a Yield-Reliability correlation model with time as an independent variable to guide the design of burn-in test parameters based on common defect models such as Poisson distribution and Negative binomial distribution.

In summary, the quality and reliability are used to qualitatively describe and reflect the manufacturing capacity and stability of product when the design level is certain. In reliability engineering practice, the quality index shows generality because it includes many product characteristics, and the reliability index is more targeted. The manufacturing quality of products is always quantitatively described by manufacturing defects (De-fects) combined with yield. From the time axis of product life cycle, manufacturing quality can be regarded as "product quality at zero time". Product reliability can be regarded as quality prediction over time accumulation, which is "product quality greater than zero time". It is also a dynamic real-time product quality. Therefore, time factors and product compliance characteristics have become the key link between product manufacturing quality and reliability. How to ensure the compliance characteristics of product design requirements in the manufacturing stage, so as to quantify the initial reliability level of the front end of the time axis, is the key to product reliability analysis and early failure rate control in the manufacturing process.

1.4.4 Failure Mechanism Analysis Technology of Manufacturing Process Defects

The physical, chemical, mechanical or other processes that cause process defects are called process defect failure mechanism. At present, the research on the mechanism of process reliability is mainly concentrated in the electronic product industry. Rich experience has been accumulated due to a lot of energy invested in the research of its fault mechanism. For integrated circuit products, the physical defects produced in the processing process not only affect the product reliability, but also have an important impact on the processing yield of products. Therefore, there is a strong correlation between product reliability and processing yield.

The failure of semiconductor products is mainly caused by basic mechanical, circuit, temperature or chemical processes. Therefore, the failure research of this kind of products mainly focuses on the physical failure model. Jia [47] has studied the quantitative relationship between the failure rate and the processing yield of integrated circuit products, has divided the defects into yield defects (resulting in hardware failure), reliability defects (resulting in shortening product life) and benign defects, and has proposed a high-precision prediction model of integrated circuit product yield. Liu et al. [48] have summarized the research results of failure physical model, and have pointed out that failure physical model is not only suitable for product design, but also an economic and effective means to improve product reliability in the manufacturing process. Kuo and Kim [49] have established the relationship model by analyzing the impact of the yield of the production line on the reliability of semiconductor products. The parameters affecting the yield and reliability of semiconductor products have been classified into three categories: parameters related to design, parameters related to manufacturing process and parameters related to operation. The first two types of parameters affect the yield. All parameters affect the reliability. At the same time, the assurance methods of product reliability and user satisfaction such as wafer acceptance test, burn-in test, final test and quality control are also studied.

Most the above models are established for integrated circuit (semiconductor) products. Accurate description of the failure mechanism of electronic products is essential for their successful application. However, there is little research on the fault mechanism of widely used specific mechanical manufacturing process. In engineering practice, the analysis of the influence mechanism between mechanical manufacturing process and product reliability and the establishment of the qualitative model describing the relationship between product reliability and various variables in mechanical manufacturing process mainly rely on the experience accumulation of manufacturing practitioners and the analysis of a large number of historical data. It is urgent to study the quantitative process defect mechanism and analysis method.

1.4.5 Reliability Monitoring Technology in Manufacturing Process

Due to the lack of systematic understanding of manufacturing process reliability, the research on reliability monitoring in manufacturing process is mainly divided into the following three discrete directions.

1. **Deviation Flow Analysis Technology**

With the development of advanced manufacturing and process technology, manufacturing processes become more complex, and higher capability of manufacturing system is required. In the production of products, multi station manufacturing system (MMS) is always needed [50], and the manufacturing of products will be realized through N interacting multi station manufacturing processes (MMP), which is shown in Fig. 1.7 The "Stream of Variation" (SoV) [51] theory describes that in the production environment of multi station manufacturing system, the fluctuation in the manufacturing process is promoted by coupling in the multi process flow in two forms: manufacturing process deviation and product quality characteristic deviation, which enables the deviation of product processing characteristics to be transmitted in the path shown in Fig. 1.7. Taking the multi station machining and manufacturing of mechanical products as an example, the process size of products will be affected by various manufacturing fluctuations, such as fixtures, machine tools, etc. The process of deviation transmission is clearly described in Figs. 1.7 and 1.8. Affected by the random error *wi*, the operation of each station will produce manufacturing process fluctuations *ui* such as machining deviation and positioning deviation. At the same time, there are dimensional deviations in the products processed and transmitted from the upstream stations. These comprehensive effects are accumulated through multiple stations, and reflected by the measured value *Yi* of key stations. Taking into account the above factors and measurement error factors *vi*, the product size does not fully match the design size, as shown in Fig. 1.8.

On the one hand, the accumulation of dimensional deviation will make the tools or fixtures in the downstream processing process need to be repaired repeatedly, which will increase the manufacturing cycle. On the other hand, it will lead to the poor and failure of the total assembly due to the mismatching or loose pack burn-in of assembly parts [52].

The analysis and modeling of the common deviation flow problem in MMP have been carried out by domestic and foreign scholars in many aspects. Agrawal et al.

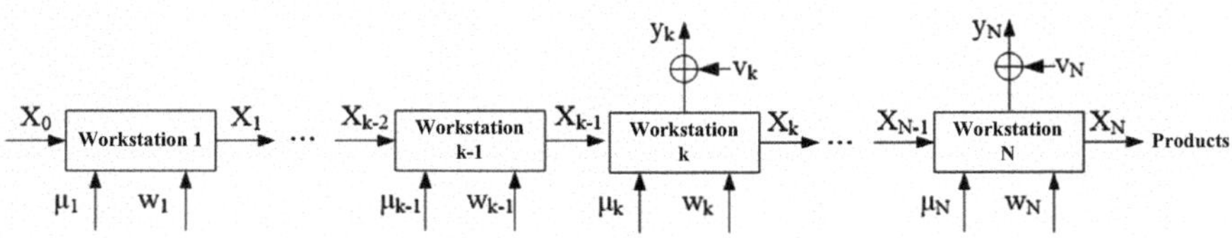

Fig. 1.7 Fluctuation transmission in multi-station manufacturing processes

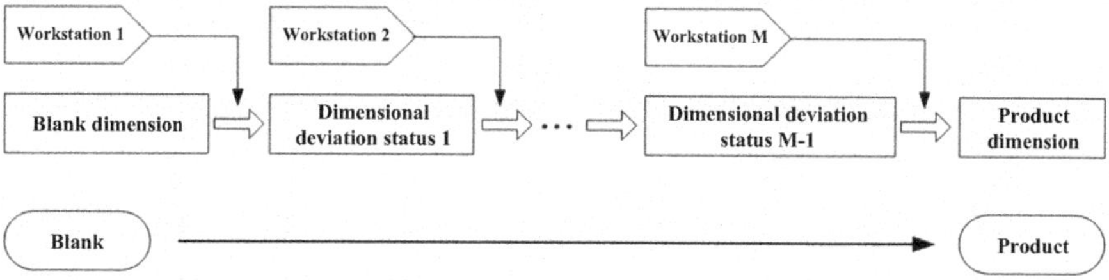

Fig. 1.8 Accumulation of deviation states in multi-station manufacturing processes

[53, 54] have modeled the deviation flow problem through autoregressive model in the late 1990s. The conclusion that the observation value *Yi* of the current station will be determined by the observation value *Yi* − 1 of the upstream station has been given. The AR(1) model including Formulas (1.2) and (1.3) has been recommended to describe it.

$$Y_i \sim N\left(\mu_i, \sigma_i^2\right) \tag{1.2}$$

$$Y_i = \alpha_i + \beta_i Y_{i-1} + e_i, \ i = 1, 2, \ldots N \tag{1.3}$$

The AR (1) model is built based on the time series autoregressive model. The upstream process and noise are considered to affect the measurement vector of the studied process. Generally, the advantage of the model is that it can analyze and predict the fluctuation of process size with many uncertain deviation factors, and it is more effective for series MMS with stable time.

As early as 1997, Hsieh et al. [55, 56], aiming at the assembly of automobile body, guided by the minimum assembly error of flexible parts of car body, the related process modeling and optimization have been carried out based on genetic algorithm, and the evaluation method of physical structure has also been presented based on elastic statics theory.

Liu et al. [57–59] have also studied the assembly capacity and assembly deviation through the finite element method for the assembly of parts. The distribution of tolerance limit in the process of flexible assembly has been focused on.

For the analysis of dimensional deviation, extreme value method and statistical method are suitable for the assembly process of rigid parts with simple mechanical structure. For the complex assembly process of complex mechanical structure, Monte Carlo simulation method is more suitable. Then, guided by Monte Carlo simulation method, Turner et al. [60–62] have measured the sensitivity of deviation by setting disturbance factors to affect product manufacturing size, and the deviation analysis method of linear assembly function under CAD model has been given. For the structure of the product, the vastness of CAD model parameters is determined by the complexity of its components, and the calculation time of deviation sensitivity has become a bottleneck in practical application.

Hu and Koren [51] have given the concept of systematic deviation flow as the basis of analysis on the problem of predicting product size characteristic deviation and then identifying potential faults.

Based on the concept of systematic deviation flow, Jin and Shi [63] have constructed the corresponding state space model in 1999 by analyzing the dimensional deviation of rigid parts moving in the plane for the deviation flow of simple and complex multi station assembly process. At the same time, the fluctuation caused by fixture positioning and tool positioning is concerned, and the difference accumulation model is proposed to measure the influence of fluctuation factors on process dimensional deviation.

Similarly, Huang et al. [64] have modeled the machined parts in an abstract and simplified way, have considered the positioning and tool deviation, and have extended the above state space model to apply to the deviation flow description and modeling of the actual multi station manufacturing process.

Zhong et al. [65] have further constructed a point distribution model to further deduce the deviation flow model of the multi station manufacturing process by integrating the deformation characteristics of flexible parts.

Then, Djurdjanovic and Ni [66] have proposed the linear deviation flow state space integrating the deviation flow concept and state space model, which can effectively control the deviation flow in multi station manufacturing process under the environment of multi-directional development and application of modern control theory.

On the one hand, The SoV technology in manufacturing process describes the correlation among multiple stations or processes, on the other hand, it quantifies the process in which the manufacturing quality deviation is transferred from the upstream process to the downstream process with the processing process, and finally reflected in the product size deviation and even the decline of product reliability. It provides theoretical support for controlling process deviation to ensure the product quality and reliability required in the design stage.

The key product characteristics which its fluctuation will have a serious impact on product safety or user experience, and the associated and key process control characteristics together constitute the key quality characteristics at the product level. Therefore, identifying and extracting the monitoring objects with the key process control characteristics as the core, and then using the relevant monitoring technology to carry out targeted control, are the two major elements of monitoring process deviation. Furthermore, it is urgent to monitor to ensure that the corresponding key product parameters are near the target value and the key process parameters with small fluctuation, so as to ensure the key product characteristics and realize the requirements of design quality and reliability.

In traditional engineering practice, the priority of product quality characteristics and even the of key quality characteristics are often determined by design experience, QFD and other quality function development tools. Based on these tools, the customer requirements are positioned, the engineering technical language is transformed, the product functional characteristics, structural characteristics and process characteristics are decomposed, and then the key process control parameters that need to be monitored are identified by the IPT product development team within the enterprise.

From the perspective of loss caused by product quality characteristic deviation, Taguchi quality loss function is always used to excavate and filter the process parameters sensitive to product reliability in the manufacturing process as key control characteristics [67]. In this idea, the fluctuation of quality characteristics can be measured by quality loss. The gentle quality loss corresponds to the quality characteristics with little fluctuation, which are non-key quality characteristics; The quality loss with large observation value corresponds to the quality characteristics with large fluctuation, which are key quality characteristics [68, 69].

For the identified key quality characteristics, the monitoring of deviation is mainly divided into the following two types: One is the single process monitoring based on traditional SPC control chart technology to realize the monitoring of the last station, or a key control characteristic which is no longer affected by downstream processing [70]; The other is the monitoring of multi station manufacturing process. Through the systematic analysis and modeling of product MMP process, the whole multi station process monitoring is realized [71, 72].

Specific to the implementation of process monitoring, on the one hand, the historical data of key control characteristics can be analyzed through the change point analysis method facing the deviation flow state space, and relevant detection statistics can be established to judge whether the process is controlled [73]; On the other hand, multi station state space model can be used to alarm when the process is unstable for real-time adjustment and control [74, 75].

2. **Censored Control Chart Technology**

The deviation of key quality characteristics is transmitted and accumulated in the manufacturing process, which forms the process product quality, affects the product reliability, and hides the crisis of product reliability degradation. Especially with the increasing requirements and complexity of product manufacturing, and limited to the restriction of detection technology, time and cost, censored quality characteristics with incomplete information widely exist in the manufacturing process, which transmit and restrict product quality and reliability in manufacturing process with the conventional quality characteristics with complete information. There are fuzziness and uncertainty in the value of censored characteristics, which is naturally related to product reliability. However, at present, due to the lack of appropriate control chart to monitor, a large number of censored characteristics are omitted or abandoned in the quality control of manufacturing process, which intensifies the degradation of product reliability in manufacturing process and results in the products reliability lower than the requirements of reliability design index. Considering the importance and particularity of censoring characteristics, DC. Montgomery [70], a senior expert in the field of process quality control, introduced the censoring characteristics control chart technology for the first time in his book "Introduction to Statistical Quality Control" (Sixth Edition) in 2009.

The censored quality characteristic control chart technology originated in the 1990s. With the increasing demand of users for product reliability assurance, the research on process control technology for censored characteristics has also developed. The early censored characteristic data processing methods generally start with the overall distribution, and weaken the distribution distortion caused by censored data through the later correction of the control limit [76–79].

Then, according to the characteristics of censored data such as skewed distribution and incomplete value, scholars have presented different methods to monitor the censored characteristics. The research can be divided into the following three categories:

(1) Probability Limits (PL)

Liu [80] has discussed the non-normal statistical model for high-quality process assurance, and has proposed the probability limits method to determine the upper and lower limits of the control chart when the quality characteristics are skewed distribution. However, the censored characteristics itself has not been mined, and the constructed control chart can not effectively display the deviation information of censored characteristics. Based on the idea of probability limit and first-order statistics, Huang et al. [81] have discussed the construction of unbiased ARL control chart using alarm limit to monitor the percentile of Weibull distribution process when the shape parameters are known. For the second kind of censored characteristic subject to Weibull distribution, by creating order statistics conforming to uniform distribution, based on the idea of probability limit, and with the help of chi-square distribution with degree of freedom $(2r - 1)$, Guo and Wang [82] gave one-sided control. However, it depended on sample size, failure number r, error alarm rate, etc., which has great constraints.

(2) Probability Plotting (PL)

Schneider [83] have drawn a probability map for the left censored data with skewed distribution, and have calculated the control limits under different distributions by using the point mapping method. Fitting the distribution of left censored data is the premise of the above work; In view of the complex effect of competitive risk factors on the censoring mechanism, Steiner and Jones [84] have proposed a new EWMA control chart to monitor the survival data mixed with competitive risk factors; In order to improve the reliability of multi station process products, Asadzadeh and Aghaie [85] have proposed a regression adjusted CUSUM censored characteristics control chart considering fixed competitive risk factors to detect the reduction of process mean value.

(3) Conditional Expected Value (CEV)

Steiner et al. [86–88] have given the construction steps of one-sided control chart for the normal censored characteristics based on CEV to realize the rapid detection of process quality degradation. When the censoring level is constant, the CEV based Shewhart control chart for process control is given, and the CEV control chart with highly censored characteristics caused by competitive risk factors is also given; Zhang and Chen [89] have constructed an EWMA control chart for censored life data under Weibull distribution based on the conditional expectation method through the transformation of exponential distribution scale, which can be used to detect the change of process mean value; Olteanu [90] have studied the hypothesis that the potential distribution of interval censored data is two parameter Weibull distribution, the construction method of CUSUM based on Likelihood Ratio (LR), and its performance was preliminarily evaluated; Pascual and Li [91] have established the minimum distribution of samples for the second type of censored Weibull process, and have determined the control limit suitable for any type of control chart based on unbiased ARL to monitor the shape parameters of Weibull distribution, but the determination of the control limit depends on the sample size, the number of failures in the second type of censoring mechanism and the shape parameter value of the controlled process; Considering the progressive attribute of rank statistics, Li et al. [92] have constructed a general operating characteristic function through the weighted rank test for right censored reliable data. Furthermore, the control chart is constructed and the first-class error probability and the second-class error probability are shown; Although Guo et al. [93] have designed the control chart for censored Weibull distribution, they have failed to restore the distortion of sample distribution caused by censored Weibull distribution, and the obtained control chart wasn't accurate; Xin [94] have studied the control limit of censored characteristics of normal distribution; In view of the incomplete embodiment of quality information caused by incomplete or censored quality characteristics, Wang and Hryniewicz [95] used the fuzzy data principle to carry out approximate normal processing and approximate chi square distribution simulation on the above quality characteristics, and constructed CUSUM control chart containing fuzzy decision rules and ARL operation.

It can be seen from the above research and analysis that the traditional truncated characteristic control chart is mainly constructed based on conditional expected value (CEV) technology. However, this method is limited to specific control chart types, which is difficult to be widely used in manufacturing process. At the same time, it lacks the research on the relationship between censored characteristic monitoring and product reliability in the manufacturing stage, which limits its development.

3. **Burn-in Technology**

Affected by systematic or random factors such as human, machine, material, method, environment and measurement in the manufacturing process, the quality of products fluctuates, which is manifested in the deviation of quality characteristic value from the standard value, resulting in the loss of quality deviation and the frequent occurrence of early faults [32, 38, 42]. McCluskey et al. [96] have considered from the perspective of product failure mechanism that when the manufacturing process is poorly controlled, manufacturing defects will occur, which will affect the robustness of products to the high stress environment, and the accumulation of defect effects will eventually lead to the early failure of products. Cha et al. [97] have focused on heterogeneous populations with different failure modes and proposed a failure rate model with percentile and tail mixed mechanism to observe the real dynamic change characteristics of the overall failure rate.

From the perspective of reliability and fluctuation, Almudever and Rubio [98] have given five manufacturing defects that lead to early product failure in the manufacturing process of carbon nanotube field-effect transistors (CNFET). The importance of strengthening manufacturing process supervision and optimizing design technology was also emphasized. In order to reduce the early failure rate, burn-in test, as a special failure oriented accelerated test in manufacturing process, is used to screen and eliminate products with quality defects to ensure the stable performance of device in use [99], but it also brings high burn-in cost [100]. The inherent manufacturing fluctuations and deviations in the manufacturing process inevitably lead to a large number of hidden losses and abnormally high early failure rate, which is an important prior information for burn-in time optimization. However, the setting of the traditional burn-in test environment didn't consider the deviation factors and their losses in manufacturing process that lead to frequent early failures, but is based on the understanding that the burn-in time should not be too short or too long to achieve the desired reliability level. Thus, the traditional burn-in test is usually simple and arbitrary. Different from the maintenance cost of early failures, the quality deviation loss in the manufacturing process can be regarded as a kind of implicit cost, and gradually becomes explicit with the burn-in cost and the cumulative maintenance cost of failures within the service warranty period. The traditional optimization research of early failure rate mostly focuses on the optimization of burn-in test time under the established warranty strategy and minimum cost, while ignoring the constraints of manufacturing quality deviation loss resulting in high early failure rate of products. Therefore, the optimization of burn-in time is independent of the quality deviation loss in the manufacturing stage, and can not fundamentally guide the optimization scheme for reducing early failure rate. The disconnection between manufacturing process quality control and reliability verification is also resulted [42].

At present, the research on the optimal burn-in test time for early failure rate optimization focuses on the structure of the cost model and the type of warranty strategy under different constraints, such as specific average remaining life, specific task reliability, specific failure rate, maximum burn-in test capacity, etc. From the perspective of system, the burn-in time of electronic product components is regarded as a decision variable, the reliability requirements that components and systems should meet are regarded as constraints, and the total cost optimization model including component burn-in cost, burn-in maintenance cost and after-sales maintenance cost is proposed by Kuo [101]. Subsequently, Chi and Kuo [102] have taken the maximum capacity of the actual burn-in test as the constraint, and have studied the burn-in time optimization model under the constraints of comprehensive reliability and burn-in test capacity. Chandrasekaran [103] has studied the total cost model under different failure characteristics based on the linear assumption of burn-in cost and total burn-in time with the maximum average remaining life as the constraint. The optimal burn-in time was also determined. Based on the decreasing failure rate characteristics, Lawrence [104] has clearly given the upper and lower limits of burn-in time satisfying the specific average residual life constraints by using the first-order moment and the percentile method of failure distribution. Marko and Schoonmaker [105] have defined the life cycle cost as the sum of burn-in cost and after-sales failure cost, and have used exhaustive search technology to optimize the burn-in time of spare components, so as to minimize the life cycle cost. Genadis [106] has discussed three necessary prerequisites for burn-in test of electronic products: whether the reliability level of non burn-in components meets the design objectives, how much the expected reliability of components is improved after burn-in, whether the burn-in of components meets the economic requirements, and considering the impact of warranty policy. Based on the Weibull failure hypothesis in the early failure stage and the exponential failure hypothesis in the accidental failure stage, the burn-in cost model of components or systems was measured, and the optimal burn-in time was determined. Although product maintenance is involved in the above models, the warranty policy is not systematically integrated into the modeling of burn-in time optimization. Nguyen and Murthy [107] have firstly proposed the minimum cost model under two types of maintenance policies: the free replacement/repair warranty (FRW) and the rebate strategy. Based on the assumption of bathtub curve failure mode, Yun et al. [108] have analyzed the total average cost model under the cumulative free replacement warranty policy, and have weighed the increased manufacturing cost and reduced maintenance cost by adjusting the burn-in time. Wu et al. [109] have proposed a cost model under free replacement and proportional maintenance, which can optimize the burn-in time and warranty duration at the same time. Similarly, for products with bathtub curve failure characteristics, Shafiee et al. [110] have studied the different effects of burn-in, warranty and preventive maintenance strategies on the total expected cost per unit product while optimizing the early failure rate, so as to guide management decisions. Aiming at the problem of traditional burn-in failure caused by higher reliability of micro electro mechanical systems, Ye et al. [111] have established the burn-in optimization model under preventive maintenance strategy based on the degradation model. In order to ensure that the product performance indexes after burn-in meet the design requirements, Ye et al. [112] have modeled the cost of the manufacturer and the performance received by the user, and have given the burn-in optimization model under double objectives.

To sum up, the optimal aging analysis aiming at eliminating early fault characteristics is mostly based on the improvement of reliability results or benefits of aging test on warranty or maintenance. The researches on the cost of burn-in test mostly analyze the optimal aging time by using the burn-in cost model composed of failure cost, maintenance or warranty cost [112]. Although some literatures emphasize the importance of manufacturing cost to the optimization of burn-in time, they only roughly make the cost of burn-in test included into the manufacturing cost, and the root hidden loss caused by manufacturing quality deviation has not been systematically studied [108]. How to integrate the manufacturing quality deviation loss causing high early failure rate of products into the optimization of burn-in test time, and characterize the manufacturing quality deviation loss model explicitly to reflect the root quality fluctuation has become a new problem in the field of burn-in test time optimization.

1.4.6 Summary and Analysis

From the above analysis, it can be seen that manufacturing reliability is not only a systematic problem of equipment development, but also a new direction in the field of reliability system engineering. However, due to the lack of research on the fault mechanism of manufacturing process, the influence relationship between manufacturing system reliability, manufacturing process quality and product reliability has not been organically combined, and there is no systematic research on process reliability technology. In addition, the only few process reliability books are part level application books focusing on a specific process category (machinery or electronics). There is a lack of general theory and technical monographs covering the whole machine, mechanical electronics and other typical processes to guide the theoretical and technological development of manufacturing reliability and help "made in China 2025" [113].

Therefore, from the perspective of reliability system engineering technology, many of the authors' scientific achievements in the field of manufacturing and process reliability are condensed in this book at the macro level of equipment, manufacturing processes and manufacturing systems. According to the framework of modern science and technology, including theory, basic technology and applied technology, a complete cognitive theory and application technology system of manufacturing process reliability are put forward for the first time. High integration of manufacturing process reliability with design reliability technology and use and maintenance assurance technology is also realized. This book provides systematic theoretical and methodological support for recognizing and preventing product manufacturing defects from the perspective of manufacturing system and manufacturing process, and will fill the gap of reliability system engineering technology in the field of manufacturing application.

References

1. Lai C. D., M. Xie. Bathtub Shaped Failure Rate Life Distributions. Stochastic Ageing and Dependence for Reliability, 2006: 71–107.
2. D. N. P. Murthy. New research in reliability warranty and maintenance. APARM 2010, New Zealand.
3. L. B. Wang. Research on Product Infant Failure Rate Analysis and Optimization Considering Quality Variations in Manufacturing Process. Beihang University, 2015.
4. D. L. Crook. Evolution of VLSI reliability engineering. Quality and Reliability Engineering International, 1991, 7(4):221–233.
5. W. H. Schroen, J. G. Aiken, G. A. Brown. Reliability improvement by process control. 10th Annual IEEE Reliability Physics Symp., 1972, Las Vegas, Nevada.
6. A C. Pronikov et al., translated by Li Changqi and Yu Liji. Accuracy and Reliability of CNC Machine Tools. Beijing: Mechanical Industry Press, 1987.
7. Paul Barringer, P. E. Barringer and Associates. Process reliability concepts. SAE 2000 Weibull User's Conference, March 10, 2000, Detroit, Michigan.

8. C. G. Chen, G. X. Lin. Dynamic Optimization Selection of Machine Tools Considering Processing Time and Reliability. Machinery Design & Manufacture, 2003, 8:3–4.
9. X. D. Kong, Y. F. En, X. W. Zhang, et al. Microelectronics Manufacture Process Reliability Evaluation and Control. Electronic Product Reliability and Environmental Testing, 2004, 3:1–5.
10. G. H. Lin. Process reliability assessment with a Bayesian approach. International Journal of Advanced Manufacturing Technology, 2005, 25: 392–395.
11. G. C. Fu, G. Y. Shang, X. K. Shi, et al. Processing System Reliability Model Based on Products Reliability. Journal of Beijing University of Aeronautics and Astronautics, 2009, 35(1):9–12.
12. Wang Wenli, Reliability of Electronic Assembly Process. 2011, Electronic Industry Press.
13. Fan Rongrong, Reliability of Modern Electronic Assembly Process. 2012, Electronic Industry Press.
14. Patrick, D. T. O., Andre, K., 2012. Practical reliability engineering. John Wiley & Sons, Ltd, Chichester, UK.
15. Robert R. Inman, Blumenfeld DE, Huang NJ, Li JS, Li J. Survey of recent advances on the interface between production system design and quality. IIE Transactions, 2013, 45:557–574.
16. Zhenghao Gan, Waisum Wong, Juin J. Liou, Semiconductor Process Reliability in Practice, 2013, The McGraw-Hill Companies, Inc.
17. Zhang Guozhi, Yang Guang, Gong Yinghai. Reliability Analysis of Complex Systems. Harbin: Harbin Institute of Technology Press, 2009.
18. Dutuit Y, Rauzy A. Approximate estimation of system reliability via fault tree. Reliability Engineering and System Safety, 2005, 87(2):163–172.
19. Perman M, Seneqacnik A, Tuma M. Semi-Markov models with an application to power-plant reliability analysis. IEEE Transactions on Reliability, 1997, 46(4): 526–532.
20. Rausand M, Hoyland A. System reliability theory: models, statistical methods and applications (2nd edition). New York: Wiley-Interscience, 2004.
21. Carneiro JSA, Ferrarini L. Reliability analysis of power system based on generalized stochastic Petri nets. Proceedings of the 10th International Conference on Probabilistic Methods Applied to Power Systems. Rincon, Puerto Rico. 2008: 425–436.
22. Z. P. Shen, T. Zheng. Exact Algorithm for Complex System Reliability Using the GO Methodology. Journal of Tsinghua University (Science and Technology), 2002, 42(5): 569–572.
23. C. Su, G. Shen, Y. Q. Xu. Modeling Theory and It's Application for Dynamic Reliability of Manufacturing System. Machine Design & Research, 2006, 22(5):17–19.
24. Kingsley E. Abhulimen. Model for risk and reliability analysis of complex production systems: Application to FPSO/flow-Riser system. Computers and Chemical Engineering, 2009, 33:1306–1321.
25. Yi-Kuei Lin and Ping-Chen Chang. A Novel Reliability Evaluation Technique for Stochastic-Flow Manufacturing Networks With Multiple Production Lines. IEEE Transactions on Reliability, 2013, 62(1):92–104.
26. Lin Li, Jun Ni. Reliability Estimation Based on Operational Data of Manufacturing Systems. Quality and Reliability Engineering International, 2008, 24:843–854.
27. Guo-Dong Li, Shiro Masuda Daisuke Yamaguchi, Masatake Nagai. A New Reliability Prediction Model in Manufacturing Systems. IEEE Transactions on Reliability, 2010, VOL. 59(1):170–177.
28. Yi-Kuei Lin, Ping-Chen Chang. System reliability of a manufacturing network with reworking action and different failure rates. International Journal of Production Research, 2012, 50(23):6930–6944.
29. Yi-Kuei Lin, Cheng-Fu Huang, Ping-Chen Chang. System reliability evaluation of a touch panel manufacturing system with defect rate and reworking. Reliability Engineering and System Safety, 2013, 118:51–60.
30. Mokhtari, Hadi, Mozdgir, Ashkan, Abadi, Isa Nakhai Kamal. A reliability/availability approach to joint production and maintenance scheduling with multiple preventive maintenance services. International Journal of Production Research, 2012, 50(20):5906–5925.

31. Yong Chen, Jionghua Jin. Quality-reliability chain modeling for system-reliability analysis of complex manufacturing process. IEEE Transaction on Reliability, 2005, 54(3):475–488.
32. Yong Chen and Jionghua (Judy) Jin. Quality-Oriented-Maintenance for Multiple Interactive Tooling Components in Discrete Manufacturing Processes. IEEE Transactions on Reliability, 2006, 55(1):123–134.
33. J. W. Sun, L. F. Xi, E. S. Pan 等. Product Dimensional Quality Oriented Reliability Modeling and Analysis of Manufacturing Systems. Journal of Shanghai Jiaotong University, 2008, 42(7):1100–1104.
34. Zhang Faping, L. J. and S. Yan Yan. Dimensional quality oriented reliability modeling for complex manufacturing process. International Journal of Computational Intelligence Systems, 2011, 4(6): 1262–1268.
35. Garvin, D. Managing quality. New York: The Free Press, 1988.
36. Zhou, C, R. B., Chinnam, A., Korostelev. Hazard rate models for early detection of reliability problems using information from warranty databases and upstream supply chain. International Journal of Production Economics, 2012, 139(1):180–195.
37. Karim, M. R., K., Suzuki. Analysis of warranty claim data: a literature review. International Journal of Quality & Reliability Management, 2005, 22(7):667–686.
38. Jiang, R., D. N. P., Murthy. Impact of quality variations on product reliability. Reliability Engineering and System Safety, 2009, 94(2):490–496.
39. Bebbington, M., C. D., Lai, D. N. P., Murthy, R., Zitikis. Modelling N- and W-shaped hazard rate functions without mixing distributions. Journal of Risk and Reliability, 2009, 223(1):59–69.
40. Roesch, W. J., D., Littleton. Assessing the Reliability Risk of a Maverick Manufacturing Anomaly. CS Mantech Conference, Boston, 2012.
41. Bosacchi, B. Fuzzy Logic Application to Quality & Reliability in Microelectronics. The International Society for Optical Engineering, 1996, 2761:224–236.
42. Roesch, W. J. Using a new bathtub curve to correlate quality and reliability. Microelectronics Reliability, 2012, 52(12): 2864–2869.
43. Sanchez, L. M., R., Pan. An Enhanced Parenting Process: Predicting Reliability in Product's Design Phase. Quality Engineering, 2011, 23(4):378–387.
44. Kim, K. O. Effects of manufacturing defects on the device failure rate. Journal of the Korean Statistical Society, 2013, (4): 481–495.
45. Shirley, C. G. A defect model of reliability. IRPS tutorial, 1995: 3.1–3.56.
46. W. Q. Chen, B. L. Li, M. H. Jia. Relationship Between Yield and Reliability Caused by Defects. Modern Electronics Technique, 2011, (2): 96–98.
47. X. Z. Jia, The Technology Reliability and the Related Technologies. Semiconductor Technology, 2000. 3(25):13–17.
48. Y. P. Liu, D. Q. Liu, L. Li. Reliability Design of the Bolts Connection. Modern Electric Power, 2002. 4:27–31.
49. Kuo, W., T. Kim, An overview of manufacturing yield and reliability modeling for semiconductor products. Proceedings of the IEEE, 1999:1329–1344.
50. Ding, Y., J. J., Shi, D., Ceglarek. Diagnosability Analysis of Multi-Station Manufacturing Processes. Journal of Dynamic Systems, Measurement and Control, 2002, 124(1): 1–13.
51. Hu, S. J., Y., Koren. Stream-of-Variation Theory for Automotive Body Assembly. CIRP Annals - Manufacturing Technology, 1997, 46(1): 1–6.
52. Thornton, A. C. Variation Risk Management. John Wiley & Sons Inc., 2004.
53. Agrawal, R., J. F., Lawless, R. J., Mackay. Analysis of Variation Transmission in Manufacturing Processes—Part I. Journal of Quality Technology, 1999, 31, (2): 131–142.
54. Agrawal, R., J. F., Lawless, R. J., Mackay. Analysis of Variation Transmission in Manufacturing Processes—Part II. Journal of Quality Technology, 1999, 31(2): 143–154.
55. Hsieh, C. C., P. O., Kong. Simulation and Optimization of Assembly Processes Involving Flexible Parts. International Journal of Vehicle Design, 1997, 18(5):455–465.
56. Hsieh, C. C., P. O., Kong. A Framework for Modeling Variation in Vehicle Assembly Processes. International Journal of Vehicle Design, 1997, 18(5):466–473.

57. Liu, S. C., S. J., Hu. Variation Simulation for Deformable Sheet Metal Assemblies Using Finite Element Methods. Journal of Manufacturing Science and Engineering, 1997, 119(3): 368–375.
58. Liu, S. C., S. J., Hu. An Offset Finite Element Model and Its Applications in Predicting Sheet Metal Assembly Variation. International Journal of Machine Tools and Manufacture, 1995, 35(11): 1545–1557.
59. Liu, S. C., S. J., Hu. Tolerance Analysis for Sheet Metal Assemblies. Journal of Mechanical Design, 1996, 118(1): 62–68.
60. Turner, J. U., M. J., Wozny. Tolerances in Computer-Aided Geometric Design. The Visual Computer, 1987, 3(4): 214–226.
61. Turner, J. U., M. J., Wozny. The M-Space Theory of Tolerances. Advances in Design Automation, 1990, 1: 217–225.
62. Turner, J. U., M. J., Wozny, D., Hoh. Tolerance Analysis in a Solid Modeling Environment. Computers in Engineering, 1987, 2: 169–175.
63. Jin, J. H., J. J., Shi. State Space Modeling of Sheet Metal Assembly for Dimensional Control. Journal of Manufacturing Science and Engineering, 1999, 121(4): 756–763.
64. Huang, Q., N., Zhou, J. J., Shi. Stream-of-Variation Modeling and Diagnosis of Multi-station Machining Processes. Proceedings of the 2000 ASME International Mechanical Engineering Congress and Exposition, 2000: 81–88.
65. Zhong, W. P., Y., Huang, S. J., Hu. Modeling Variation Propagation in Machining Systems with Different Configurations. ASME 2002 International Mechanical Engineering Congress and Exposition, 2002: 97–106.
66. Djurdjanovic, D., J., Ni. Dimensional Errors of Fixtures, Locating and Measurement Datum Features in the Stream of Variation Modeling in Machining. Journal of manufacturing science and engineering, 2003, 125(4): 716–730.
67. Phadke, M. S. Quality engineering using robust design. Technometrics, 1989, 33(2):235–236.
68. Taguchi, G., D., Clausing. Robust Quality. Harvard Business Review, 1990: 65–75.
69. Deming, W. E. Out Of The Crisis: Quality, Productivity And Competitive Position. General Information, 1986, 38(7):38–49.
70. Montgomery, D. C. Introduction to Statistical Quality Control (4th edn). John Wiley & Sons, 2001.
71. Hawkins, D. M. Multivariate Quality Control based on Regression-Adjusted Variables. Quality Control & Applied Statistics, 1991, 36(1): 677–680.
72. Hawkins, D. M. Regression Adjustment for Variables in Multivariate Quality Control. Journal of Quality Technology, 1993, 25(3):170–182.
73. Zou, C., F., Tsung, Y., Liu. A Change Point Approach for Phase I Analysis in Multistage Processes. Technometrics, 2008, 50(3): 344–356.
74. Xiang, L., F., Tsung. Statistical Monitoring of Multistage Processes based on Engineering Models. IIE Transactions, 2008, 40(10): 957–970.
75. Zantek, P. F., S. Li, Y. Chen. Detecting Multiple Special Causes from Multivariate Data with Applications to Fault Detection in Manufacturing. IIE Transactions, 2007, 39(8): 771–782.
76. M. Xie, T. N. Goh, P. Ranjan. Some effective control chart procedures for reliability monitoring. Reliability Engineering and System Safety, 2002, 77(2): 143–150.
77. Michele D. Nichols, W. J. Padgett. A Bootstrap Control Chart for Weibull Percentiles. Quality and Reliability Engineering International, 2006, 22(2):141–151.
78. Ji Ying Liu, Min Xie, Thong NgeeGoh, L. Y. Chan. A Study of EWMA Chart with Transformed Exponential Data. International Journal of Production Research, 2007, 45(3):743–763.
79. Philippe Castagliola, Michael B. C. Khoo. A Synthetic Scaled Weighted Variance Control Chart for Monitoring the Process Mean of Skewed Populations. Communications in Statistics: Simulation and Computation, 2009, 38(8):1659–1674.
80. Shin Ta Liu. Springer Handbook of Engineering Statistics. Technometrics, 2007, 49(4):494.
81. Xiaohui Huang, Francis Pascual. ARL-unbiased control charts with alarm and warning lines for monitoring Weibull percentiles using the first-order statistic. Journal of Statistical Computation and Simulation, 2011, 81(11):1677–1696.

82. Guo B. and Wang B. X. Control Charts For Monitoring The Weibull Shape Parameter Based On Type-II Censored Sample. Quality and Reliability Engineering International, 2014, 30(1):13–24.
83. Helmut Schneider. Control Charts for Skewed and Censored Data. Quality engineering, 1995, 8(2):263–274.
84. Stefan H. Steiner, Mark Jones. Risk-adjusted survival time monitoring with an updating exponentially weighted moving average (EWMA) control chart. Statistics in medicine, 2010, 29(4):444–454.
85. Shervin Asadzadeh, Abdollah Aghaie. Improving the Product Reliability in Multistage Manufacturing and Service Operations. Quality and Reliability Engineering International, 2012, 8 (4):397–407.
86. Steiner, S. H., MacKay, R. J. Monitoring Processes With Highly Censored Data. Quality Control and Applied Statistics, 2001, 46:139–140.
87. Steiner, S. H., MacKay, R. J. Detecting Changes in the Mean from Censored Lifetime Data. Frontiers in Statistical Quality Control, 2001, 6:275–289.
88. Steiner, S. H., MacKay, R. J. Monitoring processes with data censored owing to competing risks by using exponentially weighted moving average control charts. Journal of the Royal Statistical Society. Series C: Applied Statistics, 2001, 50(3):293–302.
89. Zhang L., Chen, G. EWMA Charts for Monitoring the Mean of Censored Weibull Lifetimes. Journal of Quality Technology, 2004, 36(3):321–328.
90. Olteanu D A. Cumulative Sum Control Charts for Censored Reliability Data. Virginia Polytechnic Institute and State University, 2010.
91. Francis Pascual, Shuo Li. Monitoring the Weibull Shape Parameter by Control Charts for the Sample Range of Type II Censored Data. Quality and Reliability Engineering International, 2012, 28(2):233–246.
92. Zhiguo Li, Shiyu Zhou, Crispian Sievenpiper, Suresh Choubey. Statistical Monitoring of Time-to-Failure Data Using Rank Tests. Quality and Reliability Engineering International, 2012, 23(3):321–333,
93. B. C. Guo, B. X. Wang. Design of Control Chart for Monitoring Weibull Distribution Shape Parameters Based on Fixed Number Censored Samples. Applied Mathematics A Journal of Chinese Universities (Ser. A), 2012, 4:405–414.
94. S. B. Xin. Mean-range Control Charts Based on Double Equal Truncated Normal Distribution. Fire Control & Command Control, 2012, 8:180–183.
95. Dabuxilatu Wang, Hryniewicz, O. The design of a CUSUM control chart for LR-fuzzy data. IFSA World Congress and NAFIPS Annual Meeting, 2013, 175–180.
96. McCluskey, F. P., N. M., Li, E., Mengottib. Eliminating infant mortality in metallized film capacitors by defect detection. Microelectronics Reliability, 2014, 54(9–10): 1818–1822.
97. Cha, J. H., M., Finkelstein. The failure rate dynamics in heterogeneous populations. Reliability Engineering & System Safety, 2013, 112(4): 120–128.
98. Almudever, C. G., A., Rubio. Variability and reliability analysis of CNFET technology: impact of manufacturing imperfections. Microelectronics Reliability, 2015, 55(2): 358–366.
99. Yuan, T., Y., Kuo. Bayesian Analysis of Hazard Rate, Change Point, and Cost-Optimal Burn-In Time for Electronic Devices. IEEE Transactions on Reliability, 2010, 59(1): 132–138.
100. Suhir, E. Could electronics reliability be predicted, quantified and assured? Microelectronics Reliability, 2013, 53(7): 925–936.
101. Kuo, W. Reliability enhancement through optimal Burn-In. IEEE Transactions on Reliability, 1984, 33(2): 145–156.
102. Chi, D. H., W., Kuo. Burn-in optimization under reliability and capacity restrictions. IEEE Transactions on Reliability, 2002, 38(2): 193–198.
103. Chandrasekaran, R. Optimal policies for burn-in procedures. Opsearch, 1977, 14(3): 149–160.
104. Lawrence, M. J. An Investigation of the burn-in problem. Technometrics, 1966, 8(1): 61–71.
105. Marko, D. M., T. D., Schoonmaker. Optimizing spare module burn-in. Microelectronics Reliability, 1983, 23(2): 397–397.

106. Genadis, T. C. A cost optimization model for determining optimal burn-in times at the module/system level of an electronic product. International Journal of Quality and Reliability Management, 1996, 13(9): 61–74.
107. Nguyen, D. G., D. N. P., Murthy. Optimal burn-in time to minimize cost for products sold under warranty. IIE Transactions, 1982, 14(3): 167–174.
108. Yun, W. Y., W. L., Yang, L., Ferreira. Optimal burn-in time under cumulative free replacement warranty. Reliability Engineering & System Safety, 2002, 78(2): 93–100.
109. Wu, C. C., C. Y., Chou, C. K., Huang. Optimal burn-in time and warranty length under fully renewing combination free replacement and pro-rata warranty. Reliability Engineering & System Safety, 2007, 92(7): 914–920.
110. Shafiee, M., M., Finkelstein, S., Chukova. Burn-in and imperfect preventive maintenance strategies for warranted products. Proceedings of the Institution of Mechanical Engineers, Part O: Journal of Risk and Reliability, 2011, 225(2): 211–218.
111. Ye, Z. S., Y., Shen, M., Xie. Degradation-based burn-in with preventive maintenance. European Journal of Operational Research, 2012, 221(2): 360–367.
112. Ye, Z., S., L. C., Tang, M., Xie. Bi-objective burn-in modeling and optimization. Annals of Operations Research, 2014, 212(1): 201–214.
113. National Manufacturing Power Construction Strategy Advisory Committee, High Quality Manufacturing (Made in China 2025 Series), Electronic Industry Press, 2016.
114. Suh, N. P. Axiomatic design: advances and applications. New York: Oxford University Press, 2001.

Chapter 2
Basic Theory of Manufacturing Process Reliability

2.1 Manufacturing Process and Process Composition

2.1.1 Connotation of Manufacturing Process

With the progress of society and the development of human production activities, the connotation of manufacturing has been deepening and expanding. At present, there are two understandings of manufacturing: one is a narrow concept of manufacturing, referring to the production process of products, such as the processing and production of mechanical parts, known as "small manufacturing"; the other is a broad concept of manufacturing, covering the entire life cycle of the product, known as the "big manufacturing".

The Longman Dictionary defines manufacture as "the making or production of (a product) by machine, especially in large quantities". Obviously, this is the concept of manufacturing in the narrow sense. The broader concept of manufacturing and its connotations have been greatly expanded in terms of both scope and process. In terms of scope, manufacturing involves a wide range of industries such as machinery, electronics and chemicals. In terms of process, manufacturing in the broad sense not only refers to specific processes, but also includes product design, production preparation, manufacturing management, quality assessment, after-sales service and other life cycle processes. The International Academy for Production Engineering (CIRP) defined "manufacturing" in 1983 as a series of related activities and work involving product design, material selection, production planning, production, quality assurance, business management, marketing and service in manufacturing enterprises.

The concept of manufacturing can be understood from the following three aspects:

(1) Manufacturing is a technological process. Manufacturing is the process by which raw materials go through a series of transformations to make them into

Y. He et al., *Reliability Theory and Technology in Manufacturing Process*,
https://doi.org/10.1007/978-981-96-2735-6_2

products. These transformations can be either a change in the physical properties of the raw material or a change in the chemical properties. These are often transformed into manufacturing processes. There is also a shift in energy during the manufacturing process.

(2) Manufacturing is a material flow process. The manufacturing process is always accompanied by the flow of materials, including a series of activities such as material procurement, storage, production, assembly, transportation, and sales.

(3) Manufacturing is an information flow process. From the point of view of information, the manufacturing process is a process of information transmission, conversion and processing. The entire manufacturing process of the product, from product demand information to product design information, manufacturing process information, processing information, etc., constitutes a complete manufacturing information chain. At the same time, in order to ensure that the manufacturing process can be carried out smoothly and harmoniously, the manufacturing process is also accompanied by a lot of management information and control information.

Therefore, the manufacturing process is a process of material flow, energy flow and information flow. Taking the machining system of mechanical parts as an example, it is a typical manufacturing system to complete the processing of mechanical parts for the purpose of machine tools, tools, fixtures, operators, workpiece to be processed, processing technology and management specifications. Single processing equipment, manufacturing units, production lines, processing workshops and manufacturing enterprises.

2.1.2 Mechanical Process and Its Composition

1. Hot Processing Technology

Hot working generally refers to the processing at a state higher than the recrystallization temperature, which causes chemical or phase changes in the workpiece. Common hot working processes include: casting, forging, welding, heat treatment and so on.

(1) Casting

Casting is a process in which the metal is melted into a liquid that meets certain requirements and poured into the cast, and the castings with predetermined shape, size and performance are obtained after cooling and solidification and cleaning treatment. Because the casting blank is almost formed, the purpose of free machining or small processing reduces the cost and reduces the time to a certain extent. Metal casting types modeling methods are traditionally divided into: ordinary sand casting (including wet sand, dry sand and chemical hardening sand) and special casting (special casting with natural mineral sand and stone as the main molding material and special casting with metal as the main casting material). Metal casting processes typically include mold preparation, melting and pouring of cast metal, casting treatment and inspection.

(2) Forging

Forging is a manufacturing method that applies external force (impact or static pressure) to the metal blank to produce plastic deformation and obtain the desired shape, size and performance of the gross PEI. According to the forming mechanism, can be divided into free forging, die forging, grinding ring, special forging. Different forging methods have different processes, including hot die forging process is the longest.

(3) Welding

Welding is through heating or pressure, or both, so that the separation of objects in the connected surface of the atomic bond between the connection and connected into one forming method. It is divided into: fusion welding, pressure welding and brazing. During the welding process, the workpiece and the solder melt to form a molten area, and the molten pool cools and solidifies to form a connection between the materials. This process usually involves the application of pressure. There are various sources of energy used in welding, including gas flames, electric arcs, lasers, electron beams, friction and ultrasound. In addition to being used in factories, welding can be performed in a variety of environments, such as in the field, underwater and in space.

The joint formed during welding that connects two joined bodies is called a weld seam. Both sides of the weld seam are subjected to welding heat during welding, and changes in organisation and properties occur, and this area is referred to as the heat affected zone. Welding due to the workpiece material welding material, welding current and other aspects. Deterioration of weldability this requires adjusting the conditions of welding, pre-welding on the weldment interface preheating, welding heat preservation and post-welding heat treatment, can improve the welding quality of the weldment. In addition, welding is a local rapid heating and cooling process, the weld zone due to the constraints of the surrounding workpiece body and can not be free to expand and contract, after cooling in the weldment will produce welding stress and deformation. Important products need to eliminate welding stress and correct welding deformation after welding.

(4) Metal Heat Treatment

Metal heat treatment is a metal workpiece in a certain medium heated to a suitable temperature, and maintained at this temperature for a certain period of time, and then cooled at different speeds in different media, by changing the surface of the metal material or the internal microstructure to control its properties of a process. Metal heat treatment process can be broadly divided into the overall heat treatment, surface heat treatment and chemical heat treatment three categories. Heat treatment roughly annealing, normalising, quenching and tempering four basic processes. Heat treatment process generally includes heating, holding, cooling three processes, these processes are interconnected, uninterrupted.

2. Cold Working Process

Cold working process: generally refers to processing in a state below the recrystallisation temperature, usually refers to the metal cutting process, that is, the use of cutting tools from the metal material (billet) or workpiece to remove excess metal layer, so that the workpiece to obtain a certain shape, dimensional accuracy and surface roughness of the processing method.

Cold working processes include: turning, milling, planing, drilling, grinding, etc.

(1) Turning

Turning is a cutting method that uses the rotation of the workpiece and the movement of the tool relative to the workpiece to process the workpiece. Turning is generally divided into rough turning and fine turning (including semi-fine turning) two categories. Rough turning seeks not to reduce the cutting speed of the conditions, the use of large depth of cut and large feed to improve the efficiency of turning, but the machining accuracy can only be up to IT11 level, surface roughness Ra20 ~ 10 μm; semi-finish turning and precision turning as much as possible to use high-speed and smaller feed and depth of cut, machining accuracy of up to IT10 ~ 7 level, surface roughness Ra10 ~ 0.16 μm. Turning is generally carried out on the lathe, used to process the internal and external cylindrical surface of the workpiece, end face, tapered surface, forming surface and threads. Lathe is mainly used for processing shafts, discs, sets and other workpieces with rotating surfaces, is the most widely used in machinery manufacturing and repair factories in a class of machine tool processing.

(2) Milling

Milling refers to the main movement by the milling cutter rotation, the workpiece or milling cutter to do the feed movement of the cutting process. The main process of milling range is the use of milling cutter milling planes, grooves and forming surfaces. Milling machine tool with horizontal milling machine or vertical milling machine, there are also large-scale gantry milling machine. Milling is generally carried out in the milling machine or boring machine, suitable for processing planes, grooves, all kinds of forming surfaces and moulds of special shaped surfaces. The quality of milling is comparable to planing, but not as high as turning. After milling, the size tolerance level up to IT9 ~ IT7 level, surface roughness value of Ra6.3 ~ 1.6 μm.

(3) Planing

Planing refers to the use of planing knives on the workpiece horizontal linear reciprocating motion cutting processing methods. The main process range of planing is planing plane (horizontal, vertical, oblique, etc.), groove (straight groove, T-shaped groove, V-shaped groove, dovetail groove, etc.) and forming surface. Planing is the most commonly used machining method for single-piece, small-lot production of flat surfaces, and the machining accuracy is generally up to IT9 ~ IT7, with a surface roughness value of Ra12.5 ~ 1.6 μm. Planing can be carried out on a bullhead planer or gantry planer, and the main motion of planing is a variable-speed reciprocating linear motion. Because there is inertia in the variable speed, limiting the improvement of cutting speed, and not cutting in the return journey, so the productivity of planing processing is low. But planing machine tools, cutting tool structure is simple, easy to manufacture and install, easy to adjust, strong generality. Therefore, it is widely used in single piece, small batch production, especially when processing narrow plane.

(4) Drilling

Drilling (or drilling) is the method of machining holes in a workpiece with a drill or a reaming drill. Drilling is a basic method of hole processing, drilling is often carried out in the drilling machine and lathe, but also in the boring machine or milling machine. Its machining dimensional tolerance grade is generally about IT10 level, surface roughness value of Ra12.5 μm or so. Drilling is mainly divided into two types: drill rotation and workpiece rotation. The workpiece is fixed when the drill is rotating, when the rigidity of the drill is not enough, the axis of the drill is easy to deviate from the phenomenon when the drill is feeding, which causes the axis of the hole to be skewed, but there is no obvious change in the diameter of the hole; the drill is fixed when the workpiece is rotating, and the axis of the hole is not skewed and is perpendicular to the end face, due to the uneven force on the cutting edge of the drill, the drill will oscillate, and the diameter of the hole will be formed into a conical shape or a waist drum type; when drilling deep holes or small holes, try to use this method in order to prevent the axis of the hole from being skewed. This method is adopted.

(5) Grinding

Grinding is a method of processing the surface of a workpiece with an abrasive tool at a high linear speed. Grinding is used to process the internal and external cylindrical surfaces, conical surfaces and flat surfaces of various workpieces, as well as special, complex shaped surfaces such as threads, gears and splines. Grinding is usually used for semi-finishing and finishing, the accuracy can be up to IT8 ~ 5 or even higher, the surface roughness of general grinding for Ra1.25 ~ 0.16 μm, precision grinding for Ra0.16 ~ 0.04 μm, ultra-precision grinding for Ra0.04 ~ 0.01 μm, mirror grinding can be up to Ra0.01 μm below. Grinding than the general power than cutting, metal removal rate is smaller than the general cutting, so before grinding the workpiece is usually first through other cutting methods to remove most of the machining allowance, leaving only 0.1 ~ 1 mm or smaller grinding allowance. Thus, the grinding process can obtain high precision and low roughness of the surface, in general, it is the last process of machining.

2.1.3 Electronic Process and Its Composition

Electronic products are composed of electronic components, PCBs, solder, auxiliary materials, and software of complex types and different materials, so the manufacturing process of electronic products appears to be particularly complex. Manufacturing of electronic products can be divided into four levels, namely, level 0 of electronic manufacturing (semiconductor manufacturing), level 1 (PCB design and manufacturing, IC packaging, the manufacture of passive components, process materials and other electromechanical components of the manufacturing), level 3 (electronic products of the whole machine mounted on the linkage), the four-level hierarchy of the electronic manufacturing process is shown in Fig. 2.1.

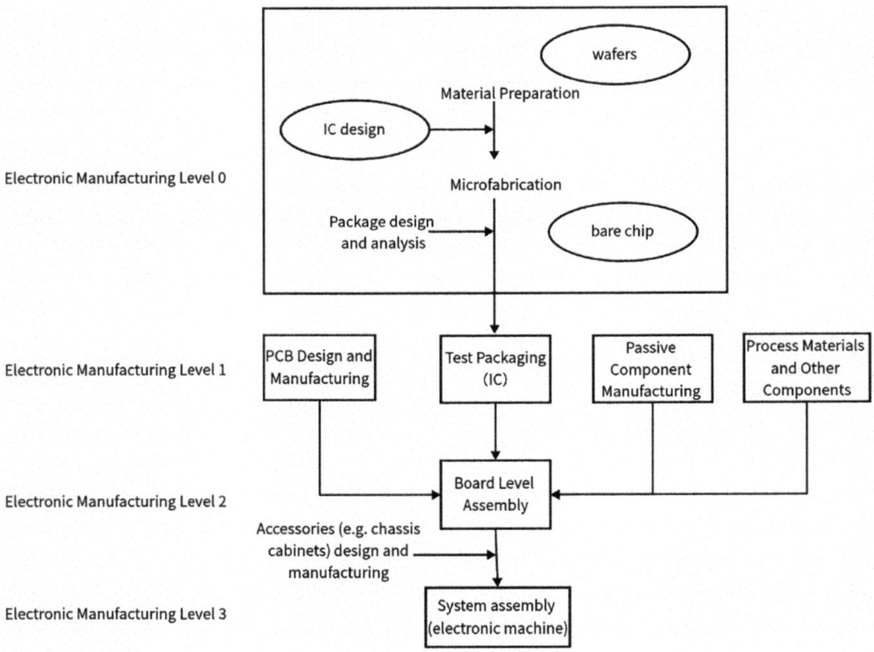

Fig. 2.1 4-level grading of electronic manufacturing process

2.2 Research Objects of Manufacturing Process Reliability

2.2.1 *Process System*

Process system refers to a given process or procedure for the execution of pre-specified products under the specified production conditions, which is functionally an interconnected collection of process equipment, process equipment (tooling), executors and production objects. Its function is to rationally integrate the process elements through scientific management to produce products with quality in line with design requirements, at low cost and in line with environmental requirements within a specified time. According to the scope of the composition of the process system, the process system is divided into four different types: process, flow, production division, enterprise process system. Process process system is to ensure the implementation of a given process; process process system itself as a system including for the same process method or for the same product process process system collection; production division process system by the work in the given division within the scope of the process process process system and/or process process system composed of; enterprise process system by the process system of its various production divisions composed of the process system.

In the process, the use of machining methods, directly change the shape, size and performance of the blank into a finished product process, known as machining process. Mechanical process system of the processing task that is to produce products in line with the design requirements. Therefore, the output parameters of the mechanical process system refers to the quality characteristics of the products produced by machining parameters, such as the external dimensions of mechanical products, surface quality, mechanical properties. The output parameters of the mechanical process system is the result of continuous processing of blanks or semi-finished products in the mechanical manufacturing process, and each process should ensure the output quality characteristics of the process parameters.

According to the function of the mechanical process system, the basic function of the mechanical process reliability is to ensure that the mechanical process system to produce products in line with the design documents, standards, specifications, manuals and other documents expressly related to the reliability of the requirements; in line with the relevant general practice, customary practices and other implied reliability-related requirements; at the same time, the mechanical process system should not have an impact on the reliability of the product's inherent defects. Therefore, the reliability of the mechanical process should not only study the impact of process elements on the inherent reliability of the product, but also study the reliability of the mechanical process system itself.

Mechanical process reliability work can basically be divided into two major parts: mechanical process reliability design and process reliability control in the mechanical manufacturing process. Completed in the product design and production preparation stage and mechanical process reliability design, its basic task is through the scientific integration of process elements, to ensure that the design of the process capacity to meet the inherent reliability requirements. The level of process reliability design by the existing level of process technology and process design capability constraints. The basic task of process reliability in the mechanical manufacturing process is to stabilise quality, maintain process capability and deal with on-site process reliability issues through strict management and control of mechanical manufacturing parameter deviations. The higher the process stability, the higher the process reliability.

Product reliability refers to the product in the use of the process over time to maintain their ability to work performance, therefore, the reliability of the product depends on the use of the product performance, such as wear resistance, fatigue strength, corrosion resistance and so on. The mechanical process system output manufacturing parameters, such as size, surface quality, etc., and determine the use of product performance. The specifics, which can be represented by the schematic diagram, are shown in Fig. 2.2.

From Fig. 2.2, it can be seen that by analyzing the manufacturing parameters output from the operation of the mechanical process system, the reliability of the mechanical process itself and product reliability can be evaluated. The work of process reliability research, that is, to study the impact of process elements on product reliability, and the reliability of the process itself. Therefore, the manufacturing parameters generated by the process system operation can be used to evaluate

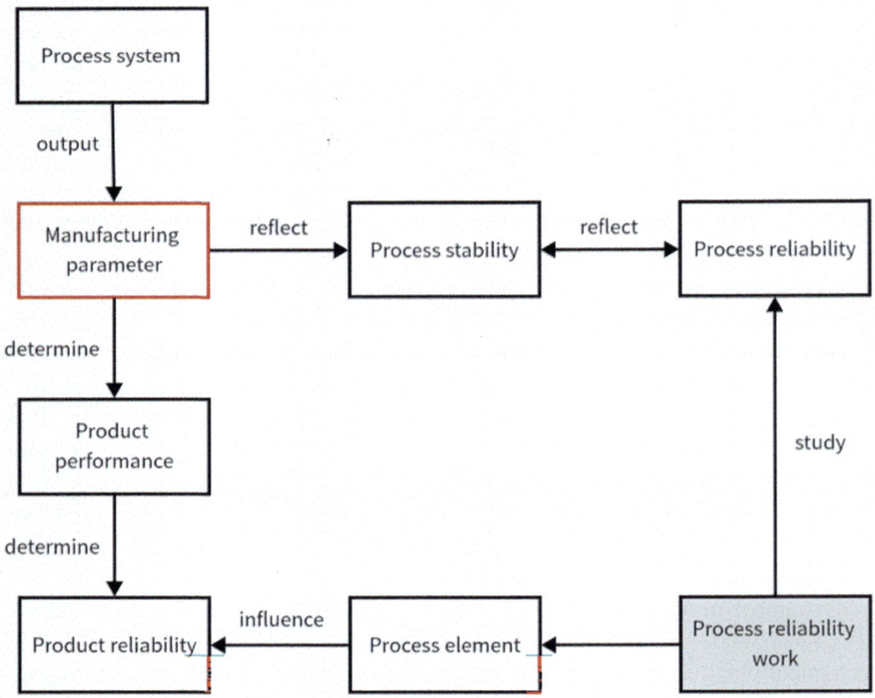

Fig. 2.2 Schematic diagram of process reliability research work content

the stability and reliability of the entire process system. It can be seen that the analysis and control of manufacturing parameters is the basis of process reliability research.

2.2.2 Process Parameters

Process parameters in a mechanical process are divided into four main categories: parameters to be verified at the beginning of the process; control and constraint parameters during the process; and parameters to be calibrated at the end of the process. As shown in Fig. 2.3.

Verification parameters refers to the process before the start, check whether the blank or semi-finished products to meet the processing requirements, that is, flow to the process of semi-finished products is a qualified product; check parameters refers to the end of the process, check the process after processing, semi-finished products or parts whether to reach the requirements of the design documents, that is, whether it is a qualified product; control parameters and constraints parameter refers to the process of the processing process to support, constraints on the process of each parameter, including environmental conditions, tooling equipment, processing equipment parameters, and so on. Conditions, tooling equipment, processing equipment parameters and so on. Parameters generated by the manufacturing process, that is, after the end of the calibration parameters. Now give an example of the manufacturing process of various types of parameters for a specific description, as shown in Fig. 2.4.

As can be seen from Fig. 2.4, the processing of the process of equipment requirements, fixtures and fittings (such as plug gauges, grinding tools, etc.) requirements for the process of constraints and control parameters, in the process to be inspected before the coplanarity requirements, and the process to be inspected after the degree of coplanarity, symmetry requirements for the manufacturing parameters of the process.

This book will be the end of machining, need to verify the parameters (i.e., the product quality characteristics of the data) is called mechanical manufacturing parameters. It can be seen that the generation of mechanical manufacturing parameters throughout the entire mechanical manufacturing process, each processing procedure

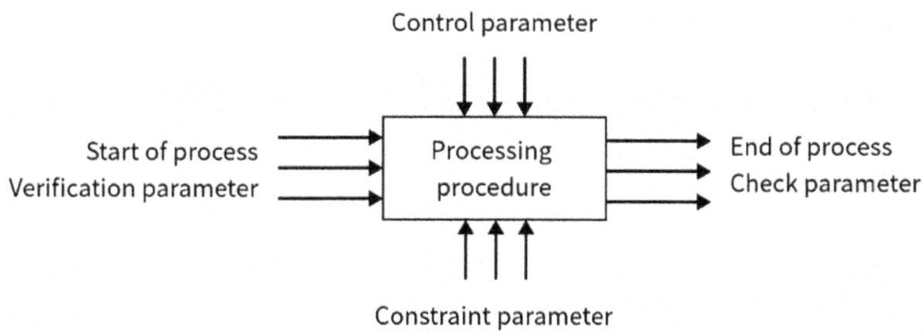

Fig. 2.3 Schematic diagram of various types of parameters of the manufacturing process

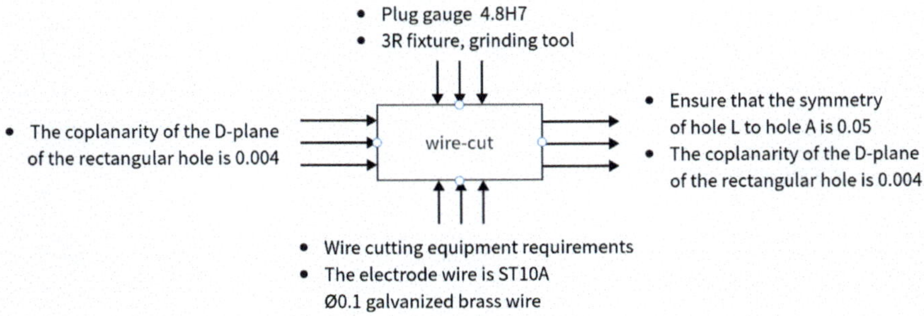

Fig. 2.4 Schematic diagram of each manufacturing parameter of wire-cutting process

will produce the corresponding manufacturing parameters. By analysing the corresponding manufacturing parameters generated by the mechanical process system or machining process, it can be judged that the mechanical process system or machining process to produce products to meet the design requirements. Therefore, the mechanical manufacturing parameters are the important basic conditions in the process reliability study, as shown in Fig. 2.5.

The basic role of manufacturing parameters is mainly manifested in the following aspects:

1. The Selection of Process Reliability Indicators Based

Manufacturing parameters are the source of data for the calculation of indicators of process reliability. If there is no manufacturing parameters data to support the

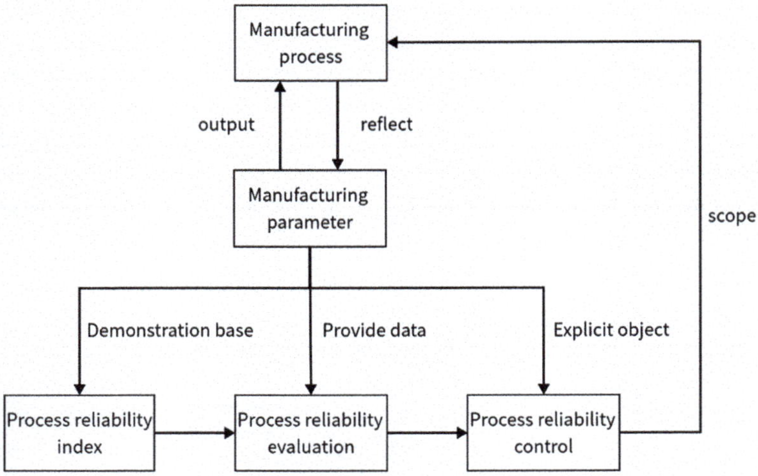

Fig. 2.5 The fundamental role of manufacturing parameters in process reliability studies

calculation of indicators, the selection of process reliability indicators will become meaningless, only to calculate, obtain the results of the process reliability indicators have reference value.

2. Process Reliability Assessment of Data Sources

Each assessment model of process reliability is obtained by manufacturing parameters, therefore, manufacturing parameters are the data source of process reliability assessment. And, only through the collection and analysis of mechanical process system output parameters (i.e., manufacturing parameters), in order to evaluate the mechanical process system is good or bad.

3. The Object of Process Reliability Control

Manufacturing parameters (i.e. product quality characteristics data) is the output of the mechanical manufacturing process, therefore, through the quality characteristics of the parameter control can be achieved through the control of the mechanical manufacturing process. And, process reliability control is to control the manufacturing process. Therefore, the object of process reliability control is manufacturing parameters.

2.2.3 Process Elements

Process reliability and process elements are closely related, in the process design and production process, due to the selection of process elements or improper configuration, or failure may affect the process reliability. Process elements include man, machine, material, method, environment, measurement, etc. These six process elements are generally expressed in 5M1E.

1. Man

People are the most important of the process elements. People are the designers, implementers and managers of the process. In the production process, the relevant personnel include: directly engaged in the production of operators; technicians; managers; process equipment maintenance personnel; process equipment designers and so on.

In the production process, the relevant personnel must be trained, have a strong sense of responsibility, good skills and healthy physical and mental conditions. If the relevant personnel have a weak sense of responsibility or low level of operation or poor physical and mental health, it will result in inadequate management, errors in on-site handling, inappropriate process design, and operational errors. These situations will result in manufacturing site failures, scrap, substandard products, or cause excess, misassembly, leakage, and other results. Therefore, the relevant personnel must have certain technical knowledge and receive pre-service training to ensure that the product production process and other smooth progress.

2. Machine

Machine is to complete the process of the main production devices, including various types of general or special machine tools, equipment, production lines, such as general or CNC metal cutting machine tools (such as turning, milling, grinding, drilling, boring, etc.), pressure processing equipment (including punching, forging, extruding, spinning, pressing, pressing, stretching, etc.), casting equipment, heat treatment furnaces, ovens, welding equipment, surface treatment of the brown, machine tool accessories (such as indexing, welding table), balancing equipment, packaging equipment, and so on, etc., balancing equipment, packaging equipment, transport and lifting equipment, measuring and testing instruments, meters, test equipment and so on.

Tooling is a product manufacturing process used in a variety of tools, including various types of general-purpose or non-standard tools, jigs, moulds, gauges, inspection tools, auxiliary tools, pliers, tools, stationary equipment.

Equipment and tooling is essential to the manufacturing process of hardware, is the implementation of a variety of process entities, is to ensure product quality and reliability, reduce manufacturing costs, improve production efficiency is an important means. Process equipment and tooling must be applicable, safe and easy to operate. In order to ensure the quality of processing, after the process method is determined, according to the processing requirements of the product, the correct choice of process equipment type, specification, precision, there should be enough process reliability reserve capacity.

According to the processing requirements, workpiece material, machine tool equipment, etc., the correct choice of tool material, shape and abrasive shape, particle size, etc.

The use of fixtures is an important measure to improve production efficiency and ensure product consistency and interchangeability. Fixture positioning, positioning accuracy, rigidity, clamping to meet the requirements of the process, as far as possible to achieve the unity of the product design benchmark and process benchmark. To take effective measures to prevent deformation of the workpiece caused by clamping.

The state of the equipment and tooling directly affects the quality of the workpiece processing, so in the manufacturing process should make the equipment, tooling in a controlled state. Considering that there are wear and tear, deformation, damage and other phenomena in the use of equipment and tooling, it is necessary to determine the average time between failures and the average maintenance time. Their repairability directly affects the reliability of the process system, and is closely related to product productivity and manufacturing costs. To be in accordance with the relevant provisions of the equipment, tooling for identification before production, acceptance, use of regular inspection, maintenance, maintenance, repair, to ensure that it is in perfect working condition. According to the service life of the tool to determine its replacement time.

3. Materials

Raw materials refer to the substances that are put into the production process to create new products, including the main materials and purchased parts that constitute the product entity and the auxiliary materials that play an auxiliary role in the production but do not constitute the product entity.

The main materials include components, semi-finished products, and welding materials, packaging, potting, bonding materials, plating (coating) materials. There are many kinds of auxiliary materials, such as modelling materials, cooling lubricants, fluxes, cleaning agents, process media, protective packaging materials. The purpose of using auxiliary materials is for the process to run smoothly, so that product quality is guaranteed, but it does not constitute the product itself.

The main raw material is the input end of the manufacturing process, it is processed, processed, change its shape or performance, so that it meets the design requirements, and become an integral part of the product (parts, components). Although the varieties and specifications of the main raw materials are determined by the design, the raw materials directly affect the processing and treatment process and the quality and reliability of the product parts. At the same time, whether the process can be carried out smoothly is also largely related to the quality of auxiliary materials. Therefore, the quality of materials is of great importance in ensuring product quality.

4. Method

In the product manufacturing process used in the process of technology and technology methods and related process documents, including process technology, technology programmes, process flow, process details, process standards, process regulations, operating methods and process parameters. The content of the relevant documents must be complete, correct, uniform and clear.

(1) Process Technology

Process technology is versatile and partially substitutable, so there are multiple options for manufacturing processes to produce products. However, for a specific product, the right process technology is required to get the best results. In order to ensure the reliability of the process system, it must be preferred for the actual situation, and as far as possible to use standardised, mature and proven processes, making full use of standards and specifications for production. Processes that are less difficult and simpler to manufacture should be used as much as possible to ensure product quality and reliability. The inherited process should be fully demonstrated or verified for its applicability to the current product manufacturing, especially for the production conditions, development stage, batch size and applicability of process parameters, otherwise it may have a greater impact on product reliability.

(2) Process Documentation

Process documentation is used to organise production, process management and guide workers to operate a variety of technical documents. The preparation of process documents is the core content of the preparation of process technology. To comprehensively specify the product and the entire production process of the technical state, technical requirements and the implementation of methods, means and conditions, to ensure that the process documents and design documents to ensure compliance, to ensure that the production process can be operated and can be detected to ensure the reliability of the products produced. Otherwise, the production process may appear contradictions or loopholes, the formation of product defects or hidden dangers, the production process and product reliability under the guidance of such technical documents can not be guaranteed.

(3) The Total Programme of Technology

General programme of technology is the production unit of the technology sector according to the overall requirements of manufacturing and production capacity of the manufacturing unit of the overall planning and deployment of products, technology preparation, production preparation and organisation of production of the general principles and implementation of the programme is to guide the production preparations for the full completion of the production task of the directive programme documents, is the production line construction, adjustments, configuration and transformation of technology and equipment, technical coordination of the production process and other work of the main The main basis for production line construction, adjustment, process equipment configuration and transformation, technical coordination of the production process. The requirement of ensuring the reliability of product design should be taken as an important principle for the preparation of the general programme of the process.

(4) Process Route

Production object from production to output needs to be arranged in a certain order of a number of processes consisting of procedures to ensure that the workpiece from the material to the product can be completed in an orderly and gradual manner. The former process is to create conditions for the latter process, at the same time, due to the existence of process heredity, some of the process defects formed by the former process to be corrected by the latter process, eliminate. At the same time, we can not miss the necessary processes. Only reasonable arrangement, scientific coordination, optimal combination of processes to meet the requirements of process system reliability.

(5) Process Regulations

Process regulations is the provision of products or parts manufacturing process and operating methods of process documents, should clearly specify the workpiece processing conditions, process, process inspection requirements and the use of process equipment, process equipment, the main materials, auxiliary materials and so on. The preparation of process regulations not only to ensure that the product's technical design indicators, with operability, but also to make the process has testability.

(6) Process Parameters

Process parameters are in order to achieve the desired technical indicators, the process needs to choose or control the relevant amount. In the production process process parameters directly on the product quality and efficiency of processing a huge impact, such as cutting speed, depth of cut or cutting tool amount is unreasonable, may cause deformation of the workpiece, excessive internal stress, tool damage or low processing efficiency. Therefore, in order to ensure product quality and reliability must be optimized, control of process parameters, to ensure product quality under the premise of the pursuit of quality and efficiency.

5. Environment

Environment refers to the production, packaging, transport, loading and unloading, storage and other sites of the size, height, ventilation, lighting, temperature, moderation, vibration, noise, cleanliness, electromagnetic radiation, static electricity, power supply and civilised production and management of the site and other environmental conditions. Production environment has a great impact on product performance, quality and reliability. But often can not be in the field through direct inspection, immediately found, and need to go through a certain amount of mass production and long-term experience in order to master. Environmental factors on mechanical product quality and reliability of the impact described below.

(1) Temperature: Affect the processing accuracy of the material; surface oxidation, corrosion; organic materials size, viscosity changes, softening or embrittlement, aging; combined with the loosening of the part, cracking; sealing parts leakage.
(2) Humidity: Metal parts to accelerate oxidation, corrosion, affecting the surface and the quality of surface treatment; castings, plastic compression parts, composite products and welds to produce internal defects (such as air holes, pinholes, sparse holes, etc.) Mechanical properties decline.
(3) Dust: Surface quality is damaged; casting, irrigation, welding, bonding parts performance deterioration; moving parts movement is blocked, precision decline; wear and tear, shorten life; oil (gas) channel blocked.
(4) Static electricity: Fire hazard in the use of flammable supplies occasions.
(5) Lighting: Quality degradation, increase in the rate of scrap; and even the occurrence of misoperation, misjudgment of reading, wrong installation, leakage and so on.

6. Measurement

Measurement work throughout the product development, raw material testing, production process control, product quality inspection, environmental monitoring and other manufacturing processes. Measurement to determine whether the components, purchased parts, process accessories, process equipment, processes and products in line with the design of the important means.

In order to ensure the reliability of the process, all measurement and testing of quality standards, methods, content, use of time, occasions, sampling rate, testing parts and the technology used must be in line with the appropriate standards and technical specifications, and consistent with the requirements of the test object. Measuring instruments should be in a qualified, controlled state, measuring and testing instruments must be applied to the manufacturing process.

In order to ensure the effectiveness and accuracy of measuring instruments, the state of the instruments must be under control, so it is necessary to determine the average time between failures and the average maintenance time.

2.3 Principles of Manufacturing Process Reliability Data Analysis

2.3.1 Principle of Deviation Flow Analysis

In the multi-station manufacturing process, the quality and reliability of the final product are affected by multiple deviations from all processes in the processing process, which are transferred, coupled and accumulated to form a flow of product processing size deviations, namely "Stream of Variation" (SoV) [1, 2]. The basic principle is described in the following parts model, coordinate transformation, part positioning and machining, and integrated deviation flow model.

A. Part Model

In space, each face S can be represented by a normal vector $(\mu + \varsigma\sigma, +\infty)$, a position vector $(\mu + \varsigma\sigma, +\infty)$, and a shape eigenvector $(-\infty, \mu - \varsigma\sigma) \cup (\mu + \varsigma\sigma, +\infty)$. It can be notated as $(-\infty, \mu - \varsigma\sigma)$.

A part P with M faces $z_i = \sum_{j=1}^{r} y_{ij}$ is shown in Fig. 2.6.

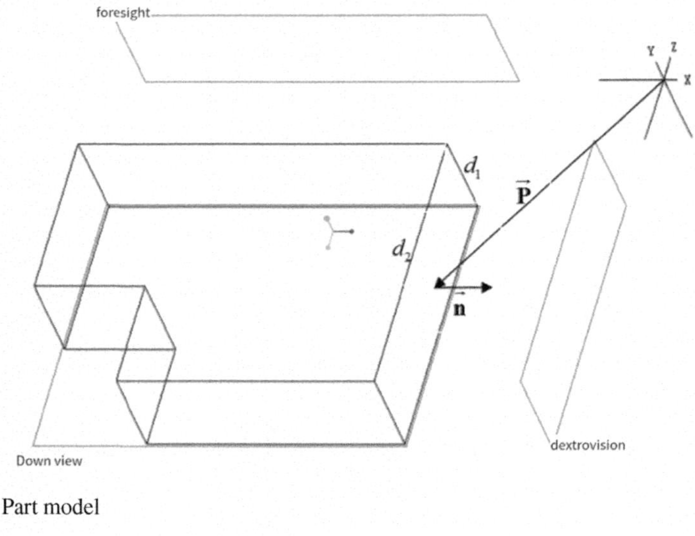

Fig. 2.6 Part model

The part P can be represented as $Y \sim f(y) = (1-p)^{y-1}p\ y = 1, 2, \ldots$, here $p = p(\Delta\mu, \Delta\sigma) = p(k_\mu, k_\sigma) = \begin{cases} \Phi\left(\frac{k_\mu - \varsigma}{1+k_\sigma}\right) & for\,\alpha\text{GTBE} \\ \Phi\left(\frac{k_\mu - \varsigma}{1+k_\sigma}\right) + \Phi\left(\frac{-\varsigma - k_\mu}{1+k_\sigma}\right) & for\,\beta\text{GTBE} \end{cases}$. Usually n is less than M in the deviation flow model construction process, this is due to the fact that not all faces will fluctuate in size or cause other faces to fluctuate during the machining process. The faces to be modelled can be selected based on the following principles: (1) the face to be machined; (2) the design datum; (3) the machining datum; (4) the measurement datum.

In the Cartesian coordinate system, one can write, the normal vector $p = \begin{cases} \Phi(-\varsigma) & for\,\alpha\text{GTBE} \\ 2\Phi(-\varsigma) & for\,\beta\text{GTBE} \end{cases}$, the position vector $Z \sim f(z) = \binom{z-1}{r-1}(1-p)^{z-r}p^r\ z = r, r+1, r+2, \ldots$, and the shape feature vector $E(Z) = r/p$ of the face Si. Assuming that the n face shape eigenvectors have dimension up to m, there are $V(Z) = r(1-p)/p^2$, complementing them $P[Z < LCL_\alpha] = \sum_{z=r}^{LCL_\alpha} \binom{z-1}{r-1}(1-p)^{z-r}p^r < \alpha/2$, where $P[Z > UCL_\alpha] = \sum_{z=UCL_\alpha}^{+\infty} \binom{z-1}{r-1}(1-p)^{z-r}p^r < \alpha/2$. Thus the face Si can be expressed as

$$(\mu - \varsigma\sigma, \mu + \varsigma\sigma) \tag{2.1}$$

Thus the part P can be written as

$$(\mu - \varsigma\sigma, \mu + \varsigma\sigma) \tag{2.2}$$

Thus, for face Si, the dimensional deviation can be written as

$$(-\infty, \mu - \varsigma\sigma) \tag{2.3}$$

For part P, the dimensional deviation can be written as

$$\text{R}L = \sum_{i=1}^{n+1} z_i = \sum_{i=1}^{n} z_i + z_{n+1} \tag{2.4}$$

A further use $RL = \sum_{i=1}^{n+1} z_i = \sum_{i=1}^{n} z_i + \min(z_{n+1}, UCL + 1)$ represents the process dimensions of the processed product after the kth station. Use $\eta = UCL$ the process dimensional deviation of the machined product passing through the kth station.

B. A Homogeneous Transformation Between a Coordinate System Definition and a Coordinate System

During machining, a chi-square transformation between the global coordinate system and the part coordinate system is involved. Definitions:

1. Global coordinate system: Refers to the coordinate system in which the machining apparatus, such as a lathe, is located;
2. Part coordinate system: The coordinate system of the part itself when the part is designed using CAD and other aids, or when the part is dimensioned.

This book agrees that for the size vector X, if the subscript G marking is considered to be the coordinates of the global coordinate system in the vector, otherwise it is a vector in the part coordinate system.

Let X be the vector representation of the part in the part coordinate system, and X_G be the vector representation of the part in the global coordinate system, then there are

$$\delta = LCL \tag{2.5}$$

$$F(\tau) = \sum_{z=r}^{\tau} f(z) = \sum_{z=r}^{\tau} \binom{z-1}{r-1} (1-p)^{z-r} p^r \tag{2.6}$$

where $\tau = r, r+1, r+2, \ldots$, there are

$$\begin{aligned} E(RL) &= E_{\{n,z\}}\left(\sum_{i=1}^{n} z_i + \min(z_{n+1}, \eta+1) | z_i \in [\delta, \eta], \right. \\ &\quad \left. z_{n+1} \in (-\infty, \delta) \cup (\eta, +\infty), i = 1, 2, \ldots n\right) \\ &= E_{\{n,z\}}\left(\sum_{i=1}^{n} z_i | z_i \in [\delta, \eta], i = 1, 2, \ldots n\right) \\ &\quad + E_{\{n,z\}}(\min(z_{n+1}, \eta+1) | z_{n+1} \in (-\infty, \delta) \cup (\eta, +\infty)) \\ &= E_{\{n,z\}}\left(\sum_{i=1}^{n} z_i | z_i \in [\delta, \eta], i = 1, 2, \ldots n\right) \\ &\quad + E_{\{z\}}(\min(z, \eta+1) | z \in (-\infty, \delta) \cup (\eta, +\infty)) \end{aligned}$$

$$\sum_{i=1}^{n} z_i.$$

where $z_i \in [\delta, \eta]$, has

$$\begin{aligned} & E_{\{n,z\}}\left(\sum_{i=1}^{n} z_i | z_i \in [\delta, \eta], i = 1, 2, \ldots n\right) \\ &= \sum_{c=1}^{+\infty} p_c E_{\{z\}}\left(\sum_{i=1}^{c} z_i | z_i \in [\delta, \eta], i = 1, 2, \ldots n\right) \end{aligned}$$

$$= \sum_{c=1}^{+\infty} p_c c E_{\{z\}}(z|z \in [\delta, \eta]) = E_{\{z\}}(z|z \in [\delta, \eta]) \sum_{c=1}^{+\infty} p_c c$$

$$p_c = [F(\eta) - F(\delta - 1)]^c [1 - F(\eta) + F(\delta - 1)]$$

$$\begin{aligned} E_{\{z\}}(z|z \in [\delta, \eta]) &= \sum_{z=\delta}^{\eta} zp\{z|z \in [\delta, \eta]\} \\ &= \sum_{z=\delta}^{\eta} z \frac{f(z)}{F(\eta) - F(\delta - 1)} \end{aligned}$$

where $\min(z_{n+1}, UCL + 1)$ is the abbreviation of $z_{n+1} \in (-\infty, \delta) \cup (\eta, +\infty)$.

$E_{\{z\}}(\min(z, \eta + 1)|z \in (-\infty, \delta) \cup (\eta, +\infty)) = \sum_{z=r}^{\delta-1} zp\{z|z \in (-\infty, \delta)\} + \sum_{z=\eta+1}^{+\infty} (\eta + 1)p\{z|z \in (\eta, +\infty)\}$ is the abbreviation of $p\{z|z \in (\eta, +\infty)\} = 1 - F(\eta)$ and the rest are analogous.

$E(RL) = \sum_{z=\delta}^{\eta} z \frac{f(z)}{F(\eta)-F(\delta-1)} \sum_{c=1}^{+\infty} [F(\eta) - F(\delta - 1)]^c \, [1 - F(\eta) + F(\delta - 1)]c$ $+ \sum_{z=r}^{\delta-1} z \frac{f(z)}{F(\delta-1)} + (\eta + 1)[1 - F(\eta)] = \frac{1}{1-F(\eta)+F(\delta-1)} \sum_{z=\delta}^{\eta} zf(z) + \frac{1}{F(\delta-1)} \sum_{z=r}^{\delta-1} zf(z) + (\eta + 1)[1 - F(\eta)]$ is the unit matrix. It is important to note $\tilde{S}$.

Use k to mark the kth station, i.e., $\tilde{\boldsymbol{P}} = \begin{pmatrix} \boldsymbol{P} & \boldsymbol{P_0} \\ \boldsymbol{0} & 1 \end{pmatrix}$ denotes the coordinate chi-square transformation that occurs at the kth station.

C. Part Positioning and Machining

The part undergoes a series of turning operations to remove surface material and gradually form a finished product.

Prior to machining, a fixture is used to hold the part in place. This can be thought of as a process of coordinate system transformation. During the positioning process, the part is fixed to the machine tool through translational operation $\boldsymbol{P}$ and rotational operations $\boldsymbol{P_0} = (\boldsymbol{I} - \boldsymbol{P})\boldsymbol{e}$, and its coordinates are transform from the part coordinate system to the global coordinate system. This process can be expressed as

$$\boldsymbol{e} = (1, 1, \ldots, 1)^T \tag{2.7}$$

The operation performed at the kth station can be expressed as

$$\boldsymbol{\pi}(0) = (\boldsymbol{\pi}, \pi_0) \tag{2.8}$$

where $\boldsymbol{\pi} = (\pi_1, \pi_2, \ldots, \pi_m)$, $\pi_i \geq 0$, $\sum \pi_i = 1$ is the newly machined face at the kth station.

$RL = \inf\{n\colon n \geq 0, X_n \in S_0\}$ characterises the machining process. There are

$$E(RL) = \boldsymbol{\pi}(\boldsymbol{I} - \boldsymbol{P})^{-1}\boldsymbol{e} \tag{2.9}$$

where $\text{v} = (\text{C}_\alpha, \text{Z}_\alpha, \text{C}_\gamma, \text{Z}_\gamma)$,

where $S = \{\boldsymbol{\nu} = (C_\alpha, Z_\alpha, C_\gamma, Z_\gamma): 0 \le C_\alpha < r_\alpha, C_\alpha \in Integers;\ C_\alpha \le Z_\alpha \le UCL_\alpha, Z_\alpha \in Integers;\ 0 \le C_\gamma < r_\gamma, C_\gamma \in Integers;\ C_\gamma \le Z_\gamma \le UCL_\gamma, Z_\gamma \in Integers\}$ is the unit matrix. $C_\alpha \le Z_\alpha$ characterises whether the face S_i is machined at station k or not.

$C_\gamma \le Z_\gamma$ describes all faces that are machined at station k, and further, ***P*** describes the faces that are machined before station k. Notation:

$$m = \frac{1}{4}(r_\alpha + 1)(2UCL_\alpha - r_\alpha + 2)(r_\gamma + 1)(2UCL_\gamma - r_\gamma + 2) \tag{2.10}$$

D. SoV Model

At station k, the sources of ***P*** deviations exist: (1) datum deviation $\boldsymbol{\pi}(0)$; (2) fixture deviation e_k^f; (3) machining deviation $\mu + \varsigma\sigma$; (4) other noise $\mu - \varsigma\sigma$.

When the part is positioned, the existence of datum deviation and fixture deviation makes the positioning error occur, recorded as $\mu + \varsigma\sigma$, the positioning deviation $\frac{r}{Z} = p = p(\Delta\mu, \Delta\sigma) = p(k_\mu, k_\sigma) = \begin{cases} \Phi\left(\frac{k_\mu - \varsigma}{1+k_\sigma}\right) & \alpha\text{GTBE} \\ \Phi\left(\frac{k_\mu - \varsigma}{1+k_\sigma}\right) + \Phi\left(\frac{-\varsigma - k_\mu}{1+k_\sigma}\right) & \beta\text{GTBE} \end{cases}$ only affects the machining operation at station k, i.e., it affects the newly formed face $\Phi\left(\frac{k_\mu - \varsigma}{1+k_\sigma}\right) = \frac{r'}{Z_\alpha}, \Phi\left(\frac{-\varsigma - k_\mu}{1+k_\sigma}\right) = \frac{r''}{Z_\gamma}$ at station k.

The machining deviation $\mu + \varsigma\sigma$ can be expressed as

$$X_G^O(k) = R_{PG}^O(k)X^O(k) + T^O(k) \tag{2.11}$$

$$B(k)x_G^u(k) = B(k)\left[X_G^u(k) - X_G^O(k)\right] \tag{2.12}$$

It should be noted that the part is fixed on the lathe during machining and the lathe coordinate system coincides with the global coordinate system, so the above machining deviations are calculated in the global coordinate system. The superscript o indicates the normal value of the dimension. When machining is completed, there is under the global coordinate system:

$$X(k) = A(k)[R_{PG}(k)X(k-1) + T(k)] + B(k)X_G^u(k) \tag{2.13}$$

Under the part coordinate system, the product process dimensions are

$$\begin{aligned} X(k) &= R_{GP}(k)\left[X_G(k) - T(k)\right] \\ &= R_{GP}(k)\left\{A(k)\left[R_{PG}(k)X(k-1) + T(k)\right] + B(k)X_G^u(k)\right\} - R_{GP}(k)T(k) \\ &= R_{GP}(k)A(k)R_{PG}(k)X(k-1) + R_{GP}(k)B(k)X_G^u(k) \\ &\quad + R_{GP}(k)A(k)T(k) - R_{GP}(k)[A(k) + B(k)]T(k) \\ &= R_{GP}(k)A(k)R_{PG}(k)X(k-1) + R_{GP}(k)B(k)\left[X_G^u(k) - T(k)\right] \\ &= A(k)X(k-1) + R_{GP}(k)B(k)\left[X_G^u(k) - T(k)\right] \end{aligned} \tag{2.14}$$

Thus, after machining at station k, the total introduced deviation can be written as T_0 and knowing that station k has no effect T_0, i.e., it is then given as

$$
\begin{aligned}
x(k) &= \Delta X(k) = X(k) - X^o(k) \\
&= A(k)X(k-1) + R_{GP}(k)B(k)\big[X_G^u(k) - T(k)\big] - X^o(k) \\
&= A(k)X(k-1) + R_{GP}(k)B(k) \\
&\quad \big[X_G^u(k) - X_G^o(k) + X_G^o(k) - T(k)\big] \\
&\quad - [A(k) + B(k)]X^o(k) \\
&= A(k)\big[X(k-1) - X^o(k-1)\big] \\
&\quad + R_{GP}(k)B(k)\big[X_G^u(k) - X_G^o(k)\big] \\
&\quad + R_{GP}(k)B(k)X_G^o(k) - R_{GP}(k)B(k)T(k) - B(k)X^o(k) \\
&= A(k)x(k-1) + R_{GP}(k)B(k)x_G^u(k) \\
&\quad + R_{GP}(k)B(k)\big[R_{PG}^o(k)X_G^o(k) + T^o(k)\big] \\
&\quad - R_{GP}(k)B(k)T(k) - B(k)X^o(k) \\
&= A(k)x(k-1) + R_{GP}(k)B(k)x_G^u(k) \\
&\quad + \big[R_{GP}(k)B(k)R_{PG}^o(k) - B(k)\big]X^o(k) \\
&\quad - R_{GP}(k)B(k)\big[T(k) - T^o(k)\big]
\end{aligned}
\tag{2.15}
$$

Defining rotational deviation T_0, translational deviation f_1, Eq. (2.15) can be reduced to

$$
\begin{aligned}
x(k) &= A(k)x(k-1) + R\_GP(k)B(k)x\_G\hat{\ }u(k) \\
&\quad + \Delta R\_GP(k)B(k)\Delta T(k)
\end{aligned}
\tag{2.16}
$$

Equation (2.16) shows that after the end of station k, the process dimensional f_2 deviation of the product is affected by the combined effect of the dimensional deviation state $x(k-1)$ at the end of machining in the previous station, the rotational deviation $f_o : \vec{n} = [a, b, c]^T$ and the translational deviation $f_L : \vec{P} = \left[x_o, y_o, z_o\right]^T$ at the time of fixation of the part, and the machining deviation $f_D : \vec{D} = \left[d_1, d_2, \ldots, d_j\right]^T$ at the time of machining.

On the other hand, noting $S = \{f_o, f_L, f_D\}$ the sources of deviation of station k, where $S_1, S_2, \ldots, S_M$ is the mounting deviation and $P = \{S_1, S_2, \ldots, S_n\}$ is the machining deviation, there are

$$
x(k) = A(k)x(k-1) + B(k)\mathbf{u}_k + H(k)x(k-1)
\tag{2.17}
$$

Comparing Eq. (2.16) with Eq. (2.17), there is:

$$B(k)u\_k + H(k)x(k-1) = R\_GP(k)B(k)x\_G^\wedge u(k) + \Delta R(k)X^\wedge o(k) - R\_GP(k)B(k)\Delta T(k) \quad (2.18)$$

Noting $\vec{P}_i = \left[p_{ix}, p_{iy}, p_{iz}\right]^T$, introducing other noise $\vec{D}_i = \left[d_{i1}, d_{i2}, \ldots, d_{ij}\right]_{j\times 1}^T$, there is

$$x(k) = A(k)x(k-1) + B(k)\mathbf{u}_k + w_k \quad (2.19)$$

Equation (2.19) characterises the transfer of the product's dimensional deviation state x(k − 1), passed down from the previous station, to the dimensional deviation state x(k), subject to positioning deviations at the kth station as well as machining deviations. That is, the deviation flow model of the process dimensions.

2.3.2 RQR Chain Analysis Model

A. RQR Chain Conceptual Model

The decline mechanism of manufacturing process product reliability is mainly manufacturing system anomalies lead to manufacturing process quality degradation, followed by the accumulation of manufacturing process deviations caused by the decline in product reliability of such a chain deviation transmission relationship [3]. Specifically, the RQR chain of Manufacturing System Reliability (RMS), Manufacturing Process Quality (QP) and Manufacturing Product Inherent Reliability (RI), of which. Manufacturing Process Quality Q is the core of the RQR chain transmission, as shown in the upper part of Fig. 2.7. Manufacturing process man, machine, material, method, environment, measurement and other deviations are the source of fluctuations in product reliability.

As shown in the lower part of Fig. 2.7, the product by the manufacturing process quality deviation (Quality Variations) will form the impact of defects, that is, the manufacturing process quality of the manufacturing process of product reliability of the QR role; at the same time, the process quality deviation mainly from the manufacturing system of the impact of the anomalies, that is, the reliability of the manufacturing system on the manufacturing process quality of the RQ role.

In engineering practice, this situation often occurs: the manufacturing process of product quality inspection qualified, but put into use, the product reliability can not meet the design reliability requirements, this is mainly because the product defects will be exposed in the use of the formation of faults, resulting in a decline in product reliability. Therefore, based on the specific role of defects on the product, it can be distinguished into explicit defects (Exteriordefect) and implicit defects (Interiordefect) two categories. As shown in Fig. 2.7, manufacturing process product reliability will be affected by both explicit and implicit defects, resulting in a decline in reliability.

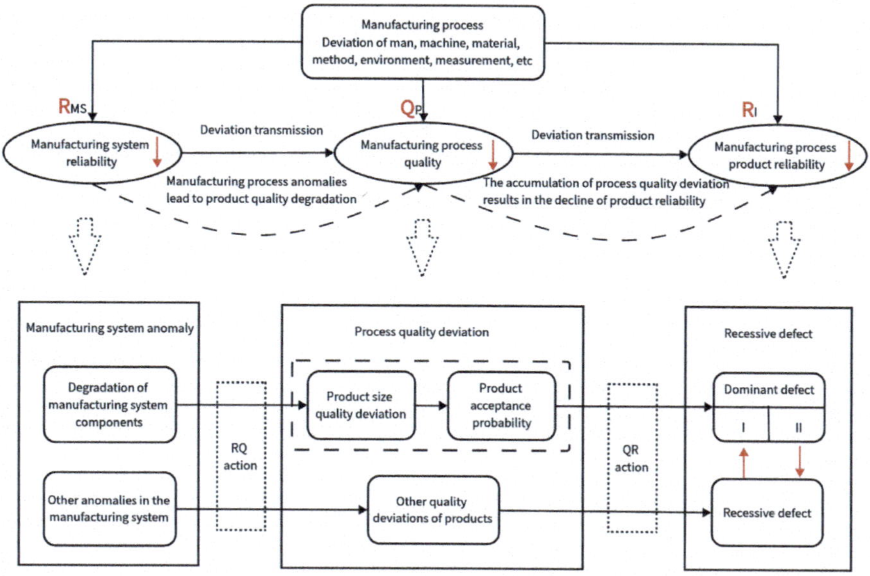

Fig. 2.7 Conceptual model of RQR chain

(1) Obvious defects. Visible defects consist of two parts, visible defects I because in the manufacturing stage will cause product quality unqualified and in the product manufacturing process of the inspection link is directly detected, can not be transferred to the use of the product stage, such as dimensional overshoot and geometric deformation, etc.; visible defects II through the product dimensions of the tolerance range of the different degrees of fluctuations in the manufacturing process of the product reliability, specifically, the dimensional deviation mainly originates from the degradation of manufacturing system components, i.e., the RQ role, and is ultimately reflected in the probability of product qualification that indicates the degree of product qualification.
(2) Hidden defects. Hidden defects mainly originate from the effects of quality deviations other than explicit dimensional deviations, i.e. QR role. It is difficult to be detected at the manufacturing stage, and will eventually manifest itself over time through the use of the user, resulting in product reliability problems, such as false welds in the welding process and potential cracks in the assembly process, etc. Accordingly, visible defects II can be regarded as a special kind of hidden defects.

There is also a relationship of mutual influence and transformation between explicit and hidden defects. In the process of product use, with the passage of time, the hidden defects, such as cracks, which are not detected in the manufacturing stage, will be gradually expanded and transformed into visible defects I, resulting in product reliability problems; at the same time, the visible defects II in the process of using the product will cause the problem of the mutual cooperation between the components,

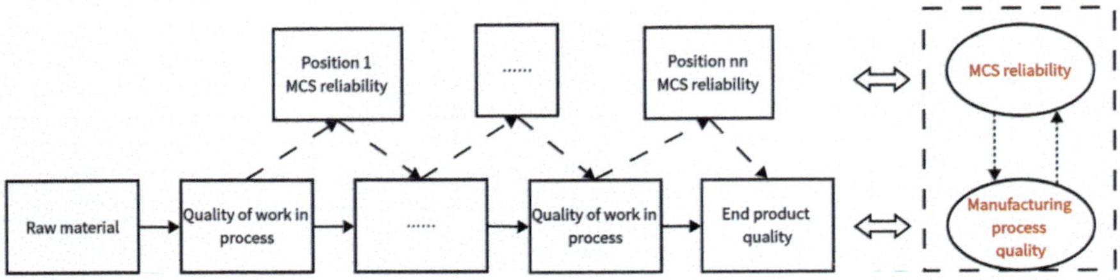

Fig. 2.8 interaction effect of manufacturing system reliability R and manufacturing process quality Q

which will lead to the generation of more hidden defects, such as residual stresses, and so on.

B. Interaction Effect of Manufacturing System Reliability R and Manufacturing Process Quality Q

As shown in Fig. 2.7, the RQ role describes the impact of manufacturing system reliability on product quality deviation, of which this section focuses on the interaction between MSC reliability and product quality in the manufacturing process. Manufacturing system components MSC as a basic factor in the RQR chain, all raw materials need to be processed, assembled, and tested by MSC (e.g., machine tools, cutting tools, etc.) to form the final product. In the QR chain, the interaction effect of the reliability of MSC and the product quality of the manufacturing process exists in the transfer accumulation between different stations in the manufacturing stage [4] (as shown in Fig. 2.8). Specifically, the interaction effect is reflected in the fact that the product quality at the current station will be affected by abnormalities in machining or positioning due to component wear, which will result in quality deviations, and at the same time, the wear and failure rates of components at the downstream station will be exacerbated by the product quality deviations at the upstream station. Eventually, the product quality deviation will be transferred and accumulated between different stations, which will be reflected in the dimensional deviation of the product.

Take the assembly process of BIW (body in white) as an example. The fixture plays a crucial role in ensuring the quality of the product, and the accuracy of the positioning of Key Product Characteristics (KPCs) reflects the quality of the assembled product. For simplicity, in the process of BIW positioning through the "two pins on one side" method, only the positioning pins are analysed as the object of concern, and the deviation of the actual positioning from the designed positioning is measured by selecting the corresponding KPCs points, which is used as the standard to measure the product quality. The interaction effect between the MSC reliability and the product quality is reflected as follows The wear of the positioning pins will affect the positioning deviation of the KPCs points, i.e., the MSC will affect the quality of the product; at the same time, the positioning deviation of the KPCs will also aggravate the wear of the pins, thus affecting the reliability of the MSC.

C. Analysis of the Relationship Between Manufacturing Process Quality and Manufacturing Process Product Reliability Based on Defects

The product is manufactured by multiple stations, in order to simplify the problem, it is assumed that the defects of each station are independent of each other, i.e., the system reliability model is a series model. Therefore, based on the influence of explicit defects II and implicit defects on product reliability, the manufacturing process product reliability is defined as the probability that explicit defects II and implicit defects do not cause failures during the product's use time t (This is defined $S_i = \left[n_{ix}, n_{iy}, n_{iz}, p_{ix}, p_{iy}, p_{iz}, d_{i1}, d_{i2}, \ldots, d_{im}\right]^T_{(6+m)\times 1}$ as the ideal state of reliability after the completion of the product's manufacturing, which is related to the influencing factors of the manufacturing stage only, and has nothing to do with the actual use of the product, regardless of the use mode of the product's use stage.) and environmental stresses:

$$X = \left[S_1^T, S_2^T, \ldots, S_n^T\right]^T \tag{2.20}$$

where $\Delta S_i = \left[\Delta n_{ix}, \Delta n_{iy}, \Delta n_{iz}, \Delta p_{ix}, \Delta p_{iy}, \Delta p_{iz}, \Delta d_{i1}, \Delta d_{i2}, \ldots, \Delta d_{im}\right]^T_{(6+m)\times 1}$ and $x = \Delta X = \left[\Delta S_1^T, \Delta S_2^T, \ldots, \Delta S_n^T\right]^T$ are the reliability based on explicit defects II and implicit defects, respectively. The value of $X(k)$ is equal to the probability of conformity of the product, which focuses on the cumulative effect of deviations in the last manufacturing station, and is calculated by defining the effect of fluctuations of the product dimensions within the conformity range on the reliability of the product in the manufacturing process. The value of $x(k)$ can be calculated by the model $X_G = R_{PG}X + T$ of the effect of hidden defects on the reliability of the product between the stations, because the generation of defects between the stations is independent of each other, and $X = R_{GP}(X_G - T)$ is a series model of $T \in R^{n(6+m)\times 1}$.

2.4 Physical Basis of Failure of Electronic Product Process Reliability

Electronic product process reliability for the system reliability of electronic products to provide protection, but the industry has not yet established a sound system theory, the system reliability of electronic products can be broken down to the process level of specific reliability indicators. For the solder joints and circuit board plating through-hole process reliability life prediction has been a more complete theoretical study. The following describes the process reliability life prediction model for solder joints and circuit board plated through holes respectively.

2.4.1 Thermal Fatigue Modelling of Solder Joints

Solder joints are used in electronic packaging technology not only for electrical and mechanical connections, but also to provide thermal dissipation channels for chips. With the development of chip packaging from through-hole mounting with pins to surface mounting with pins, until the current pinless solder ball array surface mounting. Electronic packaging is towards the solder ball array pitch is getting smaller and smaller, the solder ball size is getting smaller and smaller, for the chip size is getting bigger and bigger, the number of solder ball more and more direction, which makes the reliability of the solder joints has become more and more prominent, the study shows that 70% of the failure of the electronic device is caused by the failure of the package and the assembly, and in the failure of the electronic package and assembly, the failure of the solder joints is the main reason.

The main reason for the thermal fatigue failure of solder joints is due to the different coefficients of thermal expansion of the materials around the solder joints, which leads to the mismatch of thermal strains generated by various materials during thermal expansion or contraction, and produces a stress concentration at the strain incongruity, resulting in the emergence and expansion of cracks.

Welded joint fatigue life prediction models can be broadly divided into 3 categories:

(1) Stress-based fatigue models
(2) Strain-based fatigue model
(3) Energy-based fatigue model.

A. Stress-Based Fatigue Models

Stress-based fatigue modelling is mainly applicable in the case of vibration or shock loads, and for electronic packaging devices, the loads sustained are mainly thermal cycling loads caused by temperature changes in the environment of use.

There are two types of external stresses acting on a product. One type is reversible, that is, when the stress acts on the product or material, its parameters will change, and when the stress disappears, the product returns to its original state. Another type of action is irreversible, that is, when the stress disappears, the consequences of the stress still exists or partially exists, so that each time the role of stress will bring damage to the product, these damages cumulative more than a certain critical value, the product will fail or failure, this model is the cumulative damage model.

Linear cumulative damage theory refers to the role of cyclic loading, fatigue damage and the number of cycles of the load relationship is linear, and fatigue damage can be linearly cumulative, each stress is independent of each other and are not related to each other, when the cumulative damage reaches a certain value, the specimen or component fatigue damage occurs. Typical of the linear damage accumulation theories is the Palmgren–Miner theory, referred to as the Miner theory.

Miner (1945) formulated the assumption that the accumulation of fatigue damage is linearly related to the number of stress cycles, as proposed by Palmgren (1924), and gave the mechanical premises:

(1) Under any equal amplitude fatigue loading, the material absorbs an equal amount of network in each stress cycle, and fatigue damage occurs when the network accumulates to a critical value;
(2) Under variable-amplitude fatigue loading, the critical network for the final destruction of the material is all equal;
(3) Under variable-amplitude fatigue loading, the network absorbed at each level of the stress cycle is independent of each other and of the order of the stress levels.

Constructing a fatigue cumulative damage theory, whether it is valid or not, must quantitatively answer the following three questions:

(1) How much damage does one load cycle cause to a material or structure?
(2) How does damage accumulate over multiple load cycles?
(3) What is the critical damage at failure?

Miner's theory answers the three questions as follows:

(1) An injury caused by a cycle:

$$\mathrm{D} = \frac{1}{\mathrm{N}} \tag{2.21}$$

where N—fatigue life corresponding to the current load level.
(2) Damage caused by n cycles under equal amplitude loading:

$$D = \frac{n}{N} \tag{2.22}$$

(3) Damage caused by n cycles under variable amplitude loading:

$$D = \sum_{i=1}^{n} \frac{1}{N_i} \tag{2.23}$$

where N_i—fatigue life corresponding to the current load level S_i. Critical fatigue damage D_{CR}.

In case of constant amplitude cyclic loading, it is obvious that fatigue damage occurs when the number of cycles n is equal to its fatigue life N, i.e.: n = N, obtained from Eq. (2.3):

$$\mathrm{D_{CR}} = 1 \tag{2.24}$$

Miner's principle is summarised as follows: under a certain equal-amplitude fatigue stress S_j (corresponding to equal-amplitude fatigue life N_j), the network absorbed by the material is equal in each stress cycle, and fatigue damage occurs when this network Δw absorbed by the material reaches a critical value W. Under the action of variable amplitude stress $S_1, S_2, \ldots, S_n$, the equal amplitude life of each stress level is N_i, the actual number of cycles is n_i, the network produced is w_i, when $\sum w_i = W$ fatigue damage occurs, and there is:

$$\sum \frac{w_i}{W_i} = \sum \frac{n_i}{N_i} = 1 \tag{2.25}$$

Miner's theory can be regarded as a linear damage, linear cumulative cycle ratio theory, and its success lies in the fact that a large number of experimental results (especially the random spectrum test) show that the mean value of critical fatigue damage is really close to 1, which is widely used in engineering because of its simplicity, while other nonlinear cumulative deterministic methods need to carry out a large number of experiments to fit a large number of parameters, and the accuracy is not better than that of Miner's theory. The main shortcomings of Miner's theory are: the damage is independent of the loading state, and the damage accumulation is independent of the loading sequence.

In 1945, Miner studied the above repeated fatigue problems and proposed the fatigue damage theory. He recorded the fatigue life (number of times) of a material under a certain stress S as N, and drew the S–N curve reflecting the relationship between the two. The relationship between the action stress caused by loading and the number of actions leading to failure can be roughly described by the S–N curve.

The curve in Fig. 2.9 depicts the average of a series of tests, where S is the stress amplitude and N is the number of actions leading to failure. The central effective range portion of the S–N curve often approximates a straight line in double logarithmic coordinates. The simplified form of the S–N curve is similar to that of Fig. 2.10.

When the slope is determined, the S–N approximate relationship is shown in the following equation:

Fig. 2.9 S–N curve

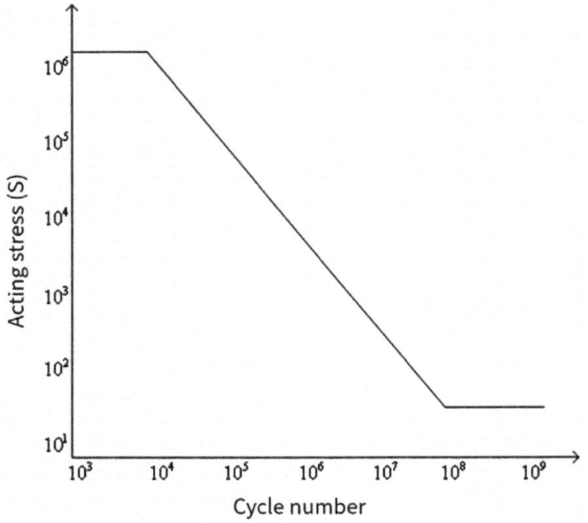

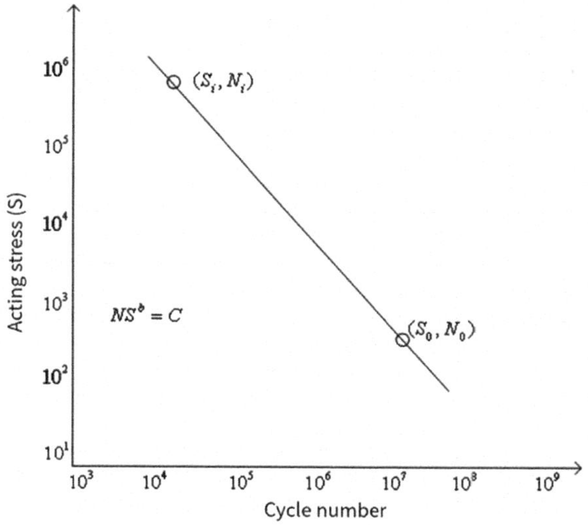

Fig. 2.10 Simplified diagram of S–N curve

$$NS^b = C \tag{2.26}$$

where b—is determined by the material (generally between 3 and 25); C—a constant.

Given b and knowing the point (N_0, S_0) on the curve, it is possible to determine any point N_i on the curve under load S_i as shown in the following equation:

$$N_i = N_0 \left[\frac{S_0}{S_i} \right]^b \tag{2.27}$$

B. Strain-Based Fatigue Model

For the more commonly used strain-based fatigue life models, the strain is usually divided into three components:

$$\Delta\gamma = \gamma_e + \gamma_p + \gamma_c \tag{2.28}$$

which γ_e is the elastic strain component, γ_p is the plastic strain component, and γ_c is the creep strain component, which are usually intertwined and produced together, and therefore it is very difficult to start a strict distinction between the three during practical application.

In accordance with the type of strain generated in the welded joint under thermal cycling conditions, strain-based fatigue model can be divided into: plastic strain-based fatigue model, creep-based fatigue model, has been both fatigue model. Plastic strain-based model that: welded joints produce fatigue damage is the main factor in the welded joints produced by the plastic deformation, while the spring deformation on the fatigue damage produced by the impact is very small, or even negligible. Therefore, the plastic strain as the main parameter to determine whether the welded joint fatigue damage. Considering the plastic strain fatigue model includes the following kinds:

(a) Coffin–Manson Model
(b) Total Strain Model
(c) Solomon Model
(d) Engelmaier Model.

These fatigue models use calculations or tests to determine the plastic shear strain applied to a welded joint, from which the fatigue life of the welded joint is derived.

The Coffin–Manson model is a widely used low-week fatigue life model:

$$\frac{\Delta\varepsilon_P}{2} = \varepsilon_f\left(2N_f\right)^c \tag{2.29}$$

where N_f—the number of thermal cycle cycles experienced by the welded joint fatigue failure; ε_P—plastic strain amplitude; ε_f—fatigue ductility coefficient, often approximated to be equal to the actual use of material true fracture toughness; c—fatigue ductility index, generally between − 0.5 and − 0.7.

Since the Coffin–Manson model only considers plastic deformation, if the effect of elastic deformation is to be considered, it can usually be combined with Basquin's equation, and the effect of elastic and plastic strains can be considered simultaneously, which is known as the Total Strain Model (TSM).

$$\frac{\Delta\varepsilon}{2} = \frac{\sigma_f}{E}\left(2N_f\right)^b + \varepsilon_f\left(2N_f\right)^c \tag{2.30}$$

where $\Delta\varepsilon$—the strain range; E—the modulus of elasticity of the solder joint material; ε_f and c—the meaning of the aforementioned; σ_f—fatigue strength coefficient; b—the fatigue strength index, also known as Basquin's index.

The improvement of the full strain model over the Coffin–Manson model is that it takes into account the effect of elastic deformation on fatigue failure.

The Solomon low-cycle fatigue model continuates the plastic shear strain with the number of fatigue cycle cycles at failure:

$$\Delta\gamma_p N_p^{\alpha} = \theta \tag{2.31}$$

where $\Delta\gamma_p$—plastic shear strain range; N_p—number of thermal cycle cycles experienced by the welded joint at failure; θ—inverse of the fatigue ductility coefficient; α—material constants.

This fatigue life model has been applied by several researchers to pinned plastic QFP packages, and flip-flop solder packages containing filler material with good results. Solomon determined the material constant in the above equation after experimental measurements to obtain the following equation:

$$\Delta\gamma_p N_p^{0.51} = 1.14 \tag{2.32}$$

i.e.:

$$N_p = 1.2928(\nabla \gamma_p)^{-1.96} \tag{2.33}$$

However, because Solomon low-week fatigue model does not take into account the effect of creep on the fatigue of solder joints, so there are still some limitations in practical use.

Another commonly used model is the Engelmaier model, shown in Eq. (2.34), which takes into account the effect of thermal cyclic loading frequency and loading temperature as well as the effect of elastic–plastic strains.

$$N_f = \frac{1}{2}\left(\frac{\Delta \gamma}{2\varepsilon_f}\right)^{\frac{1}{c}} \tag{2.34}$$

where N_f—fatigue life; ε_f—fatigue ductility coefficient, $\varepsilon_f = 0.325$ for widely used eutectic solders; c is a parameter related to the stabilised cyclic profile and is determined by Eq. (2.35):

$$c = -0.442 - 0.0006T_{sj} + 0.0147\left(1 + \frac{360}{t_H}\right) \tag{2.35}$$

where T_{sj}—average temperature of the temperature cycle (°C); t_H—high temperature holding time in the temperature cycle (minutes); $\Delta \gamma$—total shear strain range.

For the simplified first-order fatigue model, for both pinless and pinned packages, the total shear strain range consists of the overall action and localised action (Fig. 2.11):

$$\Delta \gamma = |\Delta \gamma_g| + |\Delta \gamma_l| \tag{2.36}$$

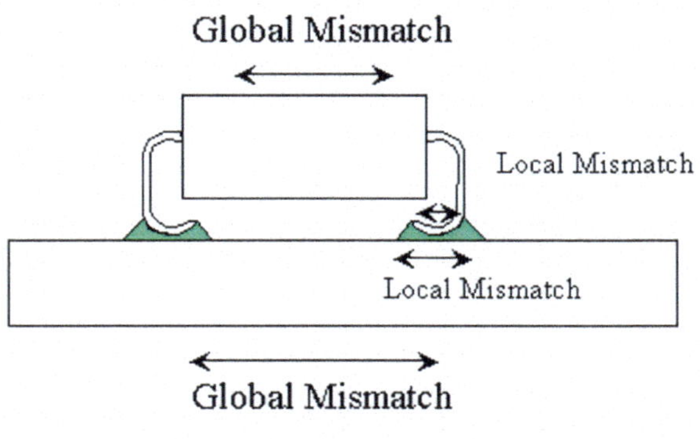

Fig. 2.11 Schematic diagram of component and board connection

(1) Overall Role

There are leaded packages:

$$\Delta\gamma_g = 0.5\mathrm{FI}\frac{\mathrm{K_D}}{(200\,\mathrm{psi})\mathrm{Ah}}(\Delta\alpha\mathrm{LT_s} - \Delta\alpha\mathrm{LT_c})^2 \tag{2.37}$$

Leadless package:

$$\Delta\gamma_g = \frac{0.5FI(\Delta\alpha LT_s - \Delta\alpha LT_c)}{h} \tag{2.38}$$

where F—user-defined calibration parameters; I—internal calibration parameters; K_D—twist factor of the lead; A—Effective connection area of solder balls; h—Symbolic height of solder balls (1/2 of actual height, typical $h = 4 \sim 5\,mils$); $\Delta\alpha LT_c = \sqrt{(L_x\alpha_{cx})^2 + (L_y\alpha_{cy})^2}\left(T_c^{\max_c^{\min}}\right)$—Ductility value of components; $\Delta\alpha LT_s = \sqrt{(L_x\alpha_{sx})^2 + (L_y\alpha_{sy})^2}\left(T_s^{\max_s^{\min}}\right)$—Ductility value of substrate. α_{cx}, α_{sx}—Linear thermal ductility coefficient of the component and the substrate in X-axis; α_{cy}, α_{sy}—Linear thermal ductility coefficient of the component and the substrate in Y-axis; T_c, T_s—Temperature of the component and the substrate.

(2) Localised Action

$$\Delta\gamma_l = \frac{\Delta T\Delta\alpha\sinh(Al_{eff})}{bA\cosh(Al_{eff})} \tag{2.39}$$

$$A = \sqrt{\frac{G}{b}\left(\frac{1}{E_l t_l} + \frac{1}{E_b t_b}\right)} \tag{2.40}$$

where G—solder ball twist factor; b—solder ball thickness; E_l—lead elastic modulus; E_b—board modulus of elasticity; t_l—lead thickness; t_b—circuit board thickness.

The fatigue model based on creep analysis focuses on the creep effect that occurs in solder joints under thermal cyclic loading. For solder joint materials, creep is generally considered to be due to both mechanisms, grain boundary slip, or matrix creep. The two creep-based fatigue models presented below are based on this premise.

Knecht and Fox proposed a simple fatigue model for matrix creep as shown in Eqs. (2.41) and (2.42):

$$N_f = \frac{C}{\Delta\gamma_{mc}} \tag{2.41}$$

$$\gamma_{mc} = \int C_0\left(\frac{\tau}{\tau_0}\right)^{nMC} dt \tag{2.42}$$

where γ_{mc}—range of shear strain due to matrix creep; nMC, C_0, τ_0—material parameters, determined by tests, in the literature of Knecht and Fox, $nMC = 7.1$, $C_0 = 890\%$.

Another creep mechanism is considered to be related to grain boundary slip, which is combined with matrix creep to establish a fatigue model as shown in Eq. (2.43):

$$N_f = \left(0.022D_{gbs} + 0.063D_{mc}\right)^{-1} \tag{2.43}$$

This model divides the creep strain into two parts, D_{gbs} and D_{mc} are the equivalent creep strain accumulated in each cycle, the former is the strain due to grain boundary slip, and the latter is the strain due to matrix creep. Syed also pointed out through the study of TSOP packages that, as the rate of change of temperature becomes larger and the overall stiffness of the package element increases, the main mechanism controlling the occurrence of creep will gradually change from grain boundary slip to matrix creep Pang also pointed out that the main mechanism to control creep will gradually change from grain boundary slip to matrix creep.

In addition Pang also proposed: Solomon plastic strain model and Knecht–Fox creep fatigue model, as shown in Eq. (2.44):

$$\frac{1}{N_f} = \frac{1}{N_p} + \frac{1}{N_c} \tag{2.44}$$

where N_p, N_c—fatigue life in Solomon's model and Knecht–Fox model.

Pang in his study applied this model to FCOA and CBGA packages to predict the fatigue life of components. After comparing it with other fatigue life models, he recommended the use of this "creep fatigue model".

C. Energy-Based Fatigue Model

Energy-based fatigue models generally consider either a hysteresis energy term or a volume-weighted average stress–strain term, and Dasgupta suggests that the total strain energy, which incorporates both stress and strain information, is an important parameter that is a good trade-off for welded joint damage. In contrast to the previous models, this energy model integrates the hysteresis return line energy as well as the effect of test conditions. Akay proposed the following fatigue model based on the total strain energy:

$$N_f = \left(\frac{\Delta W_{total}}{W_0}\right)^{\frac{1}{k}} \tag{2.45}$$

where N_f—the average number of cycle weeks experienced by the welded joint before failure; ΔW_{total}—the total strain energy; and W_0, k—the fatigue coefficient.

He applied this fatigue model to all pin package structures, and after testing he arrived at the constants summarised in the equation as: $k = -0.6342$, $W_0 = 0.1573$, but he did not address the fatigue of solder joints.

Darveaux proposed a fatigue model based on the plastic dissipation of work required for generation and crack propagation. This model uses finite element analysis to calculate the cumulative plastic work per unit volume in each cycle, and this cumulative plastic work is then used to calculate the number of cycles it takes to "induce" a crack inside the welded joint, and the number of cycles it takes for the crack to propagate inside the welded joint, which add up to the total number of cycles a welded joint will go through before it fails. The equations for crack generation and crack propagation in the Darveaux model are shown in Eqs. (2.46) and (2.47):

$$N_0 = K_1 \Delta W^{K_2} \tag{2.46}$$

$$\frac{da}{dN} = K_3 \Delta W^{K_4} \tag{2.47}$$

where K_1, K_2, K_3, K_4—material constants.

The plastic work ΔW is calculated by averaging within the welded joint cell as shown in Eq. (2.48):

$$\Delta W = \frac{\sum_i \Delta W_i V_i}{\sum_i V_i} \tag{2.48}$$

The fatigue life of the welded joint is then calculated using Eq. (2.49):

$$a = N_0 + \frac{a}{da/dN} \tag{2.49}$$

Liang considered the effect of weld joint geometry and after elastic-creep analyses of the weld joints, he proposed another energy-based fatigue model:

$$N_f = C(W_{ss})^{-m} \tag{2.50}$$

where W_{ss}—energy density under the stress–strain hysteresis return line; C, m—temperature dependent material constants.

2.4.2 Circuit Board Plated Through Hole (PTH) Failure Models

With the development of electronic technology, the printed circuit board technology puts forward more and more high requirements. For high-density packaging of electronic products, ultra-miniaturised demand, only by the miniaturisation of various types of components, pin pitch of the miniaturisation can not achieve the true meaning of high-density packaging, but also need to carry these components of the packaging substrate with which to cooperate, i.e., need to be based on the wiring pattern of the

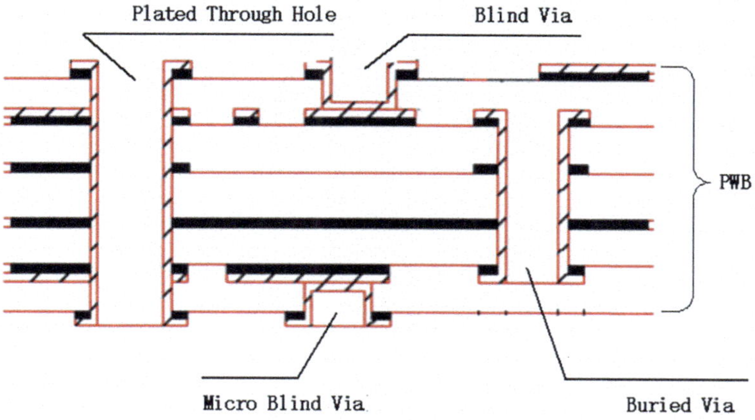

Fig. 2.12 Schematic diagram of PWB and PTH structures

miniaturisation of the substrate, the structure of the multilayer to achieve. Printed Wiring Board (PWB) refers to the entire substrate carrying electronic components, and its basic function is to carry electronic components and achieve the electrical connection between them.

Since 1961, when Hazeltine developed the multi-layer PWB manufacturing process using the plated through-hole method, it has been more than 40 years since then, and a strong industrial system has been formed. It can be said that many other types of multilayer PWBs have been developed on this basis. Figure 2.12 shows a schematic cross-section of a multilayer PWB with plated through holes.

As can be seen from Fig. 2.12, a common feature of generic PWB structures is the presence of through-holes running through the entire substrate. Plated Through Hole (PTH) refers to the through holes running through the multi-layer PWB, plated with conductive materials such as copper, nickel or solder, etc., used to provide electrical connections for the conductive metals of the different layers of the board, is an important structural component of the PWB. In order to increase the wiring freedom, there are other types of vias besides through vias, such as micro blind via, blind via, and inner buried via. In this chapter, we will mainly focus on the PTHs that run through the entire PWB type.

A typical micrograph of a PTH structure is shown in Fig. 2.13, which consists of a plated hole wall (barrel), an external pad, and an internal pad. The key geometric design parameters of PWBs related to PTHs are the diameter of the hole (Dhole), thickness of the plating (tplating), thickness of the board (tlayers), diameter of the external pad (Dpad), and spacing of neighbouring PTH holes (Lspacing), etc., as shown in Fig. 2.14.

PTH failure is mainly caused by the CTE (coefficient of thermal expansion) mismatch between the coating material and the substrate material. This mismatch is mainly manifested in the outermost pads and the thickness direction of the PWB, which is due to the fact that the CTE of the outermost pads of multilayer boards is

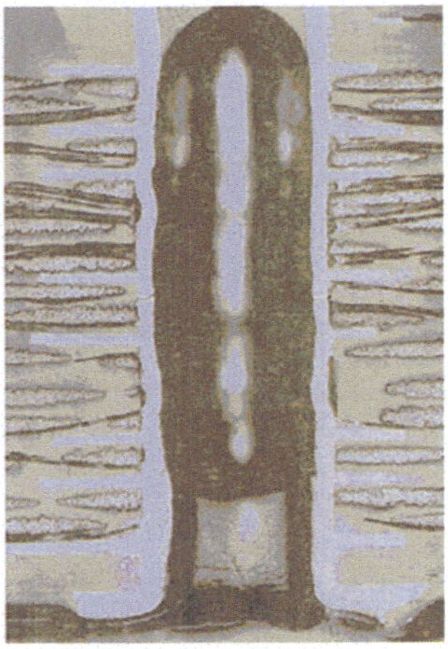

Fig. 2.13 Micrograph of PTH cross section

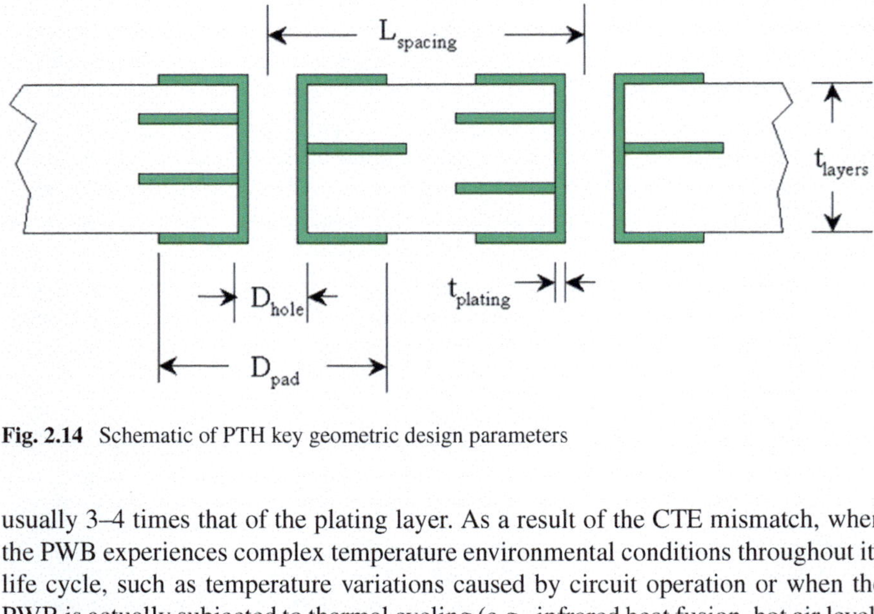

Fig. 2.14 Schematic of PTH key geometric design parameters

usually 3–4 times that of the plating layer. As a result of the CTE mismatch, when the PWB experiences complex temperature environmental conditions throughout its life cycle, such as temperature variations caused by circuit operation or when the PWB is actually subjected to thermal cycling (e.g., infrared heat fusion, hot air levelling, wave soldering during manufacture and use, and thermal cycling during use and storage), the thermal stresses generated in the PTH will cause damage to the metal plating and eventually lead to thermo-mechanical overstress failure or fatigue failure of the PTH. The thermo-mechanical behaviour of PTH structures at high and low temperatures is shown in Fig. 2.15. Table 2.1 summarises the failure of PTH structures under various thermal conditions.

Since printed circuit boards (PWBs) experience complex temperature environmental conditions throughout their life cycle, such as undergoing infrared thermal

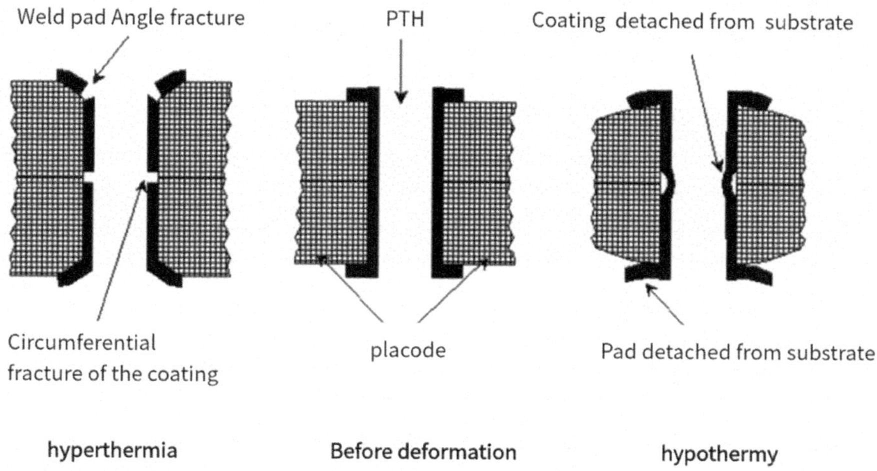

Fig. 2.15 Effect of temperature on thermo-mechanical behaviour of PTH

Table 2.1 Comparison of thermo-mechanical behaviour and failure mode analysis of PTH under different temperature loads

Hot load conditions	Stress/strain	Failure mode	Indicative parameters
High temperature	Plating is stretched, pads bend in outermost direction, failure may occur at rising edge of thermal shock	Hole wall circumferential fracture Corner failure of pads	Maximum tensile stress in the plating Maximum radial tensile stress in the pad
Low temperature	The plating is compressed and the pad is bent in the innermost direction, failure may occur at the falling edge of the thermal shock	Circumferential fracture in warped plating External pad delamination	Maximum compressive stress in warped plated layers Maximum peeling stress in the outermost pad/ substrate bonding area
Temperature cycle	Cyclic stretching and compression of the plating, failure may occur at the centre of the plating, where the pad meets the plating	Fatigue failure	Maximum plastic strain per cycle Maximum total strain per cycle

fusion, hot air levelling, wave soldering during manufacturing and use, and thermal cycling during use and storage, thermal stresses are generated in the multilayered PCBs and PTH aperture walls, and ultimately lead to thermo-mechanical stress failure or fatigue failure of the PTHs. The research content and objective of this chapter is to quantitatively assess the thermal fatigue failure of PTH under such thermal cycling conditions, so as to provide a theoretical and modelling basis for the quantitative reliability assessment or life prediction of the whole PWB and electronic products.

A. Assumptions on Which the Model Is Based

The cross-section of PTH perpendicular to the substrate is shown in Fig. 2.16. In order to facilitate the establishment of a simplified model, the following assumptions are made under the premise of ensuring that those important influences affecting the stress–strain distribution of PTH are not neglected:

(a) PTH is axisymmetric;
(b) The PTH plating thickness is uniform, i.e., defects in the plating process are not considered;
(c) The effects of pads and copper foils are not considered;
(d) No creep of the plating material and substrate material;
(e) Thermal expansion of the material is idealised, i.e. a linear relationship is satisfied $\varepsilon = \alpha \cdot \Delta T$;
(f) The temperature is the same everywhere in the PTH plating and the substrate.

B. Fundamental Equations of Elastodynamics for Axisymmetric Problems in Space

Due to the mismatch in the coefficients of thermal expansion between the substrate material and the PTH coating material (typically a 3–4 times difference), thermal stresses are generated under thermal cycling conditions and ultimately lead to fatigue fracture failure of the PTH. It has been shown that this mismatch occurs mainly in the axial direction (z-direction), which causes circumferential fracture at the centre of the plating of PTH. Therefore, PTH failure can be quantitatively assessed by determining the axial stress distribution in the PTH plating. From the geometrical structure and force characteristics of the plated through-hole, it can be seen that it belongs to the spatial axisymmetric problem, and it is usually more convenient to use the cylindrical coordinate system (r, θ, z) for solving the spatial axisymmetric

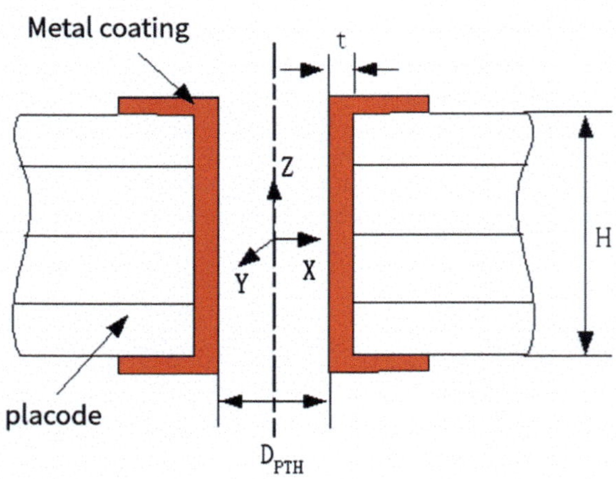

Fig. 2.16 Schematic of PTH cross-section

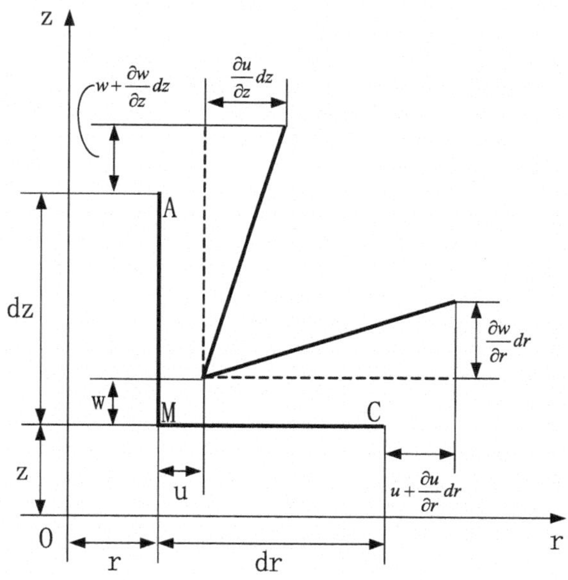

Fig. 2.17 Schematic diagram of the deformation situation of any microelement in the cylindrical coordinate system

problem in elastic mechanics. In order to establish the basic equations of elastic mechanics of PTH under thermal stress conditions, the basic equations of spatial axisymmetric problems are given in the general case first. The deformation of any microelement MABC in the cylindrical coordinate system is shown in Fig. 2.17 (taking the horizontal cross-section r–z).

From the axisymmetry of the object deformation, it can be seen that the deformation state in each radial plane should be the same, i.e., independent of the angle θ, and the deformation displacement in each radial plane is always located in that plane. That is, there are radial, tangential and axial displacement components of any point M as:

$$u = u(r, z);\ v = 0;\ w = w(r, z) \tag{2.51}$$

Combining the deformation displacement components of the microelement and the relevant definitions can be deduced to obtain the basic equations of the axisymmetric problem (only the equations related to this study are listed) as follows:

(1) Geometric equations

$$\gamma_{r\theta} = \gamma_{z\theta} = 0 \tag{2.52}$$

$$\gamma_{rz} = \frac{\partial u}{\partial z} + \frac{\partial w}{\partial r} \tag{2.53}$$

$$\varepsilon_r = \frac{\partial u}{\partial r}, \varepsilon_z = \frac{\partial w}{\partial z}, \varepsilon_\theta = \frac{u}{r} \tag{2.54}$$

(2) Physical equations

$$\varepsilon_z = \frac{1}{E}[\sigma_z - \mu(\sigma_r + \sigma_\theta)] \tag{2.55}$$

$$\gamma_{rz} = \frac{2(1+\mu)}{E}\tau_{rz} = G\tau_{rz} \tag{2.56}$$

C. Simplified Physical Model of PTH

In the actual PTH structure, the coating material and the substrate material are "bonded" together, and it is necessary to consider the effect of "adhesion" between the two materials on the stress–strain distribution. Accordingly, a simplified axisymmetric model of PTH space is constructed as shown in Fig. 2.18.

Figure 2.18 shows the deformation in any radial plane of the upper part of the symmetric structure taken along the centre plane of the substrate (only one side is taken). p denotes the plated part, and s denotes the substrate part. u and w are the radial r and axial z displacements, respectively. Under thermal stress conditions, there is some form of shear stress distributed axially between the plating and the substrate at the point of contact (i.e., at the point $r = r_0$), and it is this force τ_{rz} that contributes to the "adhesion" between the two. Assuming a very small coating thickness, the coating can be approximated as a "rigid body" in the radial direction. From the displacement continuity condition and the relationship between the action and reaction forces, the following relationship (boundary conditions) holds at the

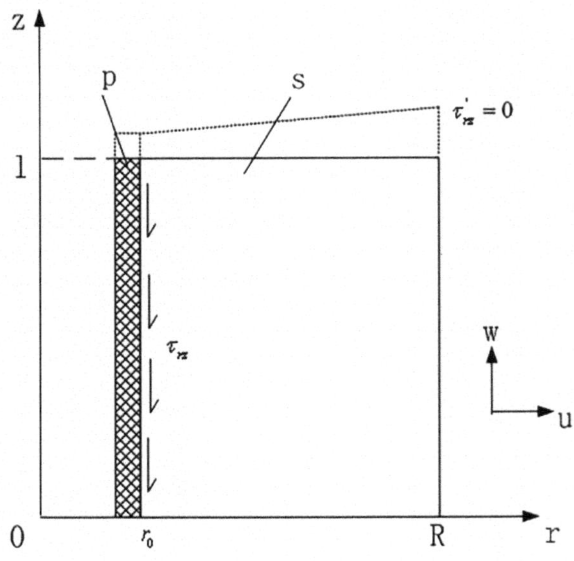

Fig. 2.18 Simplified PTH axisymmetric model

point $r = r_0$.

$$\mathrm{u}_\mathrm{p}^+ = \mathrm{u}_\mathrm{s}^-,\ \mathrm{w}_\mathrm{p}^+ = \mathrm{w}_\mathrm{s}^- \tag{2.57}$$

$$\sigma_{r_p}^+ = \sigma_{r_s}^-,\ \tau_{rz_p}^+ = \tau_{rz_s}^- \tag{2.58}$$

And since the Z-direction can be approximated as the principal stress direction of the plating force, combined with the fact that in the radial direction the plating can be approximated as a "rigid body", according to Eqs. (2.57) and (2.58) there are the following approximate physical and geometrical equations:

$$\varepsilon_z = \frac{\partial w}{\partial z} = \frac{1}{E}\sigma_z \tag{2.59}$$

$$\gamma_{rz} = \frac{\partial w}{\partial r} \tag{2.60}$$

Under thermal cycling conditions, the deformation in the coating and the substrate can be divided into two parts, i.e. the displacement due to thermal stresses $w^{thermal}$ and the displacement due to the "adhesion" w^{stress} between them. This is analysed in detail below.

D. Simplified Model of Stress–Strain Distribution of PTH

In order to determine the distribution of axial stress in the PTH coating, in the above axisymmetric model along the PTH coating axial take as shown in Fig. 2.18, and combined with the equilibrium conditions for stress analysis. In the model shown in Fig. 2.19, the upper and lower surfaces of the plating micro-element are respectively acted with axial stresses, and the lateral shear force of the substrate material acts on the plating material, which causes the "adhesion" effect between the two materials, so that the axial stresses in the PTH plating layer have a certain form of distribution. Based on the results of the preliminary FEM analysis, we can assume that the shear force $\tau_{rz} = \lambda z^2$ between the two materials is distributed along the contact surface. According to the boundary conditions, we can determine λ, i.e. the shear force τ_{rz}, and then according to the equilibrium conditions, we can finally determine the axial stress in the coating.

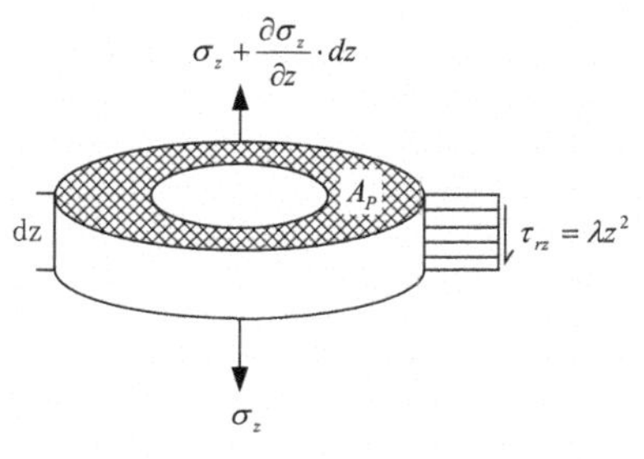

Fig. 2.19 Microelementary bodies in a simplified physical model

According to the force equilibrium condition, there is:

$$\left(\sigma_z + \frac{\partial \sigma_z}{\partial z} \cdot dz\right) \cdot A_p - \lambda z^2 \cdot 2\pi r_0 \cdot dz - \sigma_z \cdot A_p \tag{2.61}$$

where σ_z—axial stress in the PTH plating; cross sectional area of the plating $A_p \approx 2\pi r_0 \cdot t$.

Solving the differential Eq. (2.61) gives:

$$\sigma_z = \frac{\lambda}{3t} \cdot z^3 + C \tag{2.62}$$

From the boundary freedom condition $\sigma_z|_{z=l} = 0$ at the end face of the PTH plating, it is obtained as $C = -\frac{\lambda}{3t} l^3$,

$$\sigma_z = \frac{\lambda}{3t}\left(z^3 - l^3\right) \tag{2.63}$$

Equation (2.63) is the equation for the distribution of axial stress in the plating layer along the axial direction, where only λ is an unknown quantity, and λ will be determined below in conjunction with the boundary conditions.

From Eq. (2.59)

$$\varepsilon_z = \frac{\partial w_p^{stress}}{\partial z} = \frac{1}{E_p}\sigma_z = \frac{\lambda}{3tE_p}\left(z^3 - l^3\right) = \frac{\lambda l^2}{3E_p}\left(\frac{l}{t}\right)\left[\left(\frac{z}{l}\right)^3 - 1\right] \tag{2.64}$$

Solving the differential Eq. (2.64) and combining with the boundary conditions gives that

$$w_p^{stress} = \frac{\lambda l^2}{3E_p}\left(\frac{l}{t}\right)\left[\frac{1}{4}\left(\frac{z}{l}\right)^3 - 1\right] \cdot z \tag{2.65}$$

Considering the thermal stresses due to CTE mismatch between the two materials, the total displacement of the plating is given by

$$w_p^{total} = w_p^{stress} + w_p^{thermal} = \frac{\lambda l^2}{3E_p}\left(\frac{l}{t}\right)\left[\frac{1}{4}\left(\frac{z}{l}\right)^3 - 1\right] \cdot z + \alpha_p \Delta T \cdot z \tag{2.66}$$

According to the displacement continuity condition at the combination of the two materials, the same reason has the total displacement of the substrate as

$$w_s^{total} = w_s^{stress} + w_s^{thermal} = w_s^{stress} + \alpha_s \Delta T \cdot z \tag{2.67}$$

The following boundary conditions should be satisfied:

$$w_s^{stress}|_{r=r_0} + \alpha_s \Delta T \cdot z = w_p^{total} = \frac{\lambda l^2}{3E_p}\left(\frac{l}{t}\right)\left[\frac{1}{4}\left(\frac{z}{l}\right)^3 - 1\right] \cdot z + \alpha_p \Delta T \cdot z \quad (2.68)$$

$$\frac{\partial w_s^{stress}}{\partial r}|_{r=r_0} = \frac{\tau_{rz}}{G_s} = \frac{\lambda z^2}{G_s} \quad (2.69)$$

Accordingly, a function satisfying the above boundary conditions is constructed

$$w_s^{stress} = \frac{\lambda z^2 \cdot r_0}{G_s}\left(\frac{r}{r_0} - 1\right) + \left\{\frac{\lambda l^2}{3E_p} \cdot \left(\frac{l}{t}\right) \cdot \left[\frac{1}{4}\left(\frac{z}{l}\right)^3 - 1\right] + (\alpha_p - \alpha_s)\Delta T\right\} \cdot z \cdot \left[-k\left(\frac{r}{r_0} - 1\right)^2 + 1\right] \quad (2.70)$$

Deriving both sides of Eq. (2.70) with respect to r yields:

$$\frac{\partial w_s^{stress}}{\partial r} = \frac{\lambda z^2}{G_s} + \left\{\frac{\lambda l^2}{3E_p} \cdot \left(\frac{l}{t}\right) \cdot \left[\frac{1}{4}\left(\frac{z}{l}\right)^3 - 1\right] + (\alpha_p - \alpha_s)\Delta T\right\} \cdot z \cdot \left[-\frac{2k}{r_0} \cdot \left(\frac{r}{r_0} - 1\right)\right] \quad (2.71)$$

From the boundary conditions we obtain

$$\lambda = \frac{(\alpha_p - \alpha_s)\Delta T}{\frac{l}{\frac{2k}{r_0}\left(\frac{R}{r_0} - 1\right) \cdot G_s} + \frac{l^2}{4E_p} \cdot \left(\frac{l}{t}\right)} \quad (2.72)$$

Substituting the equation for λ determined by (2.72) into

$$\sigma = \sqrt{\frac{1}{2}\left[(\sigma_x - \sigma_y)^2 + (\sigma_y - \sigma_z)^2 + (\sigma_z - \sigma_x)^2 + 6\left(\tau_{xy}^2 + \tau_{yz}^2 + \tau_{zx}^2\right)\right]} \quad (2.73)$$

The axial positive stress in the plating can be finally obtained as

$$\sigma_z = \left[\left(\frac{z}{l}\right)^3 - 1\right] \cdot \frac{(\alpha_p - \alpha_s)\Delta T}{\frac{3}{2k\left(\frac{R}{r_0} - 1\right) \cdot G_s} \cdot \left(\frac{t}{l}\right) \cdot \left(\frac{r_0}{l}\right) + \frac{3}{4E_p}} \quad (2.74)$$

and there is a maximum value occurring at the centre of the PTH i.e. $z = 0$

$$\sigma_z|_{z=0} = \frac{(\alpha_s - \alpha_p)\Delta T}{\frac{3}{2k\left(\frac{R}{r_0} - 1\right) \cdot G_s} \cdot \left(\frac{t}{l}\right) \cdot \left(\frac{r_0}{l}\right) + \frac{3}{4E_p}} \quad (2.75)$$

In Eqs. (2.74) and (2.75) the subscripts p, s—the relevant parameters of the plating and substrate; σ_z—the axial stress in the plating; τ_{rz}—the shear stress in the substrate on the plating; ε_z—axial strain in the plating; w^{stress}—axial displacement due to the interaction of the substrate and the plating; $w^{thermal}$—thermal stresses axial displacement due to thermal stress; α_p, α_s—coefficient of thermal expansion of the plating and substrate material; E_p, G_s—modulus of elasticity of the plating and shear modulus of the substrate material; r_0—aperture diameter of the PTH (with plating); t—thickness of the PTH plating; R—effective radius of action of the substrate; $l = H/2$—1/2 of the thickness of the substrate (i.e., the length of the PTH); ΔT—amplitude of temperature change (°C); k—calibration coefficient, which can be determined $k = 0.25$ by comparative analysis with the results of FEM.

2.5 Physical Basis of Failure of Mechanical Product Process Reliability

Defects generated by mechanical product processing is the root cause of mechanical product failure and damage, and has an important impact on the use of product performance and reliability. The most common causes of mechanical product failure include: material organisation defects, material deformation and cracking, defects under process environment and media, and surface integrity defects.

2.5.1 Material Organisation Defects

A. Segregation of Components

The phenomenon of non-uniformity of chemical composition of metallic materials produced during solidification. Precipitation makes the organisation and properties of metal materials become uneven, corrosion resistance is reduced, in the production should try to prevent and limit in the permissible range. Metallic materials macro-regional composition of non-uniform phenomenon known as "macro segregation"; one or several grains within the composition of non-uniform phenomenon known as "micro segregation".

Microscopic segregation in alloys is caused by the uneven distribution of cytosolic/dendritic spacing in the solidification unit. Macroscopic segregation occurs when the solidified organisation changes on a macroscopic scale due to compositional differences caused by the redistribution of solute elements and impurities during solidification.

The effect of segregation defects on the elongation of alloy castings is significant, and the elongation of alloy castings tends to decrease as the level of segregation defects increases. The fracture morphology of defect-free castings and segregated castings is mainly pits and icing-sugar-like accumulation morphology compared with that of segregated castings. With the increase of segregation level, the flocculated morphology of castings increases, the fracture pits morphology decreases, and the depth of pits decreases. Macroscopic segregation defects have an effect on the tensile strength of alloy castings, and the tensile strength of alloy castings tends to decrease as the level of segregation defects increases. The fracture of non-defective castings is mainly manifested as toughness fracture, and the fracture of segregation castings is brittle fracture along the crystal.

In addition to the casting process, welding is also prone to segregation. The phenomenon of uneven distribution of chemical elements in the weld metal is called segregation. It is divided into micro segregation, regional segregation and laminar segregation three. The segregation can cause hot cracks, pores, slag and other defects in the weld.

B. Inclusionary Flaw

Inclusion refers to the presence of the chemical composition, physical properties of the material within or on the surface of the composition of the material is different from that of the base material. Inclusions destroy the continuity of the base material, resulting in the organisation of the uneven, especially for metal materials, resulting in metal materials fatigue resistance, plasticity, toughness, corrosion resistance and airtightness are adversely affected. Inclusion defects mainly include inclusions (metallic and non-metallic inclusions), inclusions (internal osmosis beans), slag pores and sand holes, etc., mainly in casting, welding, heat treatment and other thermal processing.

(1) Metal Inclusions

Castings cross-section has a different colour from the metal body of the granular material, size and shape, there are particles larger, alone in the particle cross-section, but also in the casting of a certain range of small particles in the form of the existence of. Mechanical processing due to the formation of different metal inclusions, and show a different colour. Hardness is generally higher than the metal matrix, affecting the quality of machining.

When the metal melting, add the adjustment of composition and deoxidation of the intermediate alloy is not all melted, or the alloy used in pre-furnace treatment is not completely melted, these will form inclusions, adding the wrong location may also form such defects.

(2) Non-metallic Inclusions and Slag Porosity

Non-metallic inclusions are composed of metal oxides, sulphides, phosphides, nitrides and complex compounds composed of them. The internal or surface inclusion defects in the casting are irregularly shaped holes that contain dross. The colour of the dross varies depending on the dross system. Welding inclusions in the weld is very harmful to the weld, it reduces the effective working area of the weld, reduces the weld strength, impact toughness, and reduces the corrosion resistance of the weld; especially in the shape of a sharp angle of the dross will cause stress concentration,

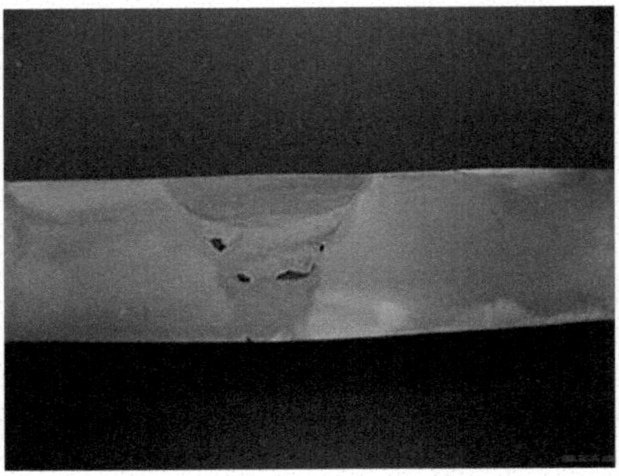

Fig. 2.20 Inclusions in the casting process

and further development of the source of cracks. Such as Fig. 2.20 for a typical casting process to produce inclusions defects.

The reason for the formation of slag pores at high temperatures, in the liquid alloy internal oxidation reaction occurs, resulting in liquid or solid oxides, and the generation of gas, solidification process oxides can not be discharged, exists in the pores that is the formation of slag pores.

Non-metallic inclusions can be divided into primary inclusions and secondary inclusions according to the formation time. Primary inclusions are formed before pouring that is, metal melting and furnace treatment process, secondary inclusions is the liquid metal in the pouring package in addition to the primary slag, pouring until the filling process is oxidised again in the formation of inclusions.

Easy oxidation alloy melting, primary and secondary oxides are many, such as melting aluminium alloy, aluminium is extremely easy to be oxidized elements, but also the alloy matrix, after melting and atmospheric contact, the surface quickly form a solid oxide film (Al_2O_3). Pouring the liquid surface is oxidised, the formation of Al_2O_3 film, and with the flow into the casting, such as too late to exclude, the casting will produce simple Al_2O_3 film inclusions. Castings on the surface and inside the liquid metal filling the formation of oxide film, damage to the continuity of the casting metal, mechanical properties are reduced, which will lead to the early failure of the component.

(3) Cold Beans, Endotoxins

Cold bean defects are bean inclusions in cavities where the metal beads are not fused to the casting matrix but are interconnected, and the chemical composition of the metal beads is the same as that of the casting body. Liquid metal splashing occurs during pouring or mould filling, forming metal beads. Liquid bead surface is oxidised, forming a layer of oxide film, if the pouring temperature is low, can not have the surface of the oxide film of the liquid bead re-fusion, casting solidification, the liquid bead surface of the oxide film reacts with the carbon in the liquid metal, generating carbon monoxide gas surrounded by liquid beads. Liquid bead solidification, the formation of holes in the metal beans that cold bean defects. Cold bean formation process is the existing bean after a hole, mostly in cast steel, cast iron.

Internal ooze, also known as internal ooze beans, similar in appearance to cold beans. In the casting of the internal hole wall, with one or more with a glossy metal beans, but the chemical composition and the casting body is not the same, is a low melting point of the melt. With the formation of cold beans for the opposite reason, the formation of holes in the gradual solidification of defects, and then in the solidification process, the existence of eutectic groups of intergranular phosphorus-containing low melting point eutectic composition of the melt, by the internal and external pressure of the extrusion. Such as cast iron eutectic transformation eutectic expansion pressure, dissolved gas precipitation pressure, external intrusion gas pressure, solidification shell contraction pressure. Under these pressures, the low melting point eutectic melt with high phosphorus content is squeezed into the hole, oozing out on the hole wall, forming metal beans that is, internal oozing bean inclusions.

(4) Trachoma

A hole that exists on the surface or inside of a casting. All or part of the hole is filled with moulding sand or core sand. The main reason is: casting sand or core sand strength is not enough, the pouring system is not set properly, etc., some of the sand falls into the casting cavity, resulting in the liquid metal pouring that is mixed or floating on the surface of the liquid, after condensation in the casting so that the formation of eyelets. Such as Fig. 2.21 for the typical casting of the surface of the eye.

C. Uneven Organisation

Manufacturing process, due to process defects will cause the processing of material organisational inhomogeneity, including grain inhomogeneity, coarse grains, overheating and overcooking and improper distribution of flow lines, resulting in material strength, stiffness or toughness and other mechanical properties and fatigue properties of the reduction, and the formation of product defects. Such defects exist in a wide range of manufacturing processes, including forging, casting.

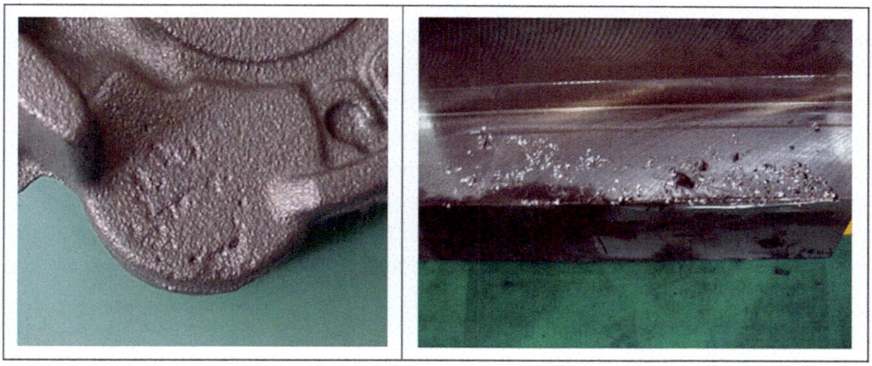

Fig. 2.21 Casting surface sand holes

(1) Uneven Grain Size

Uneven grain refers to some parts of the grain is particularly coarse, some parts are smaller. Causes include uneven deformation of the billet at various places so that the grain broken degree is not the same, or local area of the degree of deformation falls into the critical deformation zone, or high-temperature alloys local work hardening, or quenching and heating of the local grain coarse and so on. Uneven grain for forging common defects. Grain inhomogeneity will make the forging lasting performance, fatigue performance decreased significantly.

(2) Coarse Grain

Large grain is usually caused by the beginning of the forging temperature is too high and the degree of deformation is not enough, or the final forging temperature is too high, or the degree of deformation falls into the critical deformation zone. Coarse grain will make the plasticity and toughness of the forgings reduced, fatigue performance decreased significantly.

Forging aluminium and hard aluminium forging is easy to produce large grains, they are mainly distributed in the forging degree of small and large size of the larger parts, as well as the degree of deformation and deformation of the region of intense and near the fringe region. In addition, there is often a layer of coarse crystals on the surface of the forging. The reason for large grains in addition to the degree of deformation is too small, the degree of deformation is too large and deformation intense uneven caused by, heating and die forging too many times, the heating temperature is too high, the final forging temperature is too low will also produce large grains. Coarse crystals on the surface layer of the forging, one is the surface of the extrusion billet coarse crystals ring was brought into the forging; the second is the die forging when the surface of the die cavity is too rough, the abrasive tool temperature is low, lubrication is poor, so that the surface contact layer of the intense shear deformation and produce.

(3) Overheating and Overfiring

Metal in the forging heating, due to the temperature is too high, grain growth is very large or abnormal organisation, so that the performance of the phenomenon is significantly reduced, called overheating, such as Fig. 2.22 is PCrNi3Mo steel forging overheating organisation, is a typical overheating organisational morphology. Superheating temperature close to its solid phase line attachment, grain boundary oxidation and the beginning of the phenomenon of partial melting, known as overcooking, such as Fig. 2.23 is GCr15SiMn steel forgings overcooking organisation. Steel overheating will not only greatly reduce the mechanical properties of steel and the use of performance, and it is easy to produce quenching cracking.

(4) Improper Distribution of Flow Lines

The impurities, compounds, segregation, grain boundaries, etc. in the metal on the low times the tissue along the main tensile deformation direction of the fibrous distribution of the tissue, become the metal fibre tissue or flow line. Streamline is

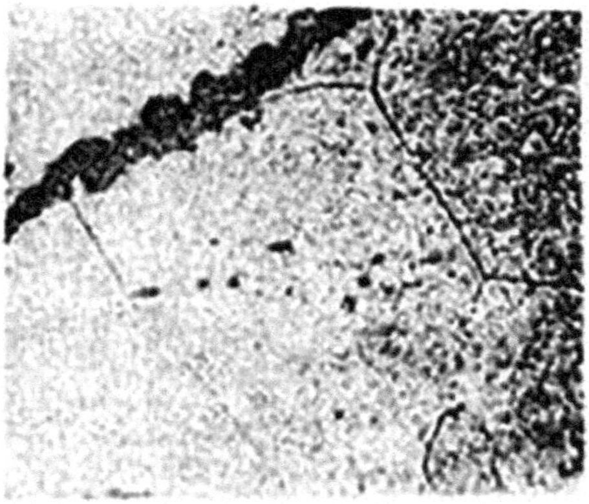

Fig. 2.22 Superheated organisation of PCrNi3Mo steel forgings

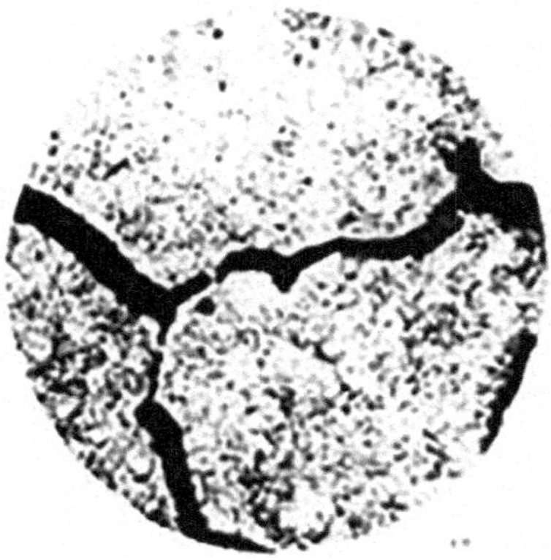

Fig. 2.23 Superheated organisation of GCr15SiMn steel forgings

not according to the forging geometry distribution of the phenomenon known as streamline not smooth or streamline distribution is not appropriate. Forging in the forging process flow line is not smooth, eddy current and through the flow. The plasticity, fatigue and corrosion resistance properties have a great impact on the strength of a smaller effect.

Fibre orientation has a greater effect on strength, especially plasticity. Because there are a large number of brittle impurities and compounds, etc. distributed along the material flow line, so the transverse specimen subjected to tensile stress strain, will be these heterogeneous plasmas as the core of the formation of microscopic holes, and continue to expand and connect into large cracks, while the longitudinal specimen is not, so the transverse performance of forgings is generally lower than the performance of the longitudinal direction.

2.5.2 *Deformation and Cracking*

Mechanical components subjected to mechanical loads, thermal loads or environmental media of a single or joint action, will produce deformation and cracking defects. These two types of defects have a certain degree of relevance, purely because of deformation and the failure process is generally slow, mostly non-catastrophic, and therefore often can not cause people to pay special attention to, but ignore the deformation defects of the supervision and prevention will lead to great losses or even catastrophic accidents, this is because the excessive deformation will eventually lead to fracture of the material. Cracking of the product material will make the zero component completely incapable of working, and sometimes cause major accidents in machinery and equipment and personal hazards, so the failure caused by cracking defects is the most dangerous failure in engineering.

A. Deformation

The change in shape and size of the material under the action of external forces is deformation. The deformation that occurs in the manufacturing process can be divided into cold working deformation and hot working deformation.

(1) Cold Working Deformation

Cold working process generally refers to the processing in the state below the recrystallisation temperature, usually refers to the metal cutting process, and the workpiece to obtain a certain shape, dimensional accuracy and surface roughness of the processing method. Including turning, milling, planing, drilling, grinding and so on. Workpiece after cold working residual part of the plastic deformation, resulting in defects or even scrap workpiece.

Bending wrinkle: In the bending process, the inner metal is subject to tangential compressive stress, which is prone to instability wrinkle or folding. When the bending radius is very small, the bending angle is very large, the sheet thickness is relatively large, wrinkled or folded phenomenon is more serious. Such as for the pipe, bending wrinkles, pipe bending processing in the deformation area of the side wall and shrinkage deformation area of the wall, wall thickness in the bend are increased. If the degree of deformation is too large, the wall tube loss of stability, causing wrinkles. Such defects occur in compression moulding process. In addition, spinning, when the spinning wheel in the high-speed rotation of the blank surface rolling, if there are other minor obstacles will cause vibration, so that the surface of the workpiece to produce wrinkles, in the latter process is not easy to eliminate, resulting in scrap.

Stretching wrinkles: Stretching is a plastic moulding process that turns a flat blank into an open hollow part. Blank in the stretching process, his surrounding edge part of the tangential compressive stress is too large, resulting in material destabilisation, so that the product along the edge of the tangential formation of uneven wrinkles, known as stretching wrinkles. After serious wrinkling, the blank is difficult to pass through the gap between the concave and convex moulds, and is pulled off, resulting in scrap.

(2) Hot Working Deformation

Thermal processing mainly includes casting, forging, welding and other machining processes, the different processes lead to different causes of deformation, the following are introduced casting deformation and welding deformation causes.

Casting deformation: Casting completely solidified after the solid state contraction stage, if the casting of the solid state contraction is impeded, the casting will produce internal stress, known as casting stress. It is the basic reason for casting deformation, the result of deformation is subject to tensile stress tends to shorten the deformation of the parts, subject to compressive stress tends to elongate the deformation of the parts, in order to make the casting in the residual stress is reduced or eliminated. According to the cause of stress, casting stress is divided into thermal stress and mechanical stress.

Thermal stress is caused by the uneven wall thickness of the casting or the cooling rate of each part is different, so that the contraction of each part of the casting is not synchronous. It still exists inside the casting after the casting is sanded, and is a kind of residual stress. Residual thermal stress is formed because the casting wall thickness is not uniform or the cooling rate of each part of the casting or the thick wall of the casting or the heart of the tensile stress, thin wall or the surface layer of the compressive stress, and with the casting of the wall thickness difference increases, the difference between the cooling rate of each part of the difference between the casting alloy line shrinkage increases and the increase in its elastic modulus, resulting in the casting of the thermal stress increases.

The basic measure to prevent the casting of thermal stress is to reduce the temperature difference between the various parts of the casting, so that it is uniformly cooled. Specifically for the selection of elasticity modulus smaller alloy as casting alloy; design casting structure, and strive to make its wall thickness uniform; the use of reasonable casting process, so that the solidification of the casting in line with the principle of simultaneous solidification. Eliminate the residual thermal stress of the casting is the method of de-stressing annealing treatment. That is, the casting is heated to the plastic state, insulation for a certain period of time, slowly cooled to room temperature, can be basically eliminated its residual casting stress.

Mechanical stress is due to the contraction of the casting by the casting or core and other mechanical obstacles to the formation of stress, as shown in Fig. 2.24. This stress is temporary, in the casting sand or mechanical obstruction will disappear after the disappearance of their own.

In order to prevent deformation, the casting wall thickness should be made as uniform as possible or make its cross-section shape symmetry, in the casting process should take corresponding measures, and strive to make it solidify at the same time; sometimes, the slender easy to deformation of the casting, in the manufacture of the model, will be made into the model and the direction of deformation is exactly the opposite of the shape in order to offset its deformation.

Welding residual deformation: After welding, the residual deformation of the weldment has plane deformation and volume deformation of two categories. Plane deformation can be divided into longitudinal contraction of the weld, weld transverse

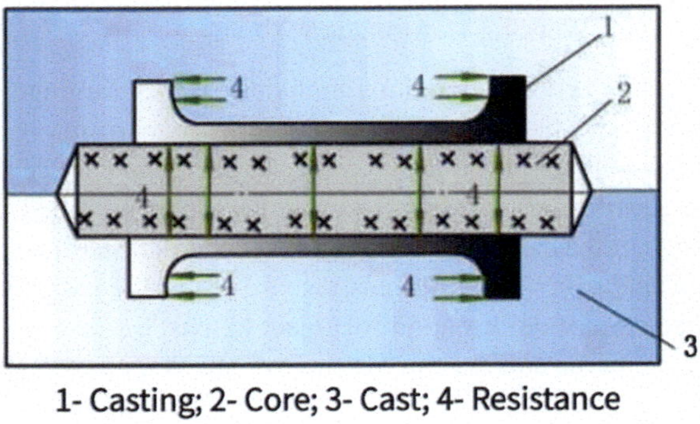

Fig. 2.24 Casting mechanical stress generation

contraction, rotational deformation. Volume deformation can be divided into angular deformation, bending deformation, torsional deformation and temperature loss wave deformation.

Commonly used for plate butt joints, plate straight butt weld welding, resulting in bending deformation are: uneven temperature distribution in the cross-section, caused by the two boards to the outside of the open bending deformation; has been welded weld transverse contraction caused by the two boards inward bending deformation; has been welded weld longitudinal contraction caused by inward bending deformation in the face; there are also due to the role of heat and the plate's out-of-face transient destabilisation deformation (wave deformation) The following are some examples of such deformations. Warp instability of the plate is due to the longitudinal residual tensile stress in the weld and caused by the compressive stress on both sides of the plate, when the value of the compressive stress is higher than the plate's critical instability of the value of the compressive stress, then the plate warping deformation occurs.

B. Cracking

Crack is one of the main defects commonly found in the processing of metal materials, and there are two main categories of causes: one is the organisational structure of the material, which is the internal cause of crack generation and expansion; the other is the stress state in which the material is located, which is the external cause of crack generation and expansion. Other factors such as deformation temperature and deformation rate affect the crack generation and expansion by influencing the material's organisational structure and stress state, which can also be attributed to the external causes of crack generation. Cracks can be caused directly by external forces, such as shear cracking in shear, expansion of the workpiece tensile cracks, etc.; can also be caused by additional stresses, additional stresses are often generated by the deformation of non-uniform or uneven flow rate.

The specific causes of cracks in different forming processes and the corresponding crack patterns are different, mainly as follows:

(1) Cracks in Castings

Cracks in castings occur when casting stresses exceed the strength limit of the metal. Typical casting cracks are hot cracks, cold cracks, cold segregation.

Hot cracking: Cracks formed at high temperatures at the end of solidification. In the ingot is not yet fully solidified or solidified and grain boundaries and dendrites still have a small amount of low melting point phase, due to the metal liquid, solid state contraction and solidification contraction is impeded, when the contraction stress exceeds the strength of the metal at that time or line shrinkage is greater than the alloy elongation when the formation of the. Such cracks mostly appear along grain boundaries, have a zigzag and irregular shape, and branch. Because it arises at higher temperatures, it is mostly oxidised in colour.

The process factors affecting thermal cracking are mainly the casting process. Certain elements in the alloy and insoluble low melting point impurities can significantly increase the tendency of thermal cracking, copper such as pouring temperature is too high will also increase the possibility of surface thermal cracking, semi-continuous casting ingot cooling intensity is greater than the iron ingot casting tendency to thermal cracking is much greater than the casting of casting in the casting to increase the speed of casting will increase the tendency of thermal cracking from the structure of ingot, the larger the cross-sectional dimensions, the more prone to thermal cracking. Figure 2.25 shows the casting process caused by thermal cracking, can be seen in the shape of the crack zigzag and with branches, can be presumed to be along the grain boundaries along the direction of the emergence of the crack.

Cold crack: It is a crack formed when the casting is in an elastic state, i.e. at low temperature. Its surface is smooth, with a metallic luster or a slightly oxidised colour, cracks occur through the grain, the shape is regular, often rounded curves or straight lines. Its mechanism for the casting solidification in the elastic state, the cooling process, the alloy will occur in the solid state phase transition caused by contraction or expansion, so that the casting volume and length changes, when this change is impeded, casting will produce casting stress. When the casting stress exceeds the tensile strength of the alloy, cracks will occur in the stress concentration area, i.e. cold cracking.

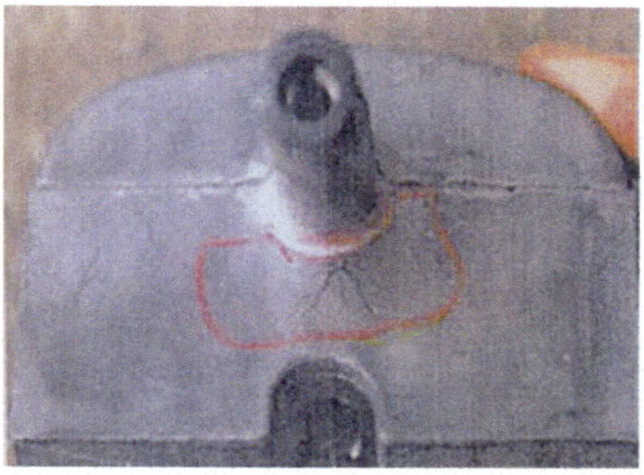

Fig. 2.25 Casting-induced thermal cracking

Residual stress is the main cause of cold cracking in castings during the machining process. There is rarely a single residual stress in the casting, generally the coexistence of residual thermal stresses and residual mechanical resistance, or then coexist with the phase change stresses of all three. Castings solidification end, the alloy has lapped into a branched crystal skeleton, in the subsequent cooling process, due to the uneven thickness of the cross-section, the cooling rate is different, there is a temperature difference caused by the casting stress, called thermal stress. Castings cooling is a solid phase change in the alloy due to the temperature to reach the phase change time is not consistent, the degree of phase change in each part is also different, the resulting stress is called phase change stress. Castings shrinkage by casting, core, pouring riser system and casting itself the structure of the obstacles generated by the stress called mechanical resistance. Three types of residual stresses lead to cold cracking.

Cold segregation: Cold segregation is a casting with incompletely fused seams that meet with rounded edges. The defect is mostly found at the wide surface of the casting away from the inner pouring channel, or the thin-walled section which is difficult to be filled and the confluence of two streams of liquid metal in the cavity. Preventive measures can be appropriate to improve the liquid metal pouring temperature and shell temperature, increase the metal hydraulic head, to prevent breakage; improve the design of the pouring system, increase the cross sectional area of the pouring system and so on. Figure 2.26 is a typical casting surface cold segregation.

(2) Forging Cracks

Cracks are usually caused by the presence of large tensile, shear or additional tensile stresses during forging. Cracks usually occur in the part of the billet where the stress is the greatest and the thickness is the thinnest. If the billet surface and internal microcracks, or billet within the existence of organisational defects, or hot processing temperature improper plasticity of the material to reduce, or deformation speed is too fast, the degree of deformation is too large, more than the material allows the plasticity of the pointer, etc., in upsetting, drawing long, punching, reaming, bending and extrusion and other processes may produce cracks.

Upsetting crack is when the height of the billet is reduced, the cross-section of the upsetting process to increase, the amount of underpressure is relatively large, the side surface of the billet is prone to longitudinal cracks or 45° direction of the cracks.

Fig. 2.26 Cold separation of casting surfaces

When upsetting, the force is along the axial direction, and the longitudinal crack on the side surface is caused by the tangential tensile stress, the tangential tensile stress arises from the uneven deformation during upsetting. The main reason for this uneven deformation is the friction between the mould and the blank, and this uneven deformation makes the appearance of the blank to be subjected to tangential tensile stress, and the closer to the surface, the greater the tangential tensile stress. When the tangential tensile stress exceeds the strength limit or tangential deformation exceeds the main degree of material allowed, it will cause longitudinal cracks. Low-plasticity materials often produce shear cracks in the 45° direction on the side surfaces due to their weak resistance to shear.

Longitudinal cracks are often caused by side surface cracks and angular cracks on the outside of the blank when the drawing process is carried out so that the cross-sectional area of the blank is reduced and the length is increased, and when long ingots and billets of low-plasticity materials (e.g., high-speed steel, etc.) are drawn on a flat anvil.

In addition, white spots are also typical forging defects, is a kind of internal defects produced in the cooling process after forging. Its morphology in the transverse low-power specimen for the fine hairline acute angle cracks, fracture for the silver-white spots. White spot is essentially a brittle sharp-edged crack, with great harm, is a very dangerous defect in martensitic and pearlitic steel. Such as Fig. 2.27 for the macro fracture on the white point morphology, you can see the silver-white spot fracture.

(3) Heat Treatment Cracks

Heat treatment cracks include cracks caused by excessive heating rate, cracks caused by surface carbonisation or decarburisation, cracks caused by overheating and over-cooking, tempering cracks, ageing cracks and so on. Among them, quenching crack is the most common type of heat treatment crack, which refers to the cracks produced during the quenching process or during the room temperature placement after quenching. Quenching cracks are macro cracks, mainly caused by macroscopic internal stress. In the manufacturing process, due to improper material selection, inaccurate quenching temperature control, quenching cold speed is not appropriate and other factors, on the one hand, increase the sensitivity of microcracks, increase the number of microcracks, reduce the material's brittle fracture resistance SK, which

Fig. 2.27 White dot morphology on macroscopic fracture

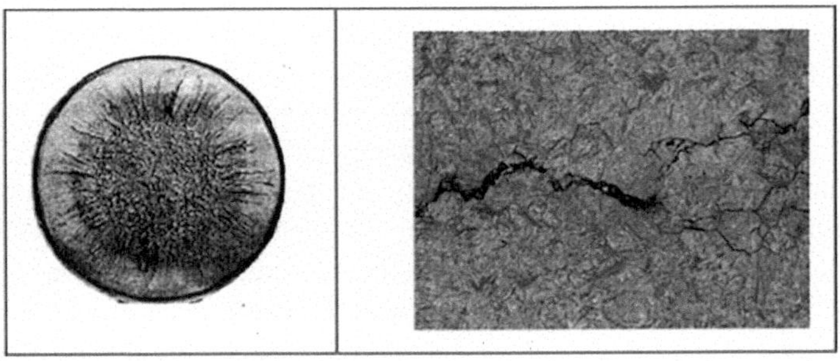

Fig. 2.28 Surface quenching transverse cracks in 9Cr2Mo steel rolls

increases the possibility of quenching crack formation; on the other hand, increase the quenching stress, so that the quenching microcracks have been formed to expand the quenching, the formation of macroscopic quenching cracks. Control of quenching cracks need to accurately control the heat treatment process, select the appropriate heating temperature, holding time, heating medium, cooling medium and cooling methods. Such as Fig. 2.28 for 9Cr2Mo steel roll surface quenching transverse cracks, and its microstructure.

(4) Welding Cracks

Welding crack is in the welding stress and other embrittlement factors together, the atomic bond of the material has been broken, the formation of new interfaces and the resulting gap. Cracks are the most dangerous defects in welded structures, often resulting in failure of the welded structure. Cracks that occur in welded structures are usually divided into five categories: hot cracks, reheat cracks, cold cracks, laminar tearing and stress corrosion cracks.

Welding thermal cracking is a kind of cracking along the crystal in the weld metal or weld heat-affected zone at high temperatures near the solid phase line. Among them, the thermal cracks occurring in the weld zone are produced in the weld crystallisation process, becoming crystalline cracks. Occurs in the heat-affected zone of thermal cracking is close to the weld of the base metal is heated to the superheat temperature, the intergranular low melting point impurities occur melting and the formation of cracks, known as liquefaction cracks.

Welding reheat cracking refers to the weldment in a certain temperature range after welding to be heated again and cracks, it is stress relief treatment cracks, post weld tempering cracks, strain aging cracks of the general term. Reheating crack formation conditions are, joints in the post-weld reheating, in the stress concentration of the coarse grain area of the residual stress relaxation, so that the grain boundary microscopic local sliding deformation of the actual plastic strain over the plastic strain capacity of the material. It was found that the process of material selection, alloy element content is more, and can make the precipitation hardening of the alloy within the crystal, only has a significant re-heating crack sensitivity.

Welding cold cracking is cracking of welded joints at temperatures near room temperature. The most common type of cold cracking is delayed cracking, i.e., cracking that occurs after a delay of some time after welding. Cold crack formation of three basic elements: the existence of hardened tissue in the welded joint, so that the joint properties of embrittlement; welded joints with high hydrogen content, resulting in hydrogen embrittlement; and hydrogen through diffusion in some welding defects in the aggregation, the formation of local high stress zone and crack initiation; welded joints exist in the role of a large welding stress. In the medium carbon steel, high carbon steel, low to medium alloy steel, tool steel, martensitic stainless steel, cast iron and titanium alloys have a tendency to produce cold cracks.

Laminar cracking exists in rolled thick steel plate corner joints, T-joints and cross joints. Due to the multi-layer welded fillet weld produced by the excessive Z stress, in the weld heat affected zone and the nearby base material caused along the direction of rolling with the development of step-like cracks. Laminar cracking is cold cracking in nature, and non-metallic impurities can often be found in the cracked platform portion. Stress corrosion cracking is a delayed cracking phenomenon that occurs under the joint action of certain specific environments and tensile stresses, and stress corrosion cracking may occur in the welded structures of low-alloy high-strength steels, austenitic steels and aluminium alloys.

2.5.3 Defects Under the Action of Process Environment Media

Mechanical product manufacturing process are in a certain environment, all the process defects are related to the environment, only sometimes the influence of the environment is not the main factor. Environmental media as the main cause of process defects, including: corrosion, hydrogen embrittlement and liquid metal embrittlement.

A. Corrode

Corrosion is the deterioration and damage suffered by a material as a result of chemical and electrochemical interaction with the surrounding medium. Therefore, corrosion damage to metal parts is in most cases a chemical process, caused by the transformation of metal atoms from a metallic state to a non-metallic state of the compound, a reaction process at the interface. According to the incentive for corrosion to occur, corrosion can basically be divided into two categories: chemical corrosion and electrochemical corrosion. The difference between the two only lies in the former is the metal and medium only chemical reaction, in the corrosion process, no current generation. And the latter in the process of corrosion in the process of current generation.

(1) Chemical Corrosion

Corrosion damage caused by purely chemical interactions between the metal and the surrounding non-electrolyte is called chemical corrosion. Its reaction course is characterised by the direct redox reaction between the atoms on the surface of the material and the oxidant in the non-electrolyte (equation below), and the corrosion products are generated on the surface where the corrosion reaction occurs. The corrosion reaction process is not accompanied by the generation of electric current.

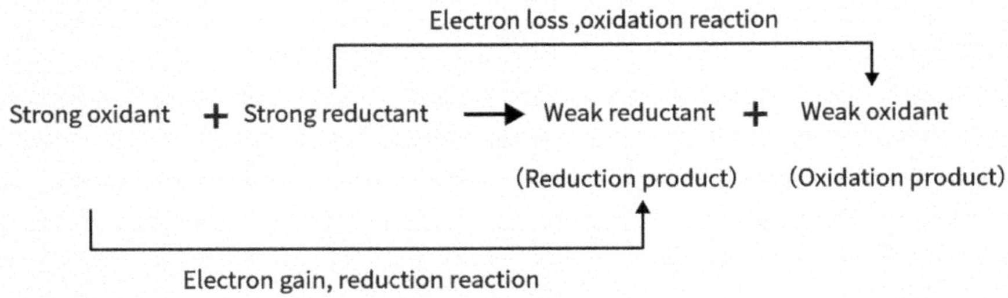

Relative to electrochemical corrosion, the occurrence of pure chemical corrosion is less, it can be divided into categories

a. Dry Gas Corrosion

Dry gas corrosion is in the metal in the dry gas (no moisture condensation on the surface) corrosion. For example, metal and dry gas O_2, H_2S, SO_2, Cl_2, etc. contact directly occurring chemical reaction.

b. Non-electrolyte Solution Corrosion

Generally refers to the corrosion of metals in non-conductive solutions, such as metal corrosion in organic liquids (such as alcohol and oil, etc.).

(2) Galvanic Corrosion

The phenomenon of corrosion of metals due to the action of a corrosion cell when the metal is in contact with an electrolyte is known as galvanic corrosion. Corrosion cell is defined as a short-circuiting primary cell that can only lead to the destruction of the metal material and cannot do useful work to the outside world. The basic process of corrosion primary cell work must include the following three aspects:

a. Anodic process: The metal M undergoes anodic dissolution and enters the solution in the form of ions, while leaving an equal amount of electrons ne the surface of the metal:

$$\left[ne^{-} \cdot M^{n+}\right] \to M^{n+} + \left[ne^{-}\right]$$

b. The cathodic process: The oxidising agent D in solution absorbs the electrons from the process at the electrode and is itself reduced:

$$D + \left[ne^{-}\right] \to \left[ne^{-} \cdot D^{n+}\right]$$

c. The two cathodic and anodic processes described above are carried out on the same metal or on different metals in direct contact with each other, and there is a current flowing in the metal circuit.

Electrochemical corrosion that may occur during the manufacturing process can be classified into the following categories according to the different environments to which they are exposed:

a. Atmospheric Corrosion

Atmospheric corrosion refers to the corrosion of metal in a moist gas. Such as water vapour, carbon dioxide, oxygen and other gases meet the metal are formed compounds. For example, oxidation and decarburisation occur in heat treatment of metals.

b. In the Electrolyte Solution Corrosion

Metal structure in natural water and acid, alkali, salt and other aqueous solutions of corrosion belongs to this category. Examples include molten salt corrosion occurring in metallurgy, corrosion occurring during pickling processes, etc.

c. Contact Corrosion (Galvanic Coupling Corrosion)

Two kinds of electrode potential different metal mutual contact is the occurrence of corrosion. Because of the two metal electrode potential is different, composed of an electric couple, so also known as galvanic coupling corrosion.

According to the way of corrosion damage, corrosion can be divided into three categories: namely, uniform corrosion (comprehensive corrosion), local corrosion and corrosion fracture. Uniform corrosion on the entire metal surface, the corrosion rate is generally the same; local corrosion is limited to a certain area of its corrosive effect, which includes spot corrosion, pustular corrosion, pitting, intergranular corrosion, crystal corrosion, selective corrosion, corrosion; corrosion rupture is in the stress (applied stresses or internal stresses) and corrosive media under the joint action of the parts or components of the final fracture. Stress corrosion process, the material first appear micro-cracks and then can be extended to macro-cracks, cracks once formed, its expansion rate than other types of corrosion faster. It caused by the damage in advance often no obvious warning and sudden brittle fracture, is the most destructive kind of corrosion. Stress corrosion fracture is brittle damage, even if the excellent ductility of the material to produce stress corrosion fracture is also brittle fracture. The fracture is flush, perpendicular to the main stress, there is no obvious signs of plastic deformation, the fracture is grey, which is usually due to a layer of corrosion products covering the fracture results, see Fig. 2.29.

B. Hydrogen Embrittlement

As hydrogen penetrates deep into the metal and causes damage, the failure of a metal part under continuous static stress below the yield limit of the material is known as hydrogen rupture failure, commonly referred to as hydrogen embrittlement. Metallic materials are susceptible to hydrogen intrusion during processing and manufacturing, as well as in the environment in which they are used. On the mechanism of hydrogen embrittlement, there is no unified understanding. The common point of various theories is: hydrogen atoms through stress-induced diffusion in the high-stress region of enrichment, only when the concentration of enriched hydrogen reaches a critical

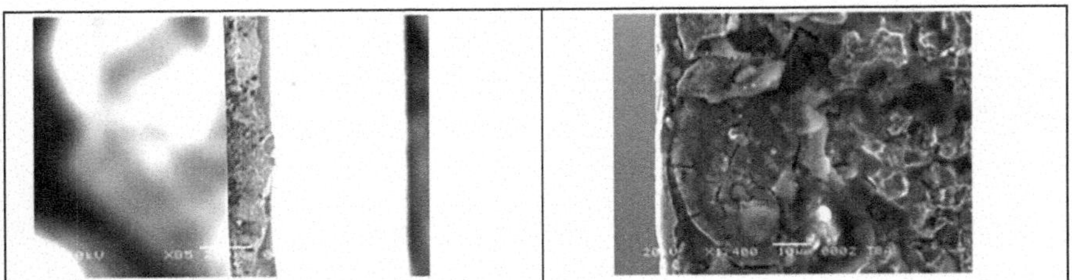

(a) The fracture morphology is low (b) Source corrosion morphology

Fig. 2.29 Fracture morphology of stress corrosion

value of Ccr, so that the material rupture stress σf reduction occurs, brittle fracture. How the enriched hydrogen acts is not clear. There are four more popular views:

(1) Hydrogen pressure theory: It is believed that supersaturated hydrogen in the metal is enriched, precipitated, and combined into hydrogen molecules at the defect location, resulting in a large internal pressure and therefore a reduction in the external stress required for crack extension. The theory can explain the existence of a gestation period, discontinuous crack extension, the effect of strain rate, etc., but it is difficult to explain the phenomenon that high lightness steels can crack even when the hydrogen partial pressure is lower than the atmospheric pressure, and it cannot explain the reversibility of reversible hydrogen embrittlement. However, this theory is recognised in the case of irreversible hydrogen embrittlement at higher hydrogen contents, such as hydrogen bulging that occurs without the action of an external force.
(2) Adsorption of hydrogen to reduce the surface energy theory: Griffith proposed that the fracture stress of the material $\sigma_f = \sqrt{\frac{2E\gamma_s}{\pi a}}$. When hydrogen is adsorbed on the crack surface, the specific surface energy γ_s decreases, thus the fracture stress decreases, causing hydrogen embrittlement. The theory explains the presence of a gestation period, the effect of strain rate, and the phenomenon of brittle fracture at low partial pressures of hydrogen, but the formula is only applicable to brittle materials. Fracture of metallic materials also requires plastic deformation work γ_p, viz. $\sigma_f = \sqrt{\frac{E(2\gamma_s+\gamma_p)}{\pi a}}$. γ_p approximately 103 times as much as γ_s, and the decrease of γ_s in hydrogen adsorption is does not have a significant effect on σ_f. In addition, O_2, CO_2, SO_2, CO, CO_2, CS_2 and other adsorption capacity than hydrogen, it should be able to cause greater brittleness, but the fact is not the case, and even a small amount of hydrogen mixed with these gases, hydrogen embrittlement has an inhibitory effect.
(3) Weak bond theory: That hydrogen into the material can make the material of the inter-atomic bonding force is reduced, the reason is that hydrogen 1s atoms into the transition group progress of the d-band, so that the d-band electron density increases, so that the s–d-band part of the overlap, and thus the inter-atomic repulsive force increases, that is, the bonding force decreases. The theory is simple, direct and easily accepted. However, the experimental evidence is not

yet sufficient, for example, the elastic modulus of the material is related to the bonding force, but the experiment did not find that hydrogen has a significant effect on the elastic modulus. In addition, aluminium alloys without a 3d band may also undergo inverse hydrogen embrittlement, so it is unlikely that 1s electrons of hydrogen enter the d band of electrons.

(4) Hydrogen promotion of local plastic deformation theory: It is believed that the essence of hydrogen cracking and general fracture process is the same, both of which are preceded by local plastic deformation and lead to cracking when they develop to the critical state, and the role of hydrogen is to promote local plastic deformation at the crack tip. Experiments show that the atomic hydrogen enriched near the crack tip by stress-induced extension acts together with the stress to promote the large-scale value-added and movement of the dislocations at the place, so that the plastic zone at the crack tip increases, and the deformation within the plastic zone increases. However, by the limitations of the immaturity of the metal fracture theory itself, the process of crack formation and expansion after a certain degree of local plastic deformation is not yet clear, and the role of hydrogen in this process also needs to be studied in depth.

Manufacturing process, hydrogen into the metal material can be summarised in the following two ways:

(1) Hydrogen Entering During Metallurgy, Welding and Heat Treatment

Due to the solubility of hydrogen in the metal material with the temperature and change, when the temperature is reduced or organisational changes, the solubility of hydrogen from large to small, hydrogen will be precipitated from the solid solution, and due to the solidification or cooling speed is faster, run out, it is residual in the body of the metal material.

(2) Hydrogen Enters in Electroplating, Pickling and Hydrogen-Emitting Corrosive Environments

This type of hydrogen usually enters during chemical or electrochemical treatment. In electroplating, the part acts as a cathode, so the deep penetration of hydrogen is unavoidable, at which time hydrogen is released at the macro-cathode or micro-cathode. Therefore, we should try to take the plating solution with less hydrogen embrittlement, or take the post-plating treatment and the use of vacuum plating, ion plating and other non-hydrogen embrittlement or less hydrogen embrittlement process. In the pickling process in addition to the metal surface oil, adhesion, oxide film and pickling solution reaction, there may be metal and pickling solution to see the chemical reaction. The hydrogen produced by the reaction in addition to the form of molecular hydrogen escapes, there are some hydrogen may enter the metal. Therefore, high-strength steel and some materials sensitive to hydrogen embrittlement generally do not allow pickling. Otherwise, the metal should be dehydrogenated as soon as possible after pickling.

Hydrogen embrittlement fracture macroscopic morphology is mainly characterized by: no macro-plastic deformation near the fracture, fracture flush, rough structure, hydrogen embrittlement fracture zone is crystalline granular, bright grey in

Fig. 2.30 hydrogen embrittlement fracture macro fracture morphology

colour, clean section, no corrosion products. Figure 2.30 shows the macro–micro fracture morphology of hydrogen embrittlement fracture of 300M steel round bar specimen.

Metallic hydrogen embrittlement fracture micro-morphology generally shows along the crystal separation, may also be through the crystal, along the via separation system along the grain boundary occurs along the grain boundary occurs along the crystal brittle fracture, icing sugar lumps. The crystal surface of the fracture is flat and free of attachments. Figure 2.31 illustrates the morphology of hydrogen embrittlement along grain fracture of a 300M steel round bar specimen.

C. Liquid Metal Embrittlement

Liquid metal embrittlement refers to the process of embrittlement of a ductile metal or alloy in contact with a liquid metal resulting in a decrease in suction and brittle fracture. In general, most liquid metal embrittlement is due to chemisorption in the liquid metal. Westwood et al. proposed that if the ratio of the maximum tensile breaking stress σ at the crack shear shortening to the maximum shear stress τ at the crack tip cross-slip surface is greater than the ratio of the true breaking stress σT to the true shear stress τT, i.e., $\sigma/\tau > \sigma T/\tau T$, then the atoms at the crack tip are separated

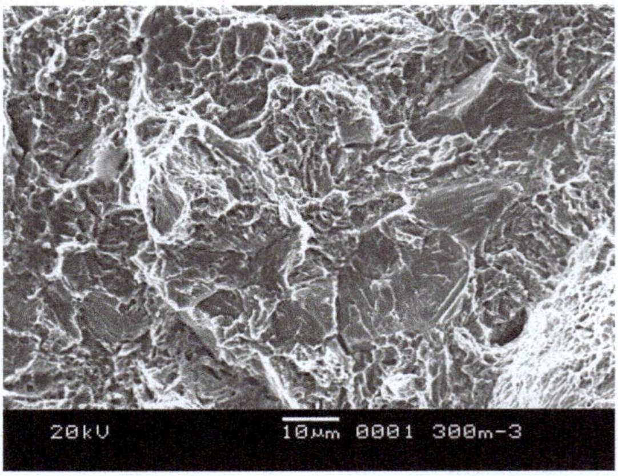

Fig. 2.31 Hydrogen embrittlement fracture along the crystal

by tension, the crack expands in a brittle manner, as the adsorption effect is shielded from the surface layer at depths greater than 10 nm, and the adsorption reduces the tensile lightness of the interatomic bonding bonds at the crack tip without affecting the slip that intersects in the plane of the crack tip, thus promoting brittle cracking rather than plastic cracking. Figure 2.32 illustrates the liquid metal embrittlement mechanism.

a. Interatomic tensile segregation at the crack tip and adsorption reduces the tensile strength between the A–A bonding bonds, but does not affect slip on the S–P surface;

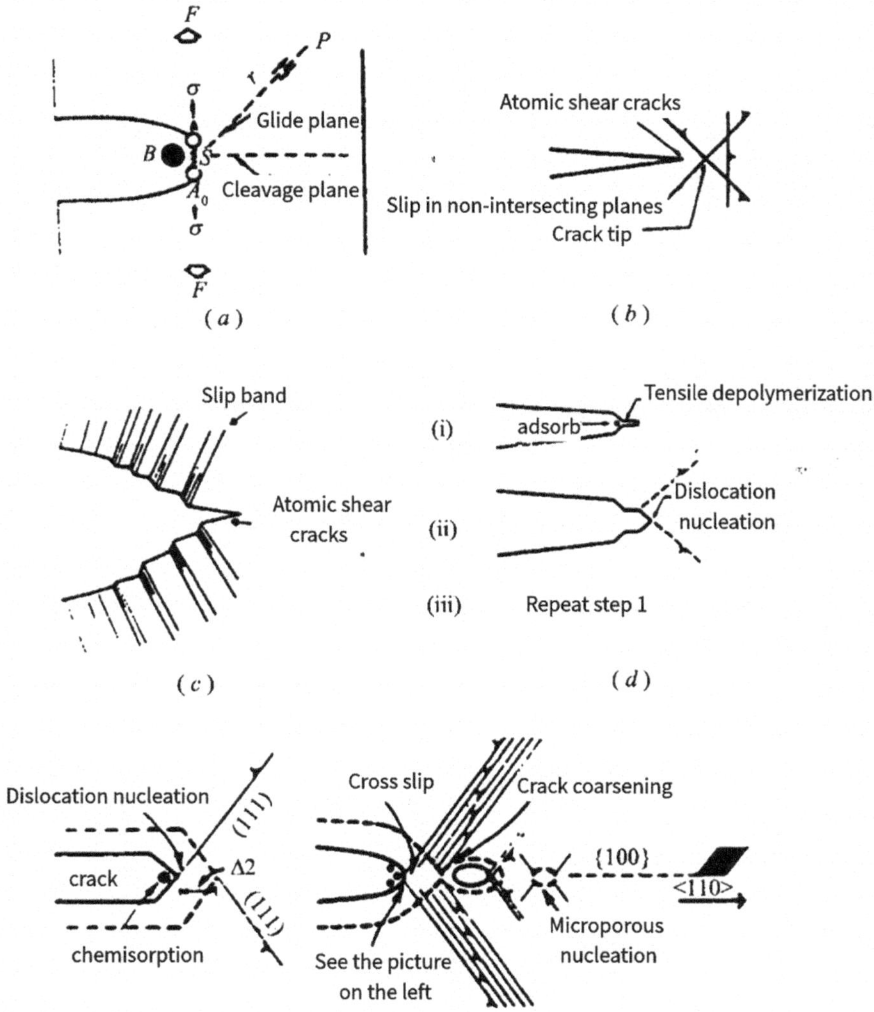

Fig. 2.32 Liquid metal embrittlement mechanism

b. Tensile depolymerisation forces acting in conjunction with dislocation loops at the crack tip disjoint;
c. The tensile depolymerisation force is accompanied by slips that widen very sharp cracks (atomic scale) to macroscopic scales;
d. Tensile depolymerisation forces alternating with dislocation dispersion;
e. Macroscopic brittle fracture occurs as a result of dislocation dispersion at the crack tip (promoting adsorption) and nucleation of a cavity in front of the crack.

There are four main pathways by which embrittlement of liquid metals occurs:

(1) Hot Dip Coating and Thermal Deformation Process

Processes leading to liquid metal embrittlement are mainly hot dip coating process and surface thermal deformation. Hot dip coating process is widely used to improve the corrosion and wear resistance of the body material, Zn, Sn, Cd, Pb and Al are often coated on the steel surface. The embrittlement of liquid metals caused by this dip coating process is divided into two categories: during the dip coating process and during the service of the component. In the latter case, the dip coating process is usually not the real cause of failure. During the dip coating process, the liquid metal is in contact with the component, and if stresses are present, this creates ideal conditions for liquid metal embrittlement.

(2) Welding Processes

Manufacturing process occurs in the liquid metal embrittlement mainly refers to the welding process in order to change the processability of the low melting point of the metal added. Such as ferrous or non-ferrous metals in the brazing material added Pb–Bi to improve the workability, it has no effect on the room temperature performance, but at high temperatures are prone to liquid metal embrittlement.

(3) Low Melting Point Metal Contamination of Melting Furnaces, etc.

Such as blades in the melting and casting process from the liquid metal cooling or machining caused by low melting point metal contamination, such as Bi–Sn positioning mould caused by surface contamination at a certain temperature when used to cause liquid metal embrittlement.

(4) Other Processes Lead to Liquid Metal Embrittlement

Such processes include bearing stall, accidental fire, over-fire and poor electrical contact caused by local low melting point metal melting, resulting in liquid metal embrittlement of structural components.

2.5.4 Surface Integrity

Surface Integrity is a general term for the geometric shape of the surface and the physical properties of the surface of the parts after processing. The former includes surface roughness, corrugation, texture, scratches, abrasion, geometric dimensions and deviations, etc.; the latter includes microstructure changes in the surface layer, plastic deformation, recrystallisation, microhardness, residual stress, intergranular corrosion, microcracks, alloying element depletion, compositional segregation, organisational non-uniformity, overheating and overcooking. Surface integrity is an important factor that affects the working life, quality of use and reliability of structural components, and is an important research topic in the machinery manufacturing industry, especially in the aerospace manufacturing industry. Common surface integrity defects in the manufacturing process are: surface morphology defects, residual stress, surface burns, work hardening and so on.

A. Surface Morphological Defects

Surface topography mainly refers to the collection of geometric features on the surface of the machined part. Surface topography not only affects the friction and wear of mechanical products, contact stiffness, fatigue strength, fit properties, transmission accuracy, sealing, detection accuracy and other mechanical properties, but also directly affects the use of mechanical products and life.

Common surface morphology defects are: surface roughness, deep grooves, scale burrs, corner radius is too small, machining accuracy does not meet.

(1) Surface Roughness

The roughness of the machined surface is expressed in terms of surface roughness. Surface roughness refers to the processing surface has a small pitch and small peaks and valleys unevenness. The smaller the surface roughness, the smoother the surface. Surface roughness and mechanical parts with the nature of the fit, wear resistance, fatigue strength, contact stiffness, vibration and noise, etc. have a close relationship, the service life of mechanical products and reliability has an important impact (Fig. 2.33).

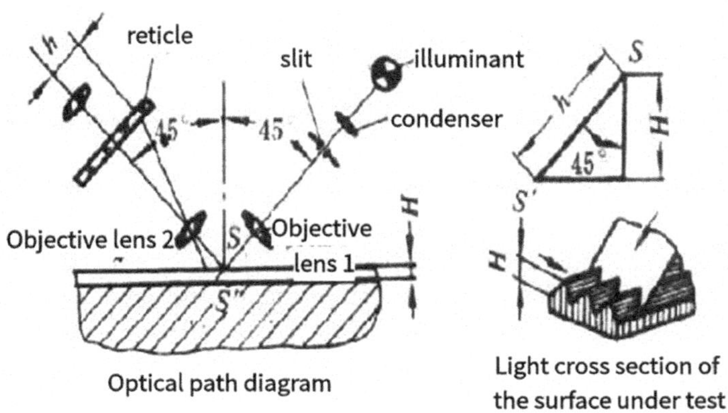

Fig. 2.33 Measuring surface roughness by the optical cut method

(2) Deep Grooves

Individual deep grooves exist on the machined surface will be a source of stress concentration in use. It leads to fatigue fracture. Low hardness of the part, high plasticity, small cutting speed or increased cutting thickness, etc., can lead to the formation of an accumulation of chipping tumours on the front face of the cutter. Due to the accumulation of chipped tumour metal in the formation process by the violent deformation and strengthening, so that its hardness is much higher than the metal being cut, it is equivalent to a rounded and blunt edge and outstretched beyond the cutting edge, and in the processed surface leaving longitudinal irregular grooves.

(3) Scale Burrs

Cutting plastic metal at low or medium cutting speed, the processing surface will often appear scaly burrs, especially for round holes using broaching method is more likely to occur, if the broaching exit burrs are not removed, it will become the use of the root cause of stress concentration.

(4) "R" Processing Is Too Small

Parts corner radius is small, especially in the shape of the cross-section of the sudden change will occur in the local stress concentration and produce micro-cracks and fatigue cracks expand into fatigue cracks, resulting in fatigue fracture.

(5) Machining Accuracy Does Not Meet

After cutting, component size, shape or location, accuracy does not meet the process drawings or design requirements. Not only directly affects the quality of workpiece assembly, but also affects the normal work of the workpiece stress state distribution, and reduces the failure resistance of the workpiece.

B. Residual Stress

Workpiece heat and cooling process, due to thermal expansion and contraction and phase change than the volume difference and volume change, due to the workpiece surface and heart of the existence of temperature difference, phase change is not simultaneous and the phase variable is different, the surface and the heart of the volume change can not be synchronised, and thus produce internal stresses, also known as residual stress.

Residual stresses exist in the workpiece as initial stresses, and when the workpiece is subjected to external loads, the result of the superposition of residual stresses may offset or increase the external stresses, thus increasing or decreasing the load carrying capacity of the workpiece. The reasonable distribution of residual stresses has a significant effect on the service behaviour of the workpiece; however, practice has shown that residual stresses, no matter how they are distributed, always have a detrimental effect on certain mechanical properties, as shown in Table 2.2.

According to the cause of internal stresses can be residual stresses can be divided into thermal stresses and organisational stresses.

(1) Thermal Stress

Table 2.2 Effect of residual stress on mechanical properties

Items	Effects
Hardness	Residual tensile stresses reduce the hardness value, residual compressive stresses increase the hardness value, the effect of tensile stresses is greater than that of compressive stresses.
Fatigue	Residual compressive stresses increase the fatigue strength of the workpiece, while residual tensile stresses reduce it.
Wear	Residual compressive stresses increase the fatigue strength of a workpiece, while residual tensile stresses decrease the fatigue strength.
Corrosion	Residual tensile stresses increase the susceptibility of the material to stress corrosion cracking.

Thermal stress refers to the internal stress caused by the temperature difference between the surface layer and the heart of the uneven expansion and contraction. The figure shows the stress change of cylindrical samples during heating and cooling.

At the beginning of heating, the surface layer of the higher temperature, thermal expansion, but by the lower temperature of the heart of the potential, the sample surface layer produces compressive stress, the heart of the tensile stress. Continue to heat up, this stress increases with the increase in the temperature difference between the heart and the surface layer, after reaching the maximum value, and then with the heart and the surface layer of the temperature difference decreases and decreases until reduced to zero, followed by stress reversal, such as Fig. 2.34a. As the yield strength of the material with the increase in temperature and reduce the internal stress exceeds the yield strength, will cause plastic deformation to relax the internal stresses.

Cooling, the situation is reversed, the beginning of the surface layer due to cooling by the heart of the limitations of the surface layer to produce tensile stresses, while the heart is subject to compressive stresses, when the stress exceeds the yield strength

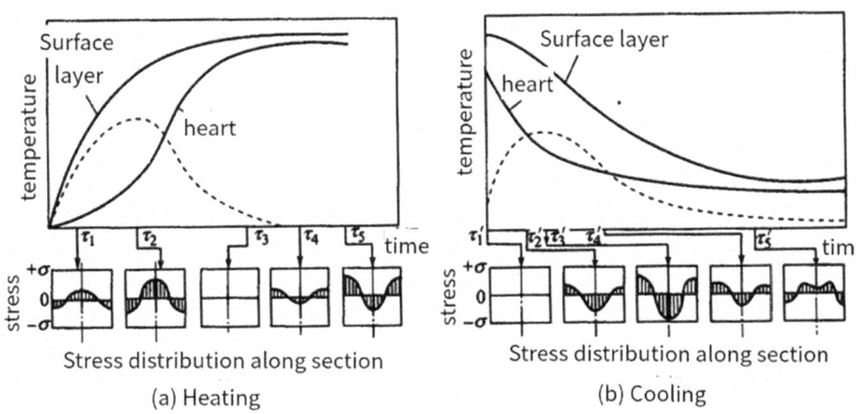

Fig. 2.34 Changes in internal stresses in a cylindrical specimen during heating and cooling

of the heart, the heart of the plastic deformation so that the internal stresses to get part of the relaxation. This internal stress as cooling continues first with the heart and surface temperature difference increases and increases, but when the surface temperature is close to room temperature or the temperature of the cooling medium, the heart to a relatively fast rate of cooling and contraction, the result is the formation of the workpiece with the cooling of the initial stage of the direction of the opposite direction of the internal stresses, the two internal stresses first offset each other, but due to the cooling of the later part of the workpiece temperature is lower, the yield strength rises, and the workpiece is not subjected to any stresses. Neither the heart or the surface will not occur at the beginning of cooling as large as the plastic deformation, if the phase change does not occur, the end of cooling, the final residual thermal stresses surface compressive stresses, the heart of the tensile stresses, such as Fig. 2.34b.

(2) Tissue Stress

Tissue stress arises from the specific volume change caused by phase transformation, also known as phase transformation stress. Figure 2.35 for cylindrical steel samples in the quenching process, the occurrence of martensitic phase transformation when the change in tissue stress. Quenching, martensite phase transition always starts at the surface and then to the heart of the expansion, the martensite phase transition occurs in the surface layer because of its volume expansion is bound to still be in the austenite heart of the tensile stress, and its own because of the limitations of the heart of the compressive stresses, compressive stresses of the peak with the phase transition to the heart of the move. Since austenite has good plasticity and very low yield strength, the phase transition stress will certainly cause plastic deformation in the austenitic state of the heart, and then when the heart temperature is reduced to the Ms point and the martensitic phase transition occurs, the accompanying volume expansion due to have been transformed into a martensite surface layer hinders the production of tissue stresses opposite to the stresses described above, and with the martensitic phase transition in the heart, the tissue stresses With the martensite phase transformation in the heart, the tissue stresses are reversed, eventually forming the residual tissue stresses of tensile stresses in the surface layer and compressive stresses in the heart.

C. Surface Burns

Surface burn is the heat generated by the processing process instantaneous several in the surface layer of the workpiece, with a very fast speed to the surface metal heated to a certain temperature, resulting in changes in the surface layer of the tissue, and the phenomenon of oxidation discolouration on the surface of the workpiece.

Titanium alloy chemical properties of active, poor thermal conductivity, viscous, these characteristics make titanium alloy more prone to surface burns. Titanium alloys in grinding, hand polishing and electrolytic processing often occur in the phenomenon of surface burns.

(1) Titanium Alloy Grinding Burns

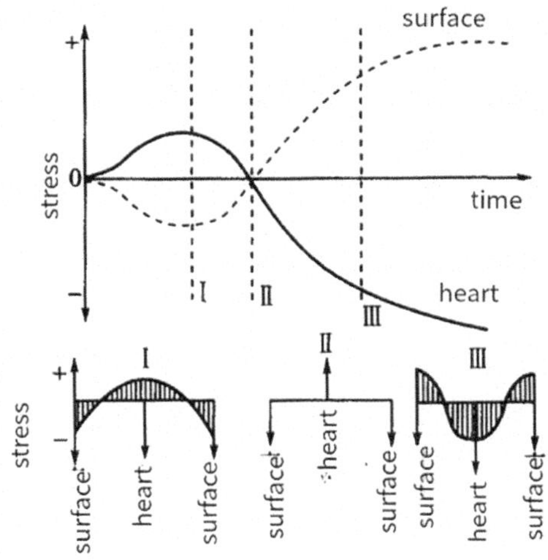

Fig. 2.35 Changes in organisational stresses during quenching of cylindrical steel samples

Titanium alloy in grinding, abrasive and parts surface contact area, abrasive chips are easy to adhere and bets the grinding wheel, resulting in a sharp decline in the cutting performance of the wheel, the grinding force increases, the local grinding temperature rises, thus forming grinding burns. When the grinding burn is serious, the surface of the part will appear net-like cracks. After corrosion, the burned area shows "white spots", as shown in Fig. 2.36a. Figure 2.36b, c for the titanium alloy compressor rotor blade grinding burns high and low tissue morphology

(2) Titanium Alloy Hand Polishing Burns

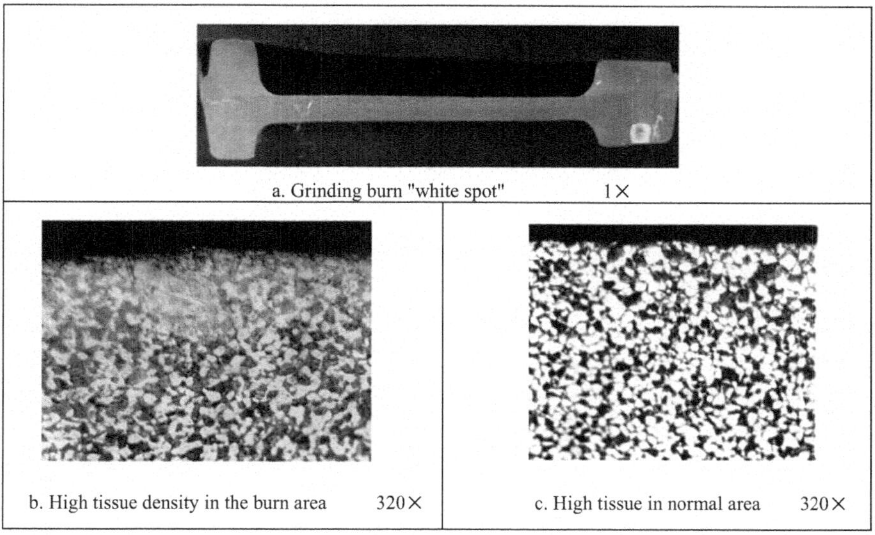

Fig. 2.36 Titanium alloy compressor rotor blade grinding burns

In the process of leaf titanium alloy sheet processing, hand polishing process is widely used. Due to the poor thermal conductivity of titanium alloy, when polishing, the heat generated at the contact between the grinding wheel and the blade surface, if not quickly conducted, will result in the surface layer of the blade caused by a small volume range of temperature increase, resulting in polishing burns.

In the corrosion inspection process, often found on the leaf "white spots". The size, shape and distribution of the "white spot" is not regular; the probability of appearing sometimes high, sometimes low, the highest up to about 30%; "white spot" of the degree of obviousness also varies, some vaguely visible, some particularly obvious. Figure 2.37a for titanium alloy rotor blade polishing burns white spot area tissue. Figure 2.37b, c for the titanium alloy compressor rotor blade polishing burns high and low tissue morphology.

(3) Titanium Alloy Electrolytic Processing Burns

Titanium alloy blades use electrolytic machining of the blade body profile, which has the advantages of high processing efficiency, no residual stress, no intergranular corrosion, etc. However, improper control of the gap between the blade (anode) and the mould (cathode) will cause short circuit burns (arc burns) in the local area, which generates extremely high temperatures in the burned area, and the titanium alloy matrix melts. The damage caused by this type of burn is much more severe than a polishing burn. In the core area of the burn, there is a melting burn phenomenon, the formation of pockmarks, high and low, black, while its surrounding yellowish or blue, after corrosion of the black pockmarks around the bright white, such as Fig. 2.38. Along the burned area dissected to prepare metallographic specimens for observation, the emergence of three obvious areas, and the burned area, the transition

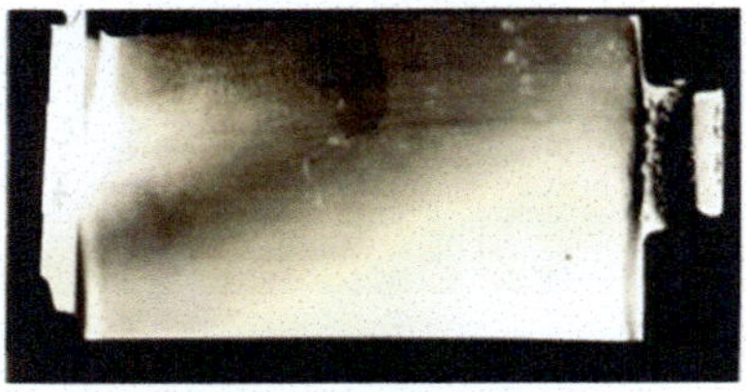

a. White spot appearance of leaf burn 0.6×

b. White spot tissue 320×

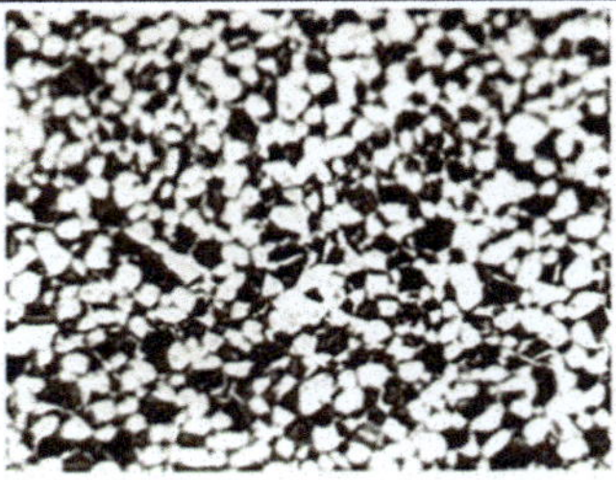

c. Matrix structure 320×

Fig. 2.37 Titanium alloy pressurised rotor blade polishing burns area and the normal body area, as shown in Fig. 2.39. It can be seen that in the core area of the burn, the temperature has exceeded the melting point of the titanium alloy, the matrix has melted to form a casting organisation, as shown in Fig. 2.40.

D. Work Hardening

Work hardening of metal materials in the recrystallisation temperature below the plastic deformation of strength and hardness increases, while the plasticity and toughness of the phenomenon of reduction, also known as cold work hardening. The reason is that the metal in plastic deformation, grain slippage, dislocation of the entanglement, so that the grain elongation, broken and fibrous, metal internal residual stress. The degree of work hardening is usually expressed by the ratio of the microhardness of the surface layer after machining and before machining and the depth of the hardened layer. Work hardening to the further processing of the metal to bring difficulties. Such as in the process of cold rolling of steel plate will be more and more hard to rolling, in the cutting process to make the surface layer of the workpiece brittle and hard to accelerate tool wear, increase the cutting force. Work hardening also has a favourable side, it can improve the strength, hardness and wear resistance of the metal. It is an important property of metallic materials.

Figure 2.41 shows a typical stress–strain curve (also known as a work-hardening curve) for a metallic single crystal, the plastic deformation portion of which is made up of three stages:

Stage I—Easy Slip Stage: Once τ reaches τ_c of the crystal, considerable deformation can be produced with little increase in stress. This stage is close to a straight line with a low slope θ_I $\left(\theta = \frac{d\tau}{d\gamma} \text{ or } \theta = \frac{d\sigma}{d\varepsilon}\right)$ i.e., work-hardening rate, typically θ_I of the order of ~ 10-4G (G is the shear modulus of the material).

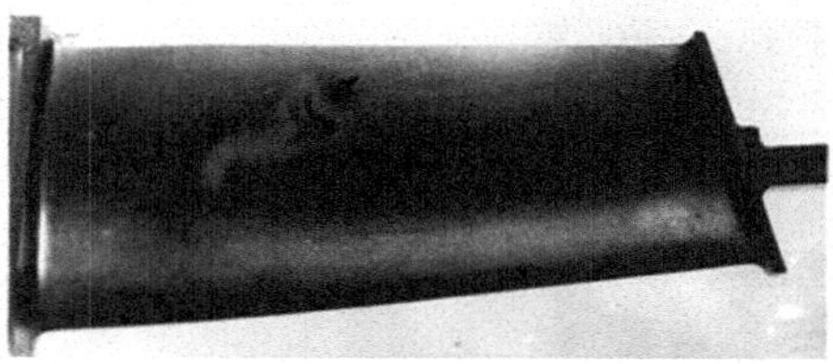

Fig. 2.38 Electrolytic burn morphology (after corrosion)

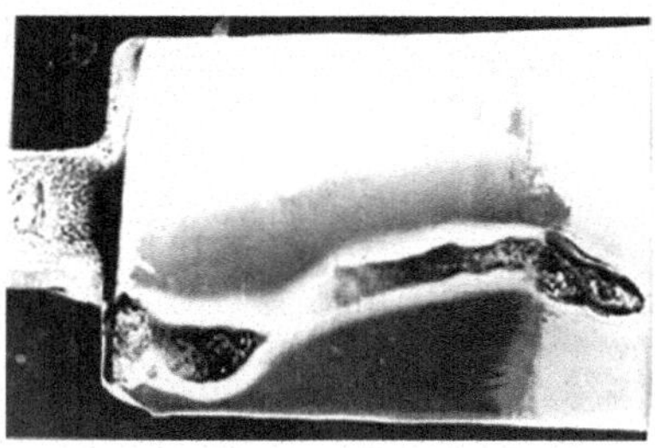

Fig. 2.39 Metallographic organisation of electrolytic burns

Fig. 2.40 Casting tissue in the burned area

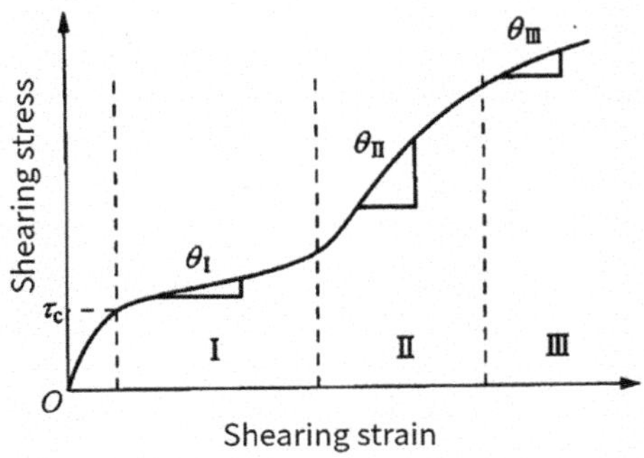

Fig. 2.41 Shear stress–shear strain curves for single crystals showing the three stages of plastic deformation

Stage II—Linear Hardening Stage: As the strain increases, the stress grows linearly, this section is also linear, and the slope is large, work-hardening is very obvious, $\theta_{II} \approx G/300$, near constant.

Stage III—Parabolic Hardening Stage: With the increase of strain, the stress rises slowly, parabolic, and θ_{III} gradually decreases.

The actual curves of various crystals vary according to the type of crystal structure, crystal site orientation, impurity content, and experimental temperature, etc. However, in general, the basic characteristics are the same, except that the length of each stage is affected by the movement, value-added, and interactions through dislocations, and a certain stage may no longer occur. The figure shows the hardening curves of three typical crystal structures of metal single crystals, in which the face-centred cubic and body-centred cubic crystals show the typical three-stage processing hardening, except that when the body-centred cubic body containing traces of impurity atoms, the yielding phenomenon described earlier will be produced due to the interactions between the impurity atoms and dislocations and will make the curves change, and the first stage of the densely tapped hexagonal metal single crystal is usually very long, far more than the other crystals structure, so much so that the specimen breaks before stage II is fully developed.

As shown in Fig. 2.42 plastic deformation of polycrystals due to the hindering effect of the crystalline junction and the coordinated fit between the grains requires that the grains can not be a single slip system action, but there must be more than one

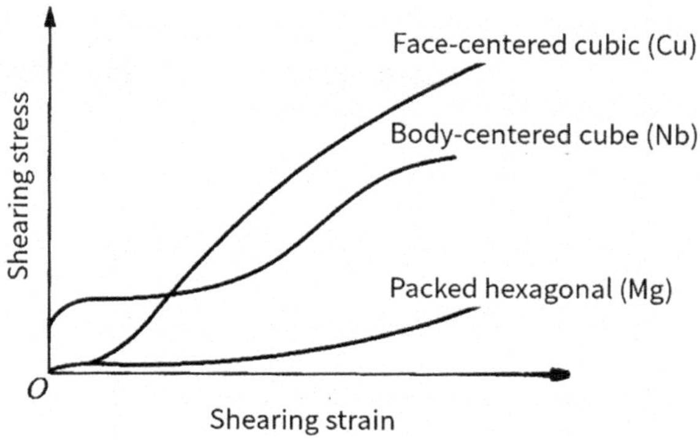

Fig. 2.42 Stress–strain curves of typical face-centred cubic, body-centred cubic and densely arranged hexagonal metallic single crystals

set of slip system at the same time, so the polycrystalline stress–strain curve does not appear in the single-crystal curve of the first stage, and its hardening curve is usually steeper.

References

1. Shi J. Stream of variation modeling and analysis for multistage manufacturing processes. 2007, CRC Press.
2. Kai Mi. Manufacturing process reliability control technology based on the correlation model of product reliability and process size. Beijing University, 2011.
3. Yin Chao. Research on the mechanism of product reliability decline in manufacturing process and risk evaluation method. Beijing University, 2014.
4. Chen Y, Jin J. Quality-reliability chain modeling for system-reliability analysis of complex manufacturing processes. IEEE Transactions on Reliability, 2005, 54(3): 475–488.

Chapter 3
Manufacturing Process Reliability Analysis Techniques

3.1 Modeling Techniques for Process System Reliability Analysis

3.1.1 Process Reliability Analysis Modeling Process

The process reliability analysis model should have a well-defined task profile, task duration, failure criteria, and environmental conditions and operating stresses encountered during the execution of the task. The process reliability modeling procedure is as follows:

(1) Determine the Task and Task Profile

Task profile is defined as: the product in the completion of the specified tasks during the period of time experienced by the events and the environment of the chronological description, is "a particular process or procedure from the beginning to the completion of the time events and the environment in which the description". A task profile should generally include: the working condition of the product; the maintenance program; the timing and sequence of product work; the timing and sequence of the product's environment (imposed and induced); and the definition of a successful or fatal failure of the task. A complex process or procedure can be modeled either with a reliability model that includes all tasks or with a corresponding reliability model based on different tasks and task profiles. For example, the reliability model can be established for all processes or parameters, but also for a process such as key processes, or certain parameters (key features) to establish a reliability model; can also be established according to different stages of the reliability model [1].

(2) Determine the Conditions of the System

A process and the completion of the process is carried out in different ways, "people, machines, materials, methods, environment, measurement" conditions. Some processes and procedures of the conditions of small changes, as the same

Y. He et al., *Reliability Theory and Technology in Manufacturing Process*,
https://doi.org/10.1007/978-981-96-2735-6_3

conditions, and some large changes in conditions, modeling should focus on consideration.

(3) Establishment of Reliability Block Diagram

Through a concise method to indicate the interrelationship between the reliability of the process or process, that is, the establishment of a reliability block diagram. The block diagram indicates the reliability value of each process or step to complete the process. In the calculation of system reliability, each box must be calculated.

(4) Establish the Corresponding Mathematical Model

The established reliability block diagram, the establishment of the corresponding mathematical model.

(5) Determine the Failure Criteria

The fault criterion of the reliability of the process system task refers to the fault that affects the completion of the process or process task, and should find out the conditions that lead to the task unsuccessful and affect the task unsuccessful process parameters and parameter limit values.

3.1.2 *Process Reliability Modeling Technique Based on Flow Balance Approach*

Consider an n-stage serial production line as shown in Fig. 3.1. Where M_i denotes the *ith* processing machine and B_i is a buffer between two neighboring machines with storage capacity b_i.

Basic assumptions:

(1) The processing time, life time and failure repair time of the machine M_i follow an exponential distribution with the parameters productivity ω_i, failure rate λ_i and repair rate μ_i, respectively.
(2) Machine M_i is not starved and machine M_n is not blocked.
(3) The machine fails only during the machining of the workpiece and does not fail during starvation or blocking.
(4) The transfer time of the workpiece is negligible.

Let the state of the machine on the production line be *ji*, i.e., machine M_i is in state *j*. *j* is categorized into the following situations:

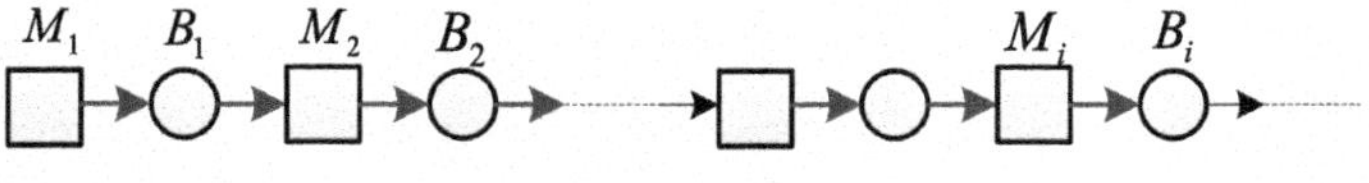

Fig. 3.1 The N-stage series production line

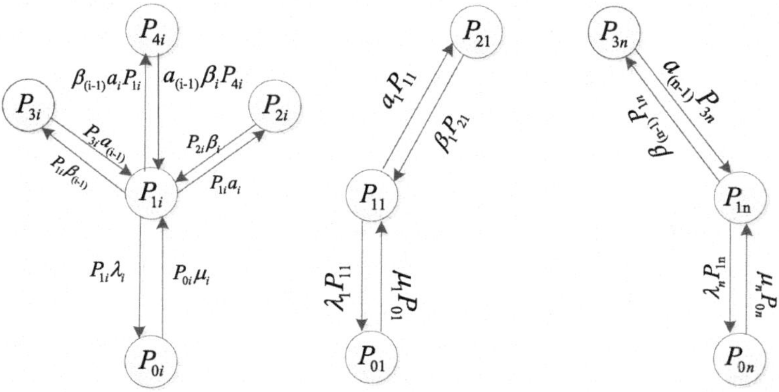

Fig. 3.2 State probability flow equilibrium of a machine

$j = 0$, the machine is faulty; $j = 1$ the machine is processing normally; $j = 2$, the machine is blocked; $j = 3$, the machine is starved; $j = 4$, the machine is both blocked and starved.

Let P_{ji} denote the probability that machine i is in state j. There are $\sum_{j=0}^{4} P_{ji} = 1\ (i = 1, 2, \ldots, n)$, where, by assumption, there are $P_{31=0}$, $P_{41=0}$ and $P_{2n=0}$, $P_{4n=0}$ a_i denotes the probability that the buffer B_i is full of workpieces, and β_i denotes the probability that there are no workpieces. Then the state probability flow transfer case of machine MI in the cascade production line has three probability flow balancing methods as shown in Fig. 3.2.

Based on Fig. 3.2, the probability of state equation for the production line is listed as follows:

$$
\begin{aligned}
\dot{P}_{01} &= -\mu_1 P_{01} + \lambda_1 P_{11} \\
\dot{P}_{11} &= \mu_1 P_{01} + \beta_1 P_{21} - (a_1 + \lambda_1) P_{11} \\
\dot{P}_{21} &= a_1 P_{11} - \beta_1 P_{21} \\
P_{01} &+ P_{11} + P_{21} = 1 \\
&\ldots \\
\dot{P}_{0i} &= -\mu_i P_{0i} + \lambda_i P_{1i} \\
\dot{P}_{1i} &= \mu_i P_{0i} + \beta_i P_{2i} - (a_i + \lambda_i + \beta_{i-1} + \beta_{i-1} a_i) P_{1i} \\
&\quad + a_{i-1} P_{3i} + a_{i-1} \beta_i P_{4i} \\
\dot{P}_{2i} &= a_i P_{1i} - \beta_i P_{2i} \\
\dot{P}_{3i} &= \beta_{i-1} P_{1i} - a_{i-1} P_{3i} \\
\dot{P}_{4i} &= \beta_{i-1} a_i P_{1i} - a_{i-1} \beta_i P_{4i} \\
P_{0i} &+ P_{1i} + P_{2i} + P_{3i} + P_{4i} = 1 \\
\dot{P}_{0n} &= -\mu_n P_{0n} + \lambda_n P_{1n} \\
\dot{P}_{1n} &= \mu_n P_{0n} - (\lambda_n + \beta_{n-1}) P_{1n} + a_{n-1} P_{3n}
\end{aligned}
$$

$$\dot{P}_{3n} = \beta_{n-1}P_{1n} - a_{n-1}P_{3n}$$
$$P_{0n} + P_{1n} + P_{3n} = 1 \quad (3.1)$$

By making $\dot{P}_{ji} = 0$ on the left side of the equation, the steady state equilibrium equation for the production line is obtained.

Where B_i has the probability $j\ (j = 0, 1, \ldots, b_i)$ of $p_{ij} = \frac{\rho_i^j(1-\rho_i)}{1-\rho_i^{b_i+1}}$ workpieces, where ρ_i is the ratio of the productivity ω_{i+1} of the ith process to the productivity ω_i of the i + 1th process, $\rho_i = \omega_i/\omega_{i+1}$ when $\omega_i = \omega_{i+1}$, $\rho= 1$, $p_{ij} = 1/(b_i + 1)$ then there are the following four state probabilities:

(1) Buffer B_i is fully full: $p_{ib_i} = \frac{\rho_i^{b_i}(1-\rho_i)}{1-\rho_i^{b_i+1}}$

(2) Buffer B_i has a storage location: $p_{i\overline{b_i}} = 1 - p_{ib_i} = \frac{1-\rho_i^{b_i}}{1-\rho_i^{b_i+1}}$

(3) Buffer B_i has no inventory: $p_{i0} = \frac{1-\rho_i}{1-\rho_i^{b_i+1}}$

(4) Buffer B_i has inventory: $p_{\overline{i0}} = 1 - p_{i0} = \frac{\rho_i\left(1-\rho_i^{b_i}\right)}{1-\rho_i^{b_i+1}}$.

When buffer B_i is fully full, all machines located in the previous process stop due to blockage; when buffer B_i is not in stock, all machines located in the subsequent process stop due to lack of material. Therefore, machine M_i can work normally only when buffer B_{i-1} is stocked and B_i has a storage location, the availability of buffer B_i is for machine M_i and M_{i+1} for machine $p_{\overline{0}}$. Therefore, the availability of buffers before and after machine M_i is $A_i = p_{\overline{(i-1)0}}p_{\overline{ib_i}}$.

3.1.3 A Stochastic Petri Net Based Process Reliability Modeling Approach

The use of Petri nets for manufacturing system modeling can clearly describe the dynamic transfer process between the various states of the system, and there are so many domestic and foreign literature is based on Petri nets complex manufacturing system modeling methods.

For the basic n machines (M_i) and n − 1 buffers (B_i) in series with a production line of a manufacturing unit, the buffer capacity of k_i, the establishment of the serial processing unit of the stochastic Petri net model, as shown in Fig. 3.3, the meaning of which is shown in Table 3.1.

Let the capacity of the ith intermediate buffer be k_i. The productivity of the previous machine is t_{4i-2} excitation rate μ_i (input rate of the buffer), and similarly the productivity of the latter machine is μ_{i+1} (output rate of the buffer), and so the steady-state solution is obtained through the equation of state transfer of the buffer, the probability of the ith buffer having storage pieces when there are $(j = 0, 1, \ldots, k_i)$ pieces of storage in the buffer is

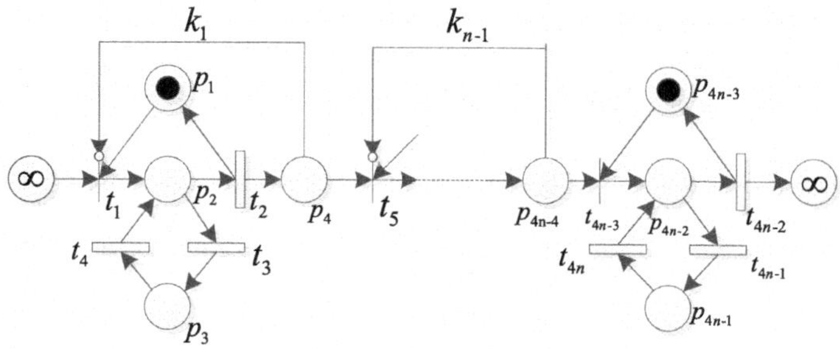

Fig. 3.3 Petri net model of basic serial repairable machining unit

Table 3.1 Variations of the model in Fig. 3.3 and the meaning of the repository

Name	Interpretation
t_{4i-3}	Providing work to machine M_i
t_{4i-2}	Machine M_i machining workpieces
t_{4i-1}	Machine M_i malfunctioned
t_{4i}	Machine M_i for maintenance
p_{4i-3}	Machine M_i is idle
p_{4i-2}	Machine M_i enabled and ready to start processing
p_{4i-1}	Machine M_i malfunctioning and ready to start repairs
p_{4i}	Buffer capacity k_i

$$p_{ij} = \frac{\rho_i^j(1 - \rho_i)}{1 - \rho_i^{k_i+1}} \tag{3.2}$$

The next analysis is the same as the flow balancing method described above.

3.2 Process FEMA Analysis Techniques

FMEA is an inductive analysis method that can be applied to the product production process, the use of the operational process, the maintenance process, the management process, etc., focusing on the production phase of the use of the process FMEA technology method [2] (Table 3.2).

3.2.1 *Purpose and Steps of a Process FMEA*

The purpose of process FMEA is to assume that the product design to meet the requirements of the premise, for the product in the production process of each process

Table 3.2 FMEA methods for product life cycle phases

	Program validation phase	Engineering development phase	Production stage	Use stage
Methodologies	Functional FMEA	• Hardware FMEA • Software FMEA • Damage mode impact analysis	Process FMEA	Statistical FMEA
Goal	Analyze and study the flaws and weaknesses of the system's functional design to provide a basis for improving the system's functional design and weighing the options	Analyze and research the defects and weaknesses of the system's hardware and software design, and provide a basis for the system's hardware and software design improvement and safeguard analysis	Analyze and study the defects and weaknesses of the production process design and its impact on the product, and provide the basis for the design improvement of the production process	Analyze and study the actual occurrence of failures, causes and their effects during the use of products, to provide the basis for providing the reliability of the use of products and for the improvement, modification or development of new products

step of all possible failure modes, causes and their impact on the product, according to the failure modes of the Risk Priority Number (RPN) value of the magnitude of the process weaknesses of the development of improvement measures, and to predict or track the effectiveness of the measures taken to reduce the value of the RPN, so that RPN to reach an acceptable level, thereby improving product quality and reliability.

Process FMEA can provide qualitative basis for determining the list of critical processes and key features, quality control plan, provide qualitative information for the development of process test outline, and provide qualitative information for determining the equipment maintenance cycle and maintenance list.

The steps of process FMEA are shown in Fig. 3.4, and the main steps of P-FMEA are as follows:

(1) Prepare a process flow diagram, listing the processes that the product undergoes and identifying the process;
(2) Analyze the functions and requirements of the process system;
(3) Identify possible process failure modes;
(4) Analyze possible failure causes and mechanisms;
(5) Analyze the impact of the failure and evaluate the impact and severity of the failure mode on the product, on the elements related to the next process, on the customer, on personnel, and on the environment and facilities;
(6) Determine the incidence, detection and risk prioritization number of the failure mode;
(7) Develop corrective actions to prevent or correct failure modes by risk priority;

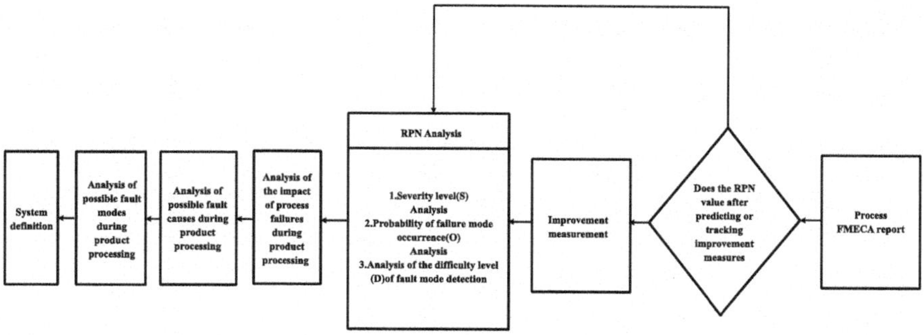

Fig. 3.4 The steps of process FMEA

(8) Anticipate or follow up to validate the risk priority number after the improvement measures taken;
(9) Formation of a P-FMEA report.

The implementation of P-FMEA is carried out by filling out the form of P-FMEA worksheet, a typical P-FMEA worksheet is shown in Table 3.3, the specific application, according to the actual situation of the form content can be supplemented, adjusted and cut.

3.2.2 *Principles to Follow When Implementing a Process FMEA*

(1) The Principle of Effectiveness

The principle of effectiveness is to make P-FMEA effective, that is, the effect of the work can affect the process design. In order to achieve the effectiveness, first of all, should be established by the process designers and reliability professionals composed of P-FMEA team, while the important results of the analysis should be asked to designers and program managers to participate in the discussion and review. Secondly, measure recommendations should be communicated to process implementers in a timely manner. Finally, the implementation of measures should be emphasized for timely tracking and analysis.

(2) Synergy Principle

Since the primary purpose of P-FMEA is to provide effective information and basis for the improvement of process design, the design of FMEA should be synchronized with the design of the process, especially in the early stages of the design of the FMEA, so that the weak links in the design can be found in a timely manner and help to optimize the design of the process. At the same time, P-FMEA should reflect the changes in design and technology, and be improved and updated as the development phase unfolds.

Table 3.3 Process FMEA worksheet

Product name: ____________ Product development/responsibility unit: ____________

Process design/responsible: ____________ Unit: ____________ Process project name: ____________

Prepared by: ____________ Participants: ____________ Review: ____________

Approval: ____________ Process/citation: ____________

Documentation: ____________ Version number: ____________

Identifier	Process/procedure name	Process (procedure) function/requirement description	Failure mode	Impact of failure				Cause/mechanism of failure	Existing controls		Severity **S**	Rate of occurrence **O**	Detection level **D**	Risk priority **RPN**	Suggested measures	Responsible unit/person and expected date of completion	Additional measures and effects						Note
				Impact on the next process	Impact on products	Impact on personnel	Impact on the environment and facilities		Detection methods	Other measures							Measures taken	Date of completion	Severity (**S**)	Rate of occurrence (**0**)	Detection level (**D**)	Risk priority (**RPN**)	

(3) "Exhaustive" Principle

P-FMEA is fully effective, the most important point is to analyze all the problems to make clear, to do everything "in mind, on record". The failure mode of the process should be exhaustive through the access to information, comparison, anticipation, recall, seminars, mutual enlightenment and other ways.

Exhaustion of the causes. For each failure mode, in as large a range as possible to list each failure reason that can be thought of, after analysis, if the cause of the failure mode is unique, that is, if the correction of the cause of the failure mode has a direct impact on the failure mode, then this part of the P-FMEA judgment process is complete.

(4) "Repeated" Principle

P-FMEA analysis is a comprehensive application of knowledge and experience within the enterprise, not once can be completed, but through the process design, process review, P-FMEA seminars, etc., and constantly revise, improve, in order to obtain a more complete prevention program, at the same time, the repetition of these processes will also promote exchanges between employees to improve the quality of the collective work of the staff and business level.

3.2.3 *Key Elements of a Process FMEA*

(1) System Definition

Process FMEA should be analyzed to define the object, the content can be summarized as a functional analysis, drawing "process flow table" and "parts–process relationship matrix".

Functional analysis is to analyze the purpose of the process being analyzed, function, role and requirements, etc.; drawing "process flow table" see table, indicating the function and requirements of each process related to the process; drawing "parts–process relationship matrix" see table, indicating that the "characteristics of parts". Drawing "Parts and components characteristics" and "process operation" of the relationship between the various processes, these are the preparatory work for the process FMEA (Tables 3.4 and 3.5).

(2) Process Defect Mode Analysis

Generally, process FMEA does not consider defects in product design. A process defect is a failure of a product or a working condition of a process system that does not meet the specified requirements due to some change in the process elements

Table 3.4 Process flow chart

Part name type of production process:
Part number department name review: Page: Total:
Equipment name/model analysts: Date of authorization to fill in the form:

Process	Importation	Output result
Process 1		
Process 2		
Process 3		
……		

Table 3.5 Component-process relationship matrix

Part name production process type:
Part number department name review: Page: Total:
Equipment name/model analysts: Date of authorization to fill in the form:

Component characteristics	Process operation			
	Process 1	Process 2	Process 3	……
Characteristic 1				
Characteristic 2				
Characteristic 3				
……				

when the product is manufactured in the specified manufacturing process. A process defect pattern is a process defect that fails to meet the requirements of the product's machining and assembly process and/or design intent (GJB 1391). The physical, chemical, biological, or other process that causes a process defect is called a process defect mechanism. A process defect mode may be the cause of a failure mode in the next (downstream) process, or it may be the consequence of a failure mode in the previous (upstream) process. An example of a typical process defect mode is shown in Table 3.6.

(3) Process Defect Cause Analysis

The cause of process defects refers to the process defects with process defect patterns corresponding to why the process defects occur, usually from the six aspects of man, machine, material, method, environment and measurement to analyze the cause of process defects. Typical examples of process defect causes are shown in Table 3.7.

(4) Process Defect Impact Analysis

The impact of process defects refers to the impact of process defects on the "customer" corresponding to the mode of process defects, "customer" refers to the next process/subsequent processes and/or the end user.

(a) For the next/subsequent process: The impact of a process defect shall be described by the process/process characteristics, see Table Table 3.8.
(b) For the end-user: The impact of process defects should be characterized in terms of product properties, see Table 3.9.

Table 3.6 Examples of typical process defect patterns

Serial number	Failure mode	Serial number	Failure mode	Serial number	Failure mode
1	Curve around	7	Size out of tolerance	13	Surface smoothness
2	Deformity	8	Position deviation	14	Not labeled
3	Crackles	9	Shape deviation	15	Mislabeling
4	Break apart	10	Short circuit	16	Handling damage
5	Whiskers	11	Open circuit	17	Dirty
6	Leaky hole	12	Surface roughness	18	Remaining remnants

Note Process defect modes should be described in physical and professional terms, rather than using visible fault phenomena

Table 3.7 Typical causes of process defects

	Serial number	Fault cause		Serial number	Fault cause
People	1	Missing parts	Legislation	11	Over and under torque
	2	Improper assembly of parts		12	Incorrect welding current, power, voltage
	3	Improper installation		13	Incorrect heat treatment time, temperature and medium
Machineries	4	Incorrect machine settings		14	Incorrect programming
	5	Incorrect tooling or fixtures		15	Poorly bonded
	6	Locator wear		16	Soldering
	7	Debris on the locator	Material	17	Excessive internal stress in the workpiece
	8	Tool wear		18	Incorrect casting gate/vent
Environment	9	Improper lubrication		19	Punch or bore a hole
	10	Unlubricated	Surveying	20	Inaccurate gauges

Table 3.8 Typical examples of the effects of process defects (for subsequent processes)

Serial number	Failure impact	Serial number	Failure impact
1	Inaccessible	6	Unable cooperate
2	Unable to turn/tap holes	7	Unmachinable surfaces
3	Mismatch	8	Causes wear and tear on the tooling process
4	Uninstallable	9	Damage to equipment
5	Connectionless	10	Hazard to the operator

(5) Risk Prioritization Number (RPN) Analysis

The evaluation parameters of the process defect pattern mainly include: the impact severity level (S) of the process defect pattern, the occurrence probability level (O) of the process defect pattern, the difficulty level (D) of the process defect pattern to be detected, and the risk-priority number (RPN). The Risk Prioritization Number RPN is the product of the Severity Level S, the Occurrence Level O and the Detection Level D, i.e.

Table 3.9 Typical examples of the effects of process defects

Serial number	Failure impact	Serial number	Failure impact
1	Excessive noise	9	Unstable work performance
2	Excessive vibration	10	Excessive wear and tear
3	Excessive resistance	11	Water seepage
4	Laborious handling	12	Oil-spill
5	Gives off a nasty odor	13	Surface defect
6	Operational irregularities	14	Excessive size, shape and position
7	Intermittent operation	15	Unscheduled maintenance
8	Not working	16	Invalidate

$$\text{RPN} = S \times O \times D$$

Risk Priority Number RPN is an evaluation of the risk level of potential failure modes of the process, reflecting a comprehensive measure of the likelihood of the occurrence of process defect modes and the severity of their consequences, the greater the RPN, the greater the hazard of the process defect mode.

In the P-FMEA work, the severity, occurrence and detection of the rating level can be subdivided, in addition to the analysis of the object of the importance of the system as a whole can be pre-determined according to the degree of a reference value (hereinafter referred to as the value of the provisions), as a judgment of whether to take improvement measures.

(a) The severity level (S) of a process defect pattern refers to the severity of the impact of a certain process defect pattern in the processing and assembly of a product, and the scoring guidelines are shown in Table 3.10.

Severity is a measure of the severity of the impact of the failure mode on the relevant parties, and is the value corresponding to the severity level of the worst potential consequences caused by the failure mode, and the impact level corresponding to different severity levels should be determined according to the actual agreement. As shown in Table 3.11 for a process FMEA in the actual agreement of the severity (S) level of the scoring guidelines, according to the process defect mode of personnel, product, task schedule, economy and the environment of the five aspects of the impact from large to small, the severity of the level of the five levels of disaster, fatal, moderate, mild and negligible, each level corresponds to a range of two values. Within the same level, the more severe takes the higher value, which can be calculated using the weighted arithmetic average method, and when the average value is not a whole number, it is rounded up according to the rounding rule.

(b) The probability of occurrence rating (O) of a process defect pattern is the likelihood of a particular process defect pattern occurring, the scoring guidelines are shown in Table 3.12, the scoring rating is a relative comparison rating and does not represent the true probability of occurrence.

Table 3.10 Scoring guidelines for severity level (S) for process defect patterns

Degree of impact	Final impact of process failure modes (for end-users)	Final impact of process failure modes (for downstream/follow-on operations)	Rating scale for severity level (S)
Disastrous	Product damage or loss of functionality	Fatalities/serious risk to operator safety and significant environmental damage	10, 9
Nasty	Product functionality is largely lost and does not work/works but performance is degraded/ end users are very dissatisfied	Danger to the safety of workers, 100% of the product may be discarded/products need to be repaired in specialized workshops and serious environmental damage	8, 7
Medium	Product works, but operational performance is degraded/end users are not satisfied, and in most cases (> 75%) the product is found to be defective	Possibility that some (< 100%) of the product will be discarded without screening/products will need to be repaired in specialized departments or off the production line and moderate environmental damage	6, 5, 4
Light	25–50% of end-users may find the product defective or without identifiable impacts	Resulting in unscheduled maintenance or repair of the product	3, 2, 1

Table 3.11 Scoring guidelines for the actual agreed upon severity (S) level in a process FMEA

Influence level	Description of the severity of the impact of failure modes	Harshness values
Disastrous	Fatalities and serious injuries; product obsolescence; facility damage; more than $1 million in damages and more than 1 month delay in progress	9–10
Deadly	Serious injuries to personnel or multiple minor injuries; loss of basic product functionality; $100,000,000–$1,000,000 in damages and delays in progress of several weeks–January	7–8
Moderate	Minor injuries to personnel; product damage reduces the use of: r. Damage to equipment can be repaired; damage to large 1–100,000 yuan delayed 3 days–1 week	5–6
Light	Mild discomfort to personnel; causing operational difficulties; no loss of basic product functionality; no loss of functionality due to damage to work equipment; loss of less than $10,000 and delay in progress of less than 3 days	3–4
Negligible	Rework does not affect schedule; reduces. Work is efficient but does not affect schedule or quality; economic losses are negligible	1–2

Table 3.12 Scoring guidelines for probability of occurrence rating (O) for process defect patterns

The possibility of process failure modes occurring	The probability of possible process failure modes occurring (P)	Rating level of probability of occurrence (O)
Very high (continuous faults)	$P_0 \geq 10^{-1}$	10
	$5 \times 10^{-1} \leq P_0 \leq 10^{-1}$	9
High (frequent faults)	$2 \times 10^{-2} \leq P_0 \leq 2 \times 10^{-2}$	8
	$1 \times 10^{-2} \leq P_0 \leq 2 \times 10^{-2}$	7
Moderate (occasional faults)	$5 \times 10^{-3} \leq P_0 \leq 1 \times 10^{-2}$	6
	$2 \times 10^{-3} \leq P_0 \leq 5 \times 10^{-3}$	5
	$1 \times 10^{-3} \leq P_0 \leq 2 \times 10^{-3}$	4
Low (rare faults)	$5 \times 10^{-4} \leq P_0 \leq 1 \times 10^{-3}$	3
	$1 \times 10^{-4} \leq P_0 \leq 5 \times 10^{-4}$	2
Extremely low (unlikely to malfunction)	$P_0 < 1 \times 10^{-4}$	1

Corresponding to the corresponding failure mode, it is a measure of the likelihood of the failure mode occurring under the existing control measures, and is determined by the "likelihood of failure" during the process implementation. The likelihood of failure is categorized into six classes: very high, high, high, medium, low, and very low. The scoring criteria for the occurrence (O) level are shown in Table 3.13.

(c) The Detected Difficulty Level (D) of a process defect pattern is the likelihood that the cause of a particular process defect pattern will be detected in the control of a product's machining process, and the Detected Difficulty Level (D) is also a relatively comparative level. In order to get a lower value of the difficulty of being detected, the product processing and assembly process needs to be improved continuously, and the scoring guidelines are shown in Table 3.14.

Detectability D is an assessment of the ease of detecting a failure when it occurs, based on available control and detection methods. The degree of detection corresponds to the likelihood that the "root cause" of the failure mode will be detected under the identified detection and control measures, and is categorized into six levels

Table 3.13 Scoring guidelines for occurrence (O) ratings

Incidence rating	Clarification	Incidence value
Extremely high	Failure is almost inevitable	10
High level	Failures occur frequently	8–9
High	Failures occur from time to time	6–7
Moderate	Failure is possible	4–5
Lower	Failures are rare	2–3
Extremely low	There's a certainty that failure won't happen	1

Table 3.14 Scoring guidelines for Detected Difficulty Level (D) for process defect patterns

Detected difficulty	Scoring criteria	Inspection method			Recommended methods for detecting difficulty	Evaluation of the difficulty level (D) of being tested Grading
		A	B	C		
Next to impossible	Unable to detect			√	Unable to detect or check	10
Very small	Current detection methods are almost impossible to detect			√	Detection through indirect inspection	9
Small	The current testing method only has a small chance of being tested measure			√	Conduct inspection by visual inspection	8
Small	The current testing method only has a very small chance of being tested measure			√	Conduct inspection through double visual inspection	7
Minor	The current testing method can detect		√	√	Conduct testing using current testing methods	6
Medium	The current detection methods can basically detect		√		Use measuring tools for testing after the product leaves the workstation	5
Slightly above average	There are many opportunities for current testing methods to be tested exit	√	√		Implement error detection in subsequent processes, or Preprocess measurement and inspection for testing	4
Upper	The current testing method is likely to detect	√	√		Errors can be detected on the spot or in subsequent processes (such as inventory, selection, setup, validation). Not accepted Defective products	3

(continued)

Table 3.14 (continued)

Detected difficulty	Scoring criteria	Inspection method			Recommended methods for detecting difficulty	Evaluation of the difficulty level (D) of being tested Grading
		A	B	C		
Very high	The current detection methods are almost certain to detect	√	√		On site detection (automatic quantity with automatic stop function) Tool. Defective product cannot pass	2
Affirm	The current testing method can definitely detect	√			The process/product is designed with error prevention measures and will not produce Defective products	1

Note Inspection method: *A* adopting error prevention measures, *B* use measuring tools for measurement, *C* person

based on the ease with which the failure can be detected, namely, extremely difficult, difficult, possible, probable, likely, and capable, and the scoring criteria for the degree of detection (D) level are shown in Table 3.15.

(6) Improvement Measures

Improvement measures are any process improvement measures that start with the reduction of the severity rating (S), occurrence probability rating (O), and detection difficulty (D) rating of a process defect pattern. Generally, regardless of the size of the process defect pattern RPN, items with a Severity Rating (S) of 9 or 10 should be addressed to reduce that risk through process design measures or product processing, assembly process control or preventive/improvement measures. In all cases where the consequences of a process defect mode could be hazardous to manufacturing/assembly personnel, preventive/improvement measures should be taken to eliminate,

Table 3.15 Scoring guidelines for detectability (D) levels

Degree of detection	Clarification	Detectability value
Extremely difficult	No test method to detect fire efficiency patterns	10
Difficult	Little likelihood of detecting a failure mode	8–9
Possible	Low probability of detecting failure modes	6–7
Likelihood	Medium likelihood of detecting failed casts	4–5
Easily	The likelihood of detecting a failure mode is very human	2–3
Capable	R must be able to detect failure modes	1

mitigate, control, or avoid the occurrence of the process defect mode. Process defect patterns for which no improvement measures are available shall be identified in the appropriate columns of the process FMEA form.

3.2.4 Process FMEA Report

The P-FMEA report should generally include the following:

(1) Overview

It mainly describes the research object and scope of P-FMEA, explains the working process of P-FMEA and the composition of the working group, as well as the main results and problems obtained.

(2) Process Description

Detailed description of the process functions, process requirements and process flow involved.

(3) Basic Rules

A description of the standards and specifications followed; the methods for determining the levels of severity, occurrence and detection, the RPN prescribed values and the basis for their determination.

(4) Work Orders

List the completed complete and accurate P-FMEA worksheets.

(5) Critical Processes/Procedures and Important Processes/Procedures Itemized Worksheet

A list of failure modes with a severity of 9 or more as critical failure modes and a list of failure modes with a severity of 7–8 as important failure modes.

(6) A List of Super-Risk and High-Risk Processes/Procedures

A list of failure modes with RPN values exceeding the specified value is listed as ultra-risky failure modes, and a list of failure modes with RPN values between 80 and 80% of the specified value is listed as high-risky failure modes.

(7) Effectiveness Evaluation

Give the implementation of process reliability improvement measures and effect evaluation.

(8) Conclusion and Recommendation

State the concluding comments on the implementation of P-FMEA. P-FMEA description.

3.2.5 *Process FMEA Example*

Illustrated with a model of a critical electronic product that involves more than 70 processes, the process defect patterns are summarized into five major categories, namely: human error, poor equipment and materials, improper process methods or parameters, and poor environment.

The evaluation criteria for severity (S) rating, occurrence (O) rating, detection (D) rating, a typical part's process defect pattern and impact analysis worksheet, and an electronic product process FMEA worksheet to determine the critical process are shown in Table 3.16, 3.17, 3.18 and 3.19 respectively.

3.3 Manufacturing Process Reliability Evaluation Technology

3.3.1 *Manufacturing Reliability Assessment Techniques Based on Evolutionary Relationships of Quality Characteristics*

The manufacturing process of a product from blank to finished product is a discrete time sequence on the process flow, and the product characteristics and process characteristics evolve along with the sequence, and the evolution law has a direct impact on the reliability of the product. The study of the evolution law of multi-stage manufacturing process includes the following two aspects: (i) the study of product characteristics in the process of inheritance, variation, superposition, decomposition, etc. between the stages; (ii) the study of product characteristics and process characteristics in the process flow of the interactive coupling, transfer of compensation and other relationships. Therefore, the following analyzes the causal relationship between the product characteristics output from each process and the process characteristics of the manufacturing process from the point of view of process dimension time series [3].

(i) Selection of Quality Characteristics

In order to quantitatively analyze the relationship between the evolution of the two types of quality characteristics in the manufacturing process, two deviation sequences characterizing the fluctuation of the characteristics are selected as the analytical indexes, i.e., the loss of deviation of the product characteristics and the loss of deviation of the process characteristics, which are denoted as Lpt(k) and Lps(k), respectively. Among them, the deviation includes two aspects: (i) the individual quality characteristic deviation of a single processed part, which reflects the quality of the output parts; (ii) the overall fluctuation deviation of a batch of products, which can characterize the stability of the manufacturing process.

Table 3.16 Evaluation guidelines for severity (S) ratings

Influence level	A statement of the seriousness of the consequences of the impact (either condition can be recognized)	Severity rating
Disastrous	(1) Death or permanent disability of personnel (2) System paralysis into product obsolescence (3) process equipment scrapping (4) Loss of > 2 million dollars (5) Delay in progress > 2 months	10
Deadly	(1) Serious injury to personnel or multiple minor injuries or serious occupational diseases (2) Loss of basic product function (3) Serious damage to process equipment (overhaul must be done before it can be reused) or most of it is scrapped (4) 50,000 < loss < 200,000 dollars (5) 1 month < delayed progress < 2 months	8 ~ 9
Adjacent	(1) Minor injuries to personnel become minor occupational diseases (2) Damaged products can only be downgraded for use or need to be returned to the factory for repair (3) Damage to process equipment that can be repaired on site (4) Rework of most of the processes (5) 10,000 < damage < 50,000 dollars (6) 10 days < delayed progress < 1 month	6 ~ 7
Light	(1) Causing serious discomfort to personnel in the work environment (2) Damage to the appearance of the product but not affecting function or performance (3) Damage to the product but no loss of basic function or repairable on site (4) Partial rework is required (5) X-art equipment damage but not loss of function temporarily may not do the repair (6) 5000 yuan < loss < 10,000 yuan (7) Delay in progress < 10 days	4 ~ 5
Faint	(1) Causing slight discomfort to personnel in the working environment (2) Slight damage to the appearance of the product, which is not easy to detect and does not affect the function or performance at all (3) Slightly damaged products but no loss of function (4) A small amount of rework is required, but does not affect the overall progress (5) Process equipment slightly damaged but no loss of function need not be repaired (6) Loss < 5000 RMB	2 ~ 3
Negligible	(1) Causes mild discomfort to personnel in the work environment (2) Rework in place (of the current process) but with no impact on the intended schedule (3) Inefficient work but neither schedule nor quality control is affected (4) Negligible economic loss within budget	1

Table 3.17 Evaluation criteria for probability of occurrence (O)

Likelihood of failure mode occurrence	Probability of possible failure modes	(Military) rank
Very high (persistent failures)	$P_O \geq 8 \times 10^{-2}$	10
High (frequent failures)	$4 \times 10^{-2} \leq P_O \leq 8 \times 10^{-2}$	9
	$2 \times 10^{-2} \leq P_O \leq 4 \times 10^{-2}$	8
	$10^{-2} \leq P_O \leq 2 \times 10^{-2}$	7
Medium (occasional failure)	$8 \times 10^{-3} \leq P_O \leq 10^{-2}$	6
	$4 \times 10^{-3} \leq P_O \leq 8 \times 10^{-3}$	5
	$2 \times 10^{-3} \leq P_O \leq 4 \times 10^{-3}$	4
Low (rarely occurring failures)	$10^{-3} \leq P_O \leq 2 \times 10^{-3}$	3
	$8 \times 10^{-4} \leq P_O \leq 10^{-3}$	2
Very low (unlikely failure)	$P_O \leq 8 \times 10^{-4}$	1

Considering that the quality loss function proposed by Taguchi is not continuous at the point of "0", and the improved quality loss function proposed by Yadav et al. can solve this problem well, Yadav's method will be used in the following analysis. First, two types of deviations are defined: expected deviation and undesirable deviation, where expected deviation means that the larger the deviation, the higher the degree of satisfaction of the characteristic, which is generally for the hopeful characteristic or the hopeful small characteristic variable; undesirable deviation means that the larger the deviation, the lower the degree of satisfaction of the characteristic, and for the hopeful quality characteristic, all deviations are undesirable deviations. Specifically, for different types of quality characteristics, the criteria for discriminating deviations are as follows:

(1) For S-type quality characteristics, d+ indicates the deviation of the actual value exceeding the maximum allowable target value, which is out of the tolerance range and is judged to be a bad deviation;
(2) For N-type quality characteristics, d+ indicates the deviation where the actual value exceeds the target value, and d− indicates the deviation where the actual value is lower than the target value, both of which are judged as undesirable deviations;
(3) For L-type quality characteristics, d− indicates the deviation of the actual value below the minimum allowable target value, which is out of the tolerance range and judged as bad deviation.

Based on the two types of deviations and the criteria for determining the deviations of quality characteristics with different expectation value attributes, the deviation loss function of the proposed characteristics is shown in Table 3.20.

Then the total deviation loss of the manufacturing process is the sum of the deviation losses of all characteristics, which has the following mathematical expression:

Table 3.18 Guidelines for evaluating detectability (D) ratings

Detectivity	Assessment criteria	Inspection methods*			Recommended detectability grading methods	Rank
		A	B	C		
Almost impossible	Undetectable			V	Unable to detect or undetectable	10
Tiny	It's almost impossible to detect with current detection methods			V	Detection by indirect vapor detection	9
Infinitesimal	Existing detection methods have only a small chance of detecting			V	Detection by S-visual inspection of the vapor	8
Small	Existing detection methods have only a small chance of detecting			V	Probe with double-g vision	7
Few	Current detection methods can detect		V	V	Detection by graphical methods	6
Moderate	Existing detection methods can essentially detect		V		Measurement is performed after the part is removed from the level	5
Upper middle class	Existing detection methods have a better chance of detecting	V	V		The error detection is carried out in the subsequent process, or the pre-process measurement is carried out to detect it	4
High	Current detection methods are likely to detect it	V	V		Errors can be detected on the spot, into detecting them in subsequent: sequences (e.g., sheds, picks, setups, verifications). Defective parts are not accepted	3

(continued)

Table 3.18 (continued)

Detectivity	Assessment criteria	Inspection methods*			Recommended detectability grading methods	Rank
		A	B	C		
High level	Existing detection methods are almost certainly capable of detecting it	V	V		On-the-spot detection (G-motivated gauge with 0-motion stop function). Defective parts cannot be passed	2
Constant	Existing detection methods are almost certainly capable of detecting it	V			Processes/products are designed to be error-proof and will not produce defective parts	1

Note Types of inspections: *A* use of error-proofing measures, *B* use of monthly gauges, *C* manual inspection

$$L(d) = \sum_{N} L(d_N) + \sum_{L} L(d_L) + \sum_{S} L(d_S) \tag{3.3}$$

Assuming that the manufacturing process of the product contains n processing procedures, and the kth process of the processing process plays a role in m_k product characteristics, denoted as $y_{k1}, \ldots, y_{km_k}$, whose target value is $y_{0k1}, \ldots, y_{0km_k}$, the corresponding variance is $v_{k1}, \ldots, v_{km_k}$, the variance target value is $v_{0k1}, \ldots, v_{0km_k}$; affect the product characteristics of the process parameters there are lk, denoted as $v_{0k1}, \ldots, v_{0km_k}$, the target value of $x_{0k1}, \ldots, x_{0kl_k}$, the corresponding variance is $\sigma_{k1}, \ldots, \sigma_{kl_k}$, the variance target value is $\sigma_{k1}, \ldots, \sigma_{kl_k}$, then the product characteristics. The product characteristics can be the lookout characteristics, lookout characteristics or lookout characteristics, manufacturing process parameters are generally lookout characteristics, all the variance belongs to the lookout characteristics.

According to the total deviation loss function formula of manufacturing process, the deviation loss sequence of product characteristics can be obtained as:

$$L_{pt}(k) = \sum_{y_k \in N} L(d_N) + \sum_{y_k \in L} L(d_L) + \sum_{y_k \in S} L(d_S) + \sum_{v_k} L(d_S) \tag{3.4}$$

The process characteristic deviation loss sequence is:

$$L_{ps}(k) = \sum_{x_k} L(d_N) + \sum_{\sigma_k} L(d_S) \tag{3.5}$$

(ii) Modeling the Evolution of Quality Characteristics

Evolution is a kind of change law on the time series, and the evolution of quality characteristics is the evolution of product characteristics and process characteristics

Table 3.19 List of important processes for a mechanical product

Contents of the process	Serial number	Process name	Failure mode	Impact of the failure on the product	Risk premium
Processed	001-01	Grinding Round	Human error	Fracture of parts	2401
			Defective material	Microcracks on the surface of the part	120
			Inappropriate process methods or parameters	Fracture of parts	120
	001-02	Ground end face	Human error	The length dimension of the part is out of tolerance and the straightness of the end face is out of tolerance	126
				Chipping on both ends of the part	126
			Equipment failure	The K degree dimension of the part is out of tolerance, and the end face is out of tolerance	U2
			Inadequate parameterization of the process	Running edges on both ends of the part	120
Processed	002-03	Ground end face	Human error	The length dimension of the part is out of tolerance, and the straightness of the ends is out of tolerance	140
			Improper equipment	Defects such as popped edges on the end face of the part	120
				The length dimension of the part exceeds the perpendicularity of the two end surfaces of the adult	112
	002-04	Point of a grindstone	Improper equipment	Parts zirconium phenomenon	120
				Parts taper angle has overshoot	144
Processed	003-02	Ground end face	Human error	The K-degree dimension of the part is out of tolerance or the perpendicularity of the two end surfaces is out of tolerance	126

(continued)

Table 3.19 (continued)

Contents of the process	Serial number	Process name	Failure mode	Impact of the failure on the product	Risk premium
			Improper equipment	Defects such as popped edges on the end face of the part	120
				Excessive deviation of the part's in-degree dimension or the perpendicularity of the two end surfaces	112

Table 3.20 Deviation loss functions

Characteristic desirability attributes	Bias loss function (math.)
Eye-level characteristics (N-type)	$L(d_N) = k\left(\exp(d_N^+)\right)^2 + \left(\exp(d_N^-)\right)^2$
Looking glass characteristics (L-type)	$L(d_L) = k\left(\exp(-d_L^+)\right)^2 + \left(\exp(d_L^-)\right)^2$
Looking down feature (S-type)	$L(d_S) = k\left(\exp(d_S^+)\right)^2 + \left(\exp(-d_S^-)\right)^2$

on the manufacturing process flow. Granger causality test can quantitatively describe the interaction of two time series, so the following will be the proposed two deviation sequences $\{L_{pt}(k)\}$ and $\{L_{ps}(k)\}$ for the Granger test respectively, the test model is as follows:

$$L_{pt}(k) = \sum_{i=1}^{q} \alpha_i L_{ps}(k-i) + \sum_{j=1}^{q} \beta_j L_{pt}(k-j) + \mu_{1k} \tag{3.6}$$

$$L_{ps}(k) = \sum_{i=1}^{s} \lambda_i L_{ps}(k-i) + \sum_{j=1}^{s} \delta_j L_{pt}(k-j) + \mu_{2k} \tag{3.7}$$

where μ1k and μ2k are random variables, which can be regarded as white noise, α, β, λ, δ are regression coefficients, and q and s denote the lag order.

In essence, the Granger causality test is a hypothesis testing problem with the original hypothesis:

$$H_0 : \alpha_1 = \alpha_1 = \cdots = \alpha_q = 0 \tag{3.8}$$

$$H_0 : \delta_1 = \delta_1 = \cdots = \delta_s = 0 \tag{3.9}$$

Granger test can test whether there is a statistically significant causal relationship between two variables, more precisely, to find out the relationship between the current value of one variable and the past value of another variable, which can quantitatively characterize the evolution of the front-end process on the current process characteristics and processed product characteristics. The essence of the model is a regression analysis of the time series [4], so in order to prevent the emergence of "pseudo-regression" phenomenon must be carried out on the time series of all variables unit root test to ensure the smoothness of the variables. In addition, Granger pointed out that if the variables are cointegrated, that is, there is a long-term correlation between the variables, then there is at least one direction of the Granger reason. In summary, the following three steps are required to determine the evolution of quality characteristics of manufacturing processes: (i) unit root test; (ii) cointegration test; and (iii) Granger causality test. The specific testing process can be implemented through Eviews software.

In engineering practice, the product quality of the output parts of the manufacturing process is directly affected by the process characteristics, according to which we consider that the sequence $\{L_{ps}(k)\}$ is the Granger cause of the sequence $\{L_{pt}(k)\}$. Then, according to whether the sequence $\{L_{pt}(k)\}$ is the Granger cause of the sequence $\{L_{ps}(k)\}$, we can get the output conclusion of this test model as follows:

(1) The sequence of product characteristic deviations $\{L_{pt}(k)\}$ and the sequence of process characteristic deviations $\{L_{ps}(k)\}$ are Granger causes for each other. It can be explained that the output product characteristics of the front-end process affect the process stability of the back-end process, which leads to changes in the fluctuation pattern of the process characteristics of the back-end process. The premise that the comprehensive process characteristics determine the product characteristics at the output end of the process, and the two deviation sequences are Granger causes for each other indicates that there is a cross-influence between the two types of characteristics, so the subsequent reliability analysis of the manufacturing process needs to consider the cross-terms.
(2) The process characteristic deviation sequence $\{L_{ps}(k)\}$ constitutes the Granger cause of the product characteristic deviation sequence $\{L_{pt}(k)\}$, while the product characteristic deviation sequence $\{L_{pt}(k)\}$ is not the Granger cause of the process characteristic deviation sequence $\{L_{ps}(k)\}$. It can be interpreted that the output product characteristics of the front-end process have no effect on the process stability of the back-end process and do not lead to changes in the fluctuation patterns of the process characteristics of the back-end process. The premise that the integrated process characteristics determine the product characteristics at the output end of the process, there is no cross-influence between the product characteristics and the process characteristics, therefore, the subsequent manufacturing process reliability analysis does not need to consider their cross terms.

The results of Granger causality test provide a model of the evolution law of the two types of characteristics of the manufacturing process, according to which the corresponding covariates are selected to construct the Cox proportional risk regression model to carry out the multifactorial analysis of the intrinsic reliability of the product, to identify the key processes and key characteristics that affect the intrinsic reliability of the manufacturing process, and to provide a basis for the subsequent improvement.

(iii) Cox Proportional Risk Regression Model

The Cox proportional risk model attributes product failure to two components, a baseline failure, which is related only to time t, and a failure due to changes or fluctuations in covariates, which is related only to the selected influences. The mathematical expression for the Cox model failure rate is given below:

$$\lambda(t|z) = \lambda_0(t)\exp\left(\omega^T z\right) \tag{3.10}$$

where z is a covariate of the manufacturing process characterizing the factors influencing the inherent reliability of the product;

λ0(t) characterizes the baseline failure rate at moment t;

λ(t|z) is the conditional failure rate, characterizing the failure rate at moment t under the condition of a certain covariate z;

ω = (ω1, ..., ωp)T, is the regression coefficient, where p is the number of covariates.

For the regression coefficients, there are the following implications:

(1) $\omega_i > 0$, the covariate is positively correlated with the product failure rate, which characterizes that the covariate is a risk factor affecting the inherent reliability of the product, and the control of this type of variable should be strengthened in the manufacturing process;
(2) $\omega_i = 0$, the covariate has no correlation with the product failure rate, which characterizes that the covariate is not a key factor affecting the inherent reliability of the product, and the control of this type of covariate can be appropriately relaxed in the manufacturing process to enhance the production efficiency;
(3) $\omega_i < 0$, the covariate is negatively correlated with the product failure rate, which characterizes that the covariate is a facilitator affecting the inherent reliability of the product, and the control of this covariate should be strengthened to enhance the reliability in the manufacturing process.

In addition, Exp(ω_i) is called the Hazards Ratio (HR) of the ith covariate, which characterizes the degree of risk of the covariate in the Cox model to the survival life of the product.

The inputs to the Cox proportional risk regression model can be represented as a triad, i.e., (Influences X, Lifespan Observations t Data Type), as shown in Table 3.21.

$$\begin{aligned} z &= (z_1, \ldots, z_{2n})^T \\ &= \left(L_{pt}(1), \ldots, L_{pt}(n), L_{ps}(1), \ldots, L_{ps}(n)\right)^T \end{aligned} \tag{3.11}$$

Table 3.21 Cox proportional risk model inputs

Brochure	x_1	x_2	…	x_m	Lifespan t	Deletion or non-deletion
1	a_{11}	a_{21}		a_{m1}	t_1	1
2	a_{12}	a_{22}		a_{m2}	t_2	0
…	…	…	…	…	…	…
n	a_{1n}	a_{2n}		a_{mn}	t_n	0

(4) When the Granger causality test shows that the sequence of product characteristic deviations {Lpt(k)} and the sequence of process characteristic deviations {Lps(k)} are mutually Granger-caused, the cross-influence of the two types of characteristics is taken into account and characterized by the cross term Lpt(k − 1) Lps(k). Then the input covariates are denoted as

$$\begin{aligned} z &= (z_1, \ldots, z_{3n})^T \\ &= z = \big(L_{pt}(1), \ldots, L_{pt}(n), L_{ps}(1), \ldots, L_{ps}(n), \\ &\quad L_{pt}(0)L_{ps}(1), \ldots, L_{pt}(n-1)L_{ps}(n)\big)^T \end{aligned} \tag{3.12}$$

(iv) Manufacturing Process Reliability Assessment

Generally, the baseline failure rate function in the Cox regression model is unknown, which brings great difficulties to the subsequent analysis and evaluation. As we know, the Weibull distribution is an important distribution in reliability analysis, and given different distribution parameters, it can portray the failure rate as increasing, constant and decreasing, so the Weibull distribution has been widely used in reliability [5], and many scholars have verified the feasibility of using this distribution as the benchmark failure rate function through a large number of examples, so without loss of generality, it is assumed that the product The benchmark failure rate function λ0(t) obeys the Weibull distribution (when λ0(t) obeys other distributions, it can also be analyzed by the same method), and its mathematical expression is as follows:

$$\lambda_0(t) = \frac{m}{\eta^m} t^{m-1} \tag{3.13}$$

Then, the proportional risk failure rate function conditional on a given covariate z can be written in the following form:

$$\lambda(t|z) = \frac{m}{\eta^m} t^{m-1} \exp\big(\omega^T z\big) \tag{3.14}$$

where m and η are the shape and scale parameters of the Weibull distribution, respectively, which can be given by combining the failure mechanism of the manufacturing process and empirical statistics.

Utilizing the relationship between the reliability function and the failure rate function, the inherent reliability function of the product under the given covariate z can be introduced, and the inference process is as follows:

$$\begin{aligned} R(t|z) &= \exp\left\{-\int_0^t \lambda(x|z)dx\right\} \\ &= \exp\left\{-\int_0^t \frac{m}{\eta^m}x^{m-1}\exp\left(\omega^T z\right)dx\right\} \\ &= \exp\left\{-\left(\frac{t}{\eta}\right)^m \exp\left(\omega^T z\right)\right\} \\ &= R_0(t)^{\exp\left(\omega^T z\right)} \end{aligned} \tag{3.15}$$

where R0(t) is the baseline reliability function for the Weibull distribution.

$$R_0(t) = e^{-\left(\frac{t}{\eta}\right)^m} \tag{3.16}$$

Analyzing manufacturing process reliability means analyzing the impact of the evolution of product characteristics and process characteristics during the manufacturing process on the inherent reliability of the output product. Product reliability is determined by the design, ensured by the manufacturing process, and manifested in the use process. According to the meaning of risk ratio of Cox model, it can be seen that

(1) When the integrated risk ratio is < 1, the inherent reliability of the delivered product is greater than the baseline reliability of the design, which means that the manufacturing process can well ensure the realization of the baseline reliability of the design, and the response manufacturing process reliability is higher;
(2) When the integrated risk ratio > 1, the inherent reliability of the delivered product is less than the benchmark reliability of the design, which means that the manufacturing process lacks the ability to ensure that the requirements of the design benchmark reliability, and the level of manufacturing process reliability needs to be improved.

According to the product failure rate function and reliability function, the failure distribution density function f(t|z) of the product under the condition of covariate z can be obtained with the following formula:

$$f(t|z) = \lambda(t|z)R(t|z) = \frac{m}{\eta^m}t^{m-1}e^{\omega^T z}R_0(t)^{\exp\left(\omega^T z\right)} \tag{3.17}$$

Then the product life expectancy that can be guaranteed by the manufacturing process is:

$$
\begin{aligned}
E(T) &= \int_0^\infty tf(t)dt \\
&= \int_0^\infty \frac{m}{\eta^m} t^m e^{\omega^T z} R_0(t)^{\exp(\omega^T z)} dt \\
&= \frac{m}{\eta^m} e^{\omega^T z} \int_0^\infty t^m R_0(t)^{\exp(\omega^T z)} dt
\end{aligned}
\tag{3.18}
$$

In the above calculations, the shape parameter m and the scale parameter η of the Weibull distribution can be determined by great likelihood estimation, such that

$$
\begin{aligned}
L(m, \eta) &= \prod_{i=1}^{N} f(t_i) \prod_{j=1}^{M} R(t_j) \\
&= \prod_{i=1}^{N} \frac{m}{\eta^m} t_i^{m-1} e^{\omega^T z_{t_i}} \prod_{j=1}^{N+M} \exp\left\{-\left(\frac{t_j}{\eta}\right)^m \exp\left(\omega^T z_{t_j}\right)\right\}
\end{aligned}
\tag{3.19}
$$

where N and M denote the number of failed samples and the number of unfailed samples, respectively, taking the logarithm of the left and right sides of the above equation yields

$$
\begin{aligned}
\ln[L(m, \eta)] &= N \ln\left(\frac{m}{\eta}\right) + \sum_{i=1}^{N} \ln\left(\frac{t_i}{\eta}\right)^{m-1} \\
&\quad + \sum_{i=1}^{N} \omega^T z_{t_i} - \sum_{j=1}^{N+M} \left(\frac{t_j}{\eta}\right)^m \exp\left(\omega^T z_{t_j}\right)
\end{aligned}
\tag{3.20}
$$

The partial derivatives of m and η are obtained separately for the above equation as follows, and the estimates of the two parameters can be obtained by joining the two equations

$$
\frac{\partial \ln[L(m, \eta)]}{\partial m} = 0 \tag{3.21}
$$

$$
\frac{\partial \ln[L(m, \eta)]}{\partial \eta} = 0 \tag{3.22}
$$

(v) Case Applications

A core device manufacturing process is complex, strict control of the processing process is still prone to low qualification rate, low processing efficiency and other issues. Therefore, it is necessary to analyze the process reliability of the core device, to find out the weak process in the manufacturing process, to ensure the qualification rate and reliability of the product.

Core device processing process is a serial process, this paper does not consider the impact of the transformation of the processing order between different processes on the early failure rate and inherent reliability of the product, product characteristics and process characteristics of the evolution of the process dimension constitutes a process of the two discrete time series, can be used to analyze the reliability of the manufacturing process of the core device using the Granger test and Cox proportional risk regression comprehensive evaluation model proposed earlier. Table 3.22 gives the process characteristics of the main processing procedures of the core device and the product characteristics related to the procedure and the deviation loss corresponding to the characteristics.

Using Eviews software, the unit root test is carried out on the series $\{L_{ps}\}$ and the series $\{L_{pt}\}$ respectively, and the test results indicate that both series are lagged first

Table 3.22 Core device processing process characteristics and output product characteristics

Serial number	Process name	Product characteristics	L_{pt}	Process characterization	L_{ps}
1	Cylindrical grinding	Straightness, cylindricity	41.378	Grinding wheel grit, feed, guide wheel speed	11.613
2	Ground end face	Parallelism, roughness	42.808	Grinding wheel grit, table flatness, table feed speed	12.586
3	Splice	Bonding strength	184.761	Curing temperature, curing time, curing pressure	20.735
4	Precision grinding of cylindrical surfaces	Roughness	25.273	Grinding wheel grit, feed, guide wheel speed	6.177
5	Ultrasonic cleaning	Surface roughness	42.574	Cleaning time, cleaning frequency	10.956
6	Slot	Power consumption, length	17.283	Laser power	3.224
7	Insulation film plating	Film thickness	31.164	Chamber temperature, coating time, sputtering power	8.699
8	Metalized film	Adhesion, film thickness	48.273	Humidity, chamber temperature, coating time, sputtering power	11.028
9	Winding assembly	Gear ratio	97.068	Winding tension	15.502

order smooth. In addition, the results of the cointegration test of the two variables indicate that there is a cointegration relationship between the variables, so the Granger test can be carried out. The results of the Granger test are shown in Table 3.23, from which it can be seen that the p-value of the F-test is greater than 0.05, so the original hypothesis is rejected under the condition of the significance level of 0.05, indicating that the two variables sequences $\{L_{ps}\}$ and $\{L_{pt}\}$ are not each other's Granger cause, which means that there is no interaction coupling between product quality and processing technology.

The test result of no interaction is due to the fact that the enterprise mostly uses CNC machine tools for processing in the production process, the processing accuracy and equipment reliability are very high, and the influence of good or bad input product quality and the size of deviation of product characteristics on the processing process is negligible. Therefore, according to the principle of covariate selection, the Cox model without cross terms should be chosen to assess the process reliability of the core device.

According to the life data of the core device, input the ternary group as shown in Table 3.21, and carry out Cox stepwise regression analysis, and the model results are shown in Table 3.24, where B is the partial regression coefficient and Exp(B) is the risk ratio HR value. The early failure rate function can be obtained as follows:

$$\lambda(t) = \lambda_0(t)\exp\big(0.196L_{ps}(5) + 0.090L_{pt}(3) + 0.067L_{pt}(6) + 0.829L_{pt}(8)\big)$$

At the significance level of 0.05, the relationship between the output quality of gluing, slotting and metalized film and the deviation of the process characteristics of ultrasonic cleaning is significant, which shows that gluing, metalized film and ultrasonic cleaning are the two key processes affecting the reliability and life of the product. Especially, the metal film plating process, which is the weak link of the core device process, has the highest HR value of the output product characteristics, which is 2.291. In addition, the HR value of the reliability of the ultrasonic cleaning

Table 3.23 Granger test results

Null hypothesis	Obs.	F-statistic	Prob.
LPS does not Granger cause LPT	7	3.71344	0.2122
LPT does not Granger cause LPS		1.23115	0.4482

Table 3.24 Cox regression results

	Regression coefficient	Wald statistic	Sig.	Risk ratio Exp(B)
$L_{ps}(5)$	0.196	7.859	0.005	1.217
$L_{pt}(3)$	0.090	31.010	0.000	1.095
$L_{pt}(6)$	0.067	7.635	0.006	1.070
$L_{pt}(8)$	0.829	30.798	0.000	2.291

process before plating also reaches 1.217. Correspondingly, in order to improve the reliability level of the core device, it is necessary to ensure that the strength of the gluing connection, cleanliness before plating, and the thickness of the film and film adhesion meet the specified requirements.

According to the process design mechanism, taking $\hat{\eta} = 180, \ \hat{m} = 12$ the baseline reliability function of the core device can be obtained as

$$R_0(t) = \exp\left\{-\left(\frac{t}{180}\right)^{12}\right\}$$

The curve of the baseline reliability function over time is shown as the solid line in Fig. 3.5. Then the reliability function R1(t|z1) under the existing process conditions is shown as the blue dashed line in Fig. 3.5, and its expression is

$$R_1(t|z_1) = R_0(t)^{\exp(\omega_1^T z_1)} = \exp\left\{-59.952\left(\frac{t}{180}\right)^{12}\right\}$$

As can be seen from Fig. 3.5, the baseline reliability level of the design is much higher than the actual reliability, indicating that fluctuations in the characteristics during the manufacturing process will accelerate the performance degradation of the product in use. In addition, the core device has a more stable performance and higher reliability in the first 100 units of time. However, after that, the reliability decreases rapidly with time, and when the use time is 120 unit time, the reliability level is only 0.6. Therefore, in order to prolong the life of the core device in the use process, it is necessary to optimize the gluing process and the process of metalized film. After the improvement and control of the metal-plated film process, the inherent reliability function R2(t|z2) is shown as the red dashed line in Fig. 3.5, and the reliability level

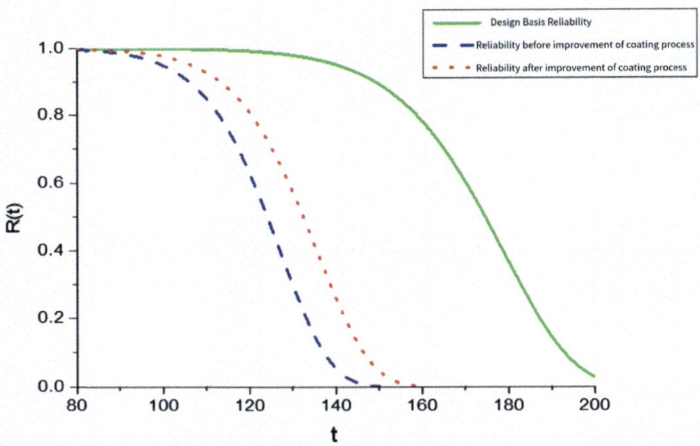

Fig. 3.5 Core device reliability function

has increased significantly, and its expression is:

$$R_1(t|z_2) = R_0(t)^{\exp(\omega_2^T z_2)} = \exp\left\{-27.598\left(\frac{t}{180}\right)^{12}\right\}$$

The concept of manufacturing process reliability is extended to study the influence of manufacturing process on inherent reliability, and the loss sequence of product characteristics on process flow and the loss sequence of process characteristics on process flow are used as indicators to analyze the evolution law between characteristics in the multi-stage manufacturing process, so as to determine the factors affecting the inherent reliability of products. In this regard, a comprehensive reliability assessment method based on Granger causality test and Cox proportional risk regression model is proposed to construct a characteristic evolution model based on the Granger test, select covariates of the Cox risk regression model based on the test results, and construct a Cox model using the product early failure data or the environmental screening test data, so as to quantitatively assess the impact of key characteristics in the manufacturing process on the intrinsic reliability of the product. reliability of the product. The comprehensive assessment method provides technical support for quantifying the evolution process between the characteristics, and also provides a basis for finding weak process links to guide the improvement of manufacturing process reliability. Finally, the feasibility of the proposed method is demonstrated through the reliability analysis of the manufacturing process of the core devices.

3.3.2 *Maturity-Based Manufacturing Risk Evaluation Techniques*

As the scale and requirements of equipment development projects continue to increase, a large number of projects suffer from schedule delays, quality degradation and cost overruns due to the lack of appropriate manufacturing technology knowledge to assist decision-making during the development phase, therefore, it is necessary to conduct a specialized manufacturing technology maturity assessment for equipment development projects. Through the manufacturing technology maturity assessment, the manufacturing risks of the equipment development project can be identified, as well as the possible impact of these risks on the development of equipment. According to the assessment results can choose the corresponding preventive measures, through the implementation of improvement measures, effectively reduce or avoid manufacturing risks, to ensure that the equipment development project successfully put into production and delivered to the troops, to accelerate the transformation of high-tech from the laboratory to the battlefield speed.

In recent years, based on the idea of technology maturity, some studies have been carried out at home and abroad in the field of manufacturing risk assessment of research and development projects, among which, the U.S. Joint Defense Manufacturing Technical Committee (JDMTP) proposes Manufacturing Readiness Level (MRL) evaluation model based on technology maturity to identify manufacturing risks by evaluating the manufacturing maturity level of key nodes of the equipment research and development projects, and to identify the manufacturing risks through the implementation of improvement measures. Maturity level to identify manufacturing risks, and initially proposed a manufacturing risk management process based on manufacturing maturity evaluation; domestic scholars have conducted qualitative research on the concept of production risk and manufacturing risk evaluation strategy; as can be seen from the above, there are fewer references to quantitative manufacturing risk evaluation models and methods in the existing literature. For this reason, this book proposes a new manufacturing risk assessment method based on manufacturing maturity and fuzzy hierarchical analysis method.

I. Manufacturing Risk Connotation and Evaluation Process

(i) Manufacturing Maturity

Manufacturing maturity is a new tool introduced and applied by the U.S. Army in recent years in program acquisition management based on Technology Readiness Level (TRL), which is used to determine whether the manufacturing technology is mature during the development of equipment and whether there is any risk in the process of technology transformation, so as to manage and control the system production to achieve the optimal in terms of quality and quantity, i.e. maximize equipment quality, reduce cost, and shorten production cycles to meet operational mission requirements.

Manufacturing maturity is an extension of the technology maturity concept and is used primarily to compensate for the difficulty of assessing the economic effectiveness of equipment production systems at the technology maturity level. Like technology maturity ratings, manufacturing maturity ratings provide systematic criteria and measures for assessing the maturity of a manufacturing technology at a given stage and allow for consistent maturity comparisons between different types of technologies. The Manufacturing Maturity Levels are categorized into 10 levels as follows:

(1) MRL Levels 1–3. A production concept or production requirements have been identified based on laboratory research;
(2) MRL Level 4. Design-to-cost (DTC) objectives have been identified and initial production ability assessment has been initiated;
(3) MRL Level 5. Yield/efficiency issues, critical quality characteristics, and S&T/specialized test equipment requirements have been identified and production planning has been initiated;
(4) MRL Level 6. Inspection/test equipment is validated in a laboratory environment and production cost impact factors and objectives are analyzed;
(5) MRL Level 7. Requirements for small-volume pilot production have been met, and the process has been validated in a production environment;

(6) MRL Level 8. The producibility risk evaluation is nearing completion and all manufacturing processes have been finalized;
(7) MRL Level 9. The system, component, or part is in small-lot pilot production, the cost of small-lot pilot production meets the production goals, and the equipment, tooling, and inspection and test equipment deliveries meet 3σ or other quality requirements;
(8) MRL Level 10. This stage is the mass production or maintenance stage, is the highest level of manufacturing maturity, the requirements of all raw materials, manufacturing processes, inspection and test equipment production capacity to meet 6σ or other quality requirements.

At each milestone decision point in the development process of an equipment project, it is necessary to assess the technical maturity of the overall project from a manufacturing perspective. The core objective is to assess the risk of manufacturability of the equipment project based on the maturity of the manufacturing and production processes, and to ensure a smooth transfer of the equipment technology program from the design department to the production plant.

(ii) Manufacturing Risk Connotations

Manufacturing risk refers to the equipment development process, due to the production system and the uncertainty of the relevant factors and their changes in the possibility of new product development failure. For example, it is difficult to realize mass production, the production cycle is too long, the process is unreasonable, equipment and instrumentation damage, backward testing methods, product quality is difficult to guarantee, poor reliability, supply system can not meet the requirements of mass production.

In order to quantify and control the manufacturing risk in the process of equipment development, you can determine the size of the probability of occurrence of manufacturing risk by evaluating the size of the deviation between the measured value of the maturity level of the production system of the key nodes and the target value; the larger the deviation, the higher the probability, and the smaller the deviation, the lower the possibility of occurrence. The size of the deviation of the maturity level of the production system also has a direct impact on the consequences of manufacturing risks, which may lead to the following forms of consequences: ① the physical quality (performance) of the equipment does not meet the expected targets, and the quality is unstable; ② the delay in the development progress, and the production and delivery of the batch can not be made on time; ③ the overspending of the development costs, the need for additional funding, and increase the investment in funding.

(iii) Manufacturing Risk Assessment Process

According to the idea of system engineering and parallel engineering, in order to effectively prevent and reduce the manufacturing risk in the process of equipment development, the manufacturing risk assessment activities should be carried out throughout the development and production process, rather than waiting until the time of commissioning to be carried out passively. Based on the manufacturing

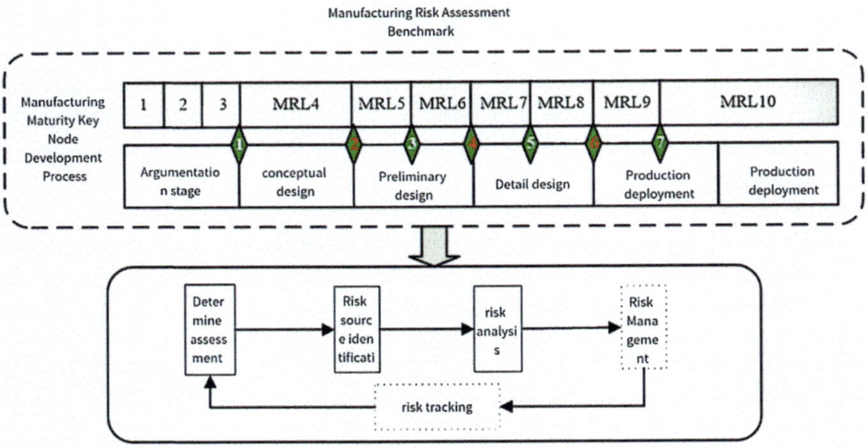

Fig. 3.6 Manufacturing risk assessment process

maturity assessment and risk management process of the U.S. Army Acquisition Program, the manufacturing risk assessment process shown in Fig. 3.6 is proposed.

As shown in Fig. 3.6, in the process of equipment development, there are a total of 7 nodes need to carry out manufacturing risk assessment, and each node has clear requirements for the maturity of the project's production system, of which the 2nd, 4th and 6th are milestone nodes, and in the U.S. Army Acquisition management process, these 3 nodes are the focus of supervision. The core of the process of single-node manufacturing risk assessment is to assess the baseline, clear manufacturing risk sources, analyze the probability of their occurrence and impact of the consequences of the final comprehensive assessment of the single-node manufacturing risk. In project manufacturing management, the transfer stage decision-making in addition to considering the current assessment of the node manufacturing risk value, but also need to consider the size of the previous node manufacturing risk value, to ensure that each stage of the manufacturing risk value is within a reasonable range of fluctuations is the ultimate goal of manufacturing risk assessment and management.

II. System of Indicators for Manufacturing Risk Assessment

In order to effectively measure the manufacturing risk of the equipment development over the assessment node, it is necessary to construct an assessment index system that can reflect the manufacturing maturity level variations and the consequences of the impact, based on the reference manufacturing maturity evaluation model, it is proposed that the risk from the following 9 aspects: technology and industrialization basis, design scheme, cost and budget, production materials, process capability and control, quality management, personnel, equipment, production management, etc. To comprehensively assess the manufacturing risk.

(1) Technology and Industrialization Base Risk

Technology and industrialization-based risk refers to the overall technical level of the country's current project design, development, production, operation, maintenance, end-of-life and other aspects of uncertainty. The main risk factors to be considered are: the maturity level of manufacturing technology, the level of industrialization base capacity, and the degree of industrial capacity and capability to meet the task.

(2) Design Program Risk

Design program risk refers to the degree to which the design program meets the demand. The main risk factors to be considered are: the level of technical status of the design scheme, the assessment of the manufacturability of key nodes, the existence of technical status acceptance standards, the implementation of production engineering and planning, and the preparation of special tooling and test equipment.

(3) Production Material Risk

Production material risk refers to the supply risk of raw materials and ancillary parts. The main risk factors to be considered are: verification of raw material reasonableness, raw material security, material list risk management, environmental factors and contractual special handling measures, and so on.

(4) Cost and Budget Risk

Cost and budget risk refers to the risk of the project to meet the requirements of the research and development funds and budget plan. The main risk factors to be considered are: cost-oriented design, production cost assessment, cost reduction measures, financial security.

(5) Process Capability and Control Risk

Process capability and control risk refers to the manufacturing process to ensure the reproducibility of key features and stability of the risk. The main risk factors to be considered are: key features and process analysis, fluctuation deviation analysis and control, process capability analysis and evaluation, process reliability analysis.

(6) Quality Management Risk

Quality management risk refers to the quality control and quality improvement risk. The main risk factors to be considered are: quality management organization and strategy, supplier quality management, quality planning, quality assurance program implementation, etc.

(7) People Risk

Personnel risk is the risk of the skills and security of a certain number of people who support the smooth running of manufacturing. The main risk factors to be considered are: participation in design and development, identification of various types of production personnel, identification of special skills of personnel, training and certification.

(8) Equipment Risk

Equipment risk refers to the risk of production line equipment capacity and capacity assurance. The main risk factors to be considered are: production line planning and construction, new equipment procurement and calibration, production line test run and verification, etc.

(9) Production Management Risk

Production management risk refers to the risk of integration and coordination of all elements of the equipment manufacturing production process. The main risk factors to be considered are: manufacturing strategy and plan, workshop scheduling plan, production line operation control, warehouse management and other situations.

III. Fuzzy Integrated Assessment of Manufacturing Risks

The assessment of the manufacturing risk of equipment development projects has a strong ambiguity. The first is the ambiguity of the concept of manufacturing risk itself, the theoretical community has not yet a unified definition, and the research on the factors and manifestations it includes has not formed a unified system. Secondly, the ambiguity of the influencing factors, due to the complexity of the risk, there are many uncertain influencing factors, and these factors may be associated with each other, at the same time, the degree of influence of these influencing factors on the loss of non-linear and dynamic, so that the assessment of them can not be effective data statistics to determine the probability of occurrence of the risk event, but also can not be directly and accurately determine the severity of the occurrence of the event.

For this reason, in order to reduce the arbitrariness of judgment and improve the scientificity of the assessment results, this book combines the fuzzy evaluation method with the AHP method, firstly, the AHP method is used in the indicator system, and the weight vectors of the various indicators are calculated according to the recursive structure of the indicators, and then the multi-level fuzzy comprehensive evaluation model is used to assess the manufacturing risk.

(i) Establishment of an Indicator System

According to the analysis of the manufacturing risk, firstly, according to the nature of the manufacturing risk is divided into nine structural indicators, and the risk factor of each type of risk is used as the analysis index of each type of structural indicators, and the set of structural indicators $O_P = \left[\underbrace{0, 0, 0, x_{op}, y_{op}, z_{op}, \overbrace{0, \ldots, 0}^{m\,term}}_{6+m\,term}\right]$ is set, and the indicator system is shown in Table 3.25.

(ii) Identification of a Collection of Rubrics

Comment set refers to all possible comments on the assessment object, according to the actual situation and the needs of manufacturing risk management, each indicator is divided into several levels according to its degree of performance, comment set $V = \{v_1, v_2, v_3, v_4, v_5\}$, where v_1, v_2, v_3, v_4, v_5 corresponds to the risk level of "high risk", "high risk", "average risk", "low risk", "low risk", respectively.

Table 3.25 Manufacturing risk assessment indicator system

Manufacturing risk assessment index system for equipment development projects	Structural indicators	Analysis indicators	Structural indicators	Analysis indicators
	Technology and industrialization Basic risk (U_1)	Manufacturing technology maturity level (U_{11}) Basic capacity level of industrialization (U_{12}) Industrial capacity and capacity (U_{13})	Quality management risks (U_6)	Quality management organization and strategy (U_{61}) Supplier quality management (U_{62}) Quality planning and assurance (U_{63})
	Design plan risk (U_2)	Technical status level (U_{21}) Manufacturability (U_{22})	Personnel risk (U_7)	Staff skills (U_{71}) Personnel supportability (U_{72})
	Production material risk (U_3)	Quality level of raw materials (U_{31}) Supporting management (U_{32})	Equipment risk (U_8)	Equipment capacity (U_{81}) Equipment capacity (U_{82})
	Cost and budget risks (U_4)	Design to cost (U_{41}) Cost assessment (U_{42}) Funding guarantee (U_{43})	Production management risks (U_9)	Manufacturing strategy and plan (U_{91}) Warehouse and workshop scheduling (U_{92}) Production line operation control (U_{93})
	Process capability and risk control (U_5)	Key characteristics and process analysis (U_{51}) Analysis and control of fluctuation deviation (U_{52}) Process capability analysis and evaluation (U_{53})		

(iii) Constructing Weight Sets

In comprehensive evaluation, the weights reflect the degree of importance of each indicator in the overall indicator system, and whether the weights are determined scientifically and reasonably directly affects the accuracy of the evaluation. Due to the different project milestone nodes, the various factors in the manufacturing risk have different degrees of influence on the assessed indicators, and there are different risk preferences, and different decision makers have different perceptions and evaluations of the same risk. Thus, the assessment of manufacturing risks for equipment development projects should reflect the degree of impact of various risk factors on equipment and minimize the differences in subjective impact among assessors. Therefore, the hierarchical analysis method can be used to rank the importance of the structural and analytical indicators, and according to the degree of influence of the indicators and their importance of the corresponding weights, so as to determine the manufacturing risk of the weight set of indicators at all levels. Let the weight set of structural indicators be $A = \{A_1, A_2, A_3, A_4, A_5, A_6, A_7, A_8, A_9\}$. The corresponding weights for each analyzed indicator are $A = \{a_{i1}, a_{i2}, \ldots, a_{ij}\}$, $i = 1, 2, 3, \ldots, 9$, j indicates the number of analyzed indicators included in structural indicator U_i.

(iv) Fuzzy Integrated Evaluation

(1) Establish the fuzzy judgment matrix. Using Delphi (Delphi method), with reference to the established set of rubrics V to evaluate the indicators at each level in turn, to get the degree of belonging to the $u_{ij} = (i = 1, 2, 3, \ldots 9)$, v_tth rubric r_{ijt}, where $r_{ijt} = m_{ijt}/m$, m is the total number of experts involved in fuzzy comprehensive evaluation, and m_{ijt} is the number of participating experts who think that the indicator m_{ijt} belongs to the evaluation level v_t. From this, we can get the following fuzzy evaluation matrix R_t:

$$Rt : Rt = \begin{bmatrix} r_{i11} & r_{i12} & r_{i13} & r_{i14} & r_{i15} \\ r_{i21} & r_{i22} & r_{i23} & r_{i24} & r_{i125} \\ \vdots & \vdots & \vdots & \vdots & \vdots \\ \vdots & \vdots & \vdots & \vdots & \vdots \\ r_{ij1} & r_{ij2} & r_{ij3} & r_{ij4} & r_{ij5} \end{bmatrix}$$

(2) First-level fuzzy comprehensive judgment. The first-level fuzzy comprehensive judgment is evaluated by all analytical indicators in the structural indicators, and according to the fuzzy evaluation matrix R_t, combined with the weight of each analytical indicator, each analytical indicator is evaluated comprehensively, and the first-level fuzzy comprehensive judgment set is obtained as (using the weighted average type of arithmetic):

$$B_t = R_t \cdot R_t = \sum_{j=1}^{n} a_{ij} r_{ij} \quad (i = 1, 2, \ldots, 9;\ t = 1, 2, \ldots, 5)$$

(3) Second-level fuzzy comprehensive judgment. The second level of fuzzy comprehensive judgment is to judge between the structural indicators, each structural indicator U_i as a factor, using B_i as a single factor evaluation of U_i, then the fuzzy comprehensive judgment matrix is:

$$R = [B_1 B_2 B_3 B_4 B_5 B_6 B_7 B_8 B_9] = (b_{it})_{9\times 5}$$

Therefore, the second level fuzzy composite judgment level can be obtained as:

$$B = A \circ R = (b_1 b_2 b_3 b_4 b_5)$$

Taking a weighted average type of arithmetic, then:

$$b_t = \sum_{i=1}^{9} A_i b_{it} \quad (t = 1, 2, \ldots 5)$$

(4) Processing of judging indicators. After fuzzy evaluation, a fuzzy vector $B = (b_1, b_2, b_3, \ldots, b_5)$ is obtained, and $b_t = (t = 1, 2, \ldots, 5)$ represents the degree of affiliation of the manufacturing risk to the tth rubric v_t. Finally, different levels of rubric v_t can be assigned the specified value β_t, and the weight of the affiliation degree b_t is used to obtain the comprehensive assessment score of manufacturing risk as:

$$\beta = \sum_{t=1}^{t} b_t^k \beta_t / \sum_{t=1}^{5} b_t^k, \text{ ordinary course } k = 1, 2.$$

IV. Arithmetic Examples

The above method is applied to a missile equipment development project undertaken by a unit as an example to assess the risk of each factor, taking the manufacturing risk assessment indicators given in Table 3.25 as the domain of the assessed factor U, and inviting 10 experts to form an assessment team to assess the manufacturing risk before commissioning. The main steps and results of the application process are briefly described below:

Step 1: Determine indicator weights. First of all, the structural and analytical indicators in Table 3.25 are weighted and scored according to the experts using the hierarchical analysis method, and after normalization, the specific weighting results are as follows:

$$\begin{aligned}
A &= \{0.05, 0.07, 0.09, 0.09, 0.2, 0.1, 0.15, 0.10, 0.15\},\\
A_1 &= \{0.3, 0.3, 0.4\},\\
A_2 &= \{0.4, 0.6\},\\
A_3 &= \{0.5, 0.5\},\\
A_4 &= \{0.4, 0.3, 0.3\},\\
A_5 &= \{0.3, 0.3, 0.4\},\\
A_6 &= \{0.35, 0.35, 0.3\},\\
A_7 &= \{0.7, 0.3\},\\
A_8 &= \{0.5, 0.5\},\\
A_9 &= \{0.3, 0.3, 0.4\}.
\end{aligned}$$

Step 2: Construct a fuzzy comprehensive judgment matrix. The assessment team scored the risk weights of each analytical indicator in Table 3.24 according to the questionnaire provided based on the set of assessment rubrics, and obtained the second-level judgment matrix after normalization, which was combined with the weights to obtain the first-level fuzzy judgment matrix, and the weight scoring table and the first-level fuzzy judgment matrix are as shown in Table 3.26.

The first level fuzzy judgment matrix is as follows:

$$R = \begin{bmatrix} 0.03 & 0.17 & 0.29 & 0.45 & 0.06 \\ 0 & 0.08 & 0.46 & 0.4 & 0.06 \\ 0.05 & 0.15 & 0.15 & 0.3 & 0.35 \\ 0 & 0.07 & 0.13 & 0.52 & 0.28 \\ 0.1 & 0.22 & 0.38 & 0.22 & 0.08 \\ 0 & 0.135 & 0.195 & 0.535 & 0.135 \\ 0 & 0.1 & 0.25 & 0.48 & 0.17 \\ 0 & 0.25 & 0.25 & 0.35 & 0.15 \\ 0.14 & 0.1 & 0.3 & 0.34 & 0.1 \end{bmatrix}$$

Step 3: The manufacturing risk is quantitatively assessed.

$B = A \circ R =$ (0.0470, 0.1464, 0.2749, 0.3798, 0.1489), and the normalization process yields: $B' = A \circ R =$ (0.047, 0.147, 0.276, 0.381, 0.149). Quantifying the set of rubrics as: $V = \{10, 8, 6, 4, 2\}$, the total manufacturing risk score $\beta = B' * V = 5.124$, i.e., the value of risk at the node is between low and average.

Step 4: Risk Cause Analysis and Control. The assessment results show that although the total manufacturing risk before the batch production of the project is not high, it can be seen from Table 3 that the risk assessment of A and the indicators is high, i.e., there is still a certain amount of risk in the control of the production process and the production management of this type of missile, and there is a need to further strengthen the contents of the work required by each analytical indicator and improve the ability of the control of the production process and the level of production management, so as to further reduce the manufacturing risk.

Table 3.26 Statistics on the evaluation results of analyzed indicators

	U_7	U_{71}	U_{72}	U_3	U_{31}		U_{32}	U_8	U_{81}	U_{82}	U_4		U_{41}	U_{42}	U_{43}	U_9	U_{91}
U11	0	1	2	5	2	U42	0	1	2	5	2	U71	0	1	1	6	2
U12	1	2	5	2	0	U43	0	0	1	7	2	U72	0	1	6	2	1
U13	0	2	2	6	0	U51	1	3	4	2	0	U81	0	3	2	4	1
U21	0	2	4	4	0	U52	1	3	2	4	0	U82	0	2	3	3	2
U22	0	0	5	4	1	U53	1	1	5	1	2	U91	1	1	4	1	1
U31	0	1	2	3	4	U61	0	1	0	7	2	U92	1	1	2	5	1
U32	1	2	1	3	3	U62	0	2	3	4	1	U93	2	1	3	4	1
U41	0	1	1	4	4	U63	0	1	3	5	1						

3.3.3 Reliability Decline Risk Evaluation Technique for Batch Production Process

I. RQR Chain-Based Risk Source Analysis of Product Reliability Decline in Manufacturing Processes

Manufacturing system is the mother of manufacturing products, manufacturing process is the formation of products, people, machines, materials, law, environment, measurement and other manufacturing factors deviation is the root cause of the decline in product reliability [6]. In engineering practice often occurs in this case, the manufacturing process, product quality inspection qualified, but put into use, product reliability can not meet the design reliability requirements. This is because in the manufacturing process manufacturing system component degradation/failure and other abnormalities will be transferred to the process quality deviation, reflected in the QR interaction effect, and the process quality deviation will cause in the manufacturing stage is difficult to detect product defects, product defects in the use of the stage of the formation of faults will be exposed to cause the decline in product reliability. Therefore, the principle of manufacturing process reliability decline is mainly manufacturing system abnormalities lead to manufacturing process quality degradation, followed by the accumulation of manufacturing process quality deviation caused by the decline in product reliability of such a chain deviation conduction relationship, as shown in Fig. 3.7. Specifically, embodied in the manufacturing system reliability R (Manufacturing System Reliability), manufacturing process quality Q (Process Quality) and manufacturing process product reliability R (Product Inherent Reliability) RQR chain, in which the manufacturing process quality Q is the core of the RQR chain conduction, but also the manufacturing process quality Q is the core of the RQR chain conduction, and the manufacturing process quality Q is the core of the RQR chain conduction. Among them, Manufacturing Process Quality is the core of RQR chain transmission and the main source of manufacturing process product reliability decline risk, as shown in the upper part of Fig. 3.7.

Reliability as a product's hidden characteristics in the manufacturing process is difficult to directly monitor, and manufacturing process quality data as the product life cycle formation process of the most abundant data, which contains a large number of reflecting the future reliability characteristics of the product as well as the potential risk of reliability decline information. The risk of decline in manufacturing process reliability mainly refers to the risk of decline in future product reliability indicators caused by manufacturing process quality deviation, the larger the manufacturing process quality deviation, the probability of decline in product reliability risk will also increase. At the same time, as shown in the lower part of Fig. 3.7, the manufacturing process quality Q as the core of the RQR chain transmission, component (such as tools, fixtures, etc.) degradation and other manufacturing system anomalies will also be reflected in the manufacturing process quality deviation, and will be transmitted through the transmission of quality deviation to the product reliability caused by the risk of decline. Key Quality Characteristics (KQCs) as the main carrier of manufacturing process quality information, is the main hand of quality analysis and

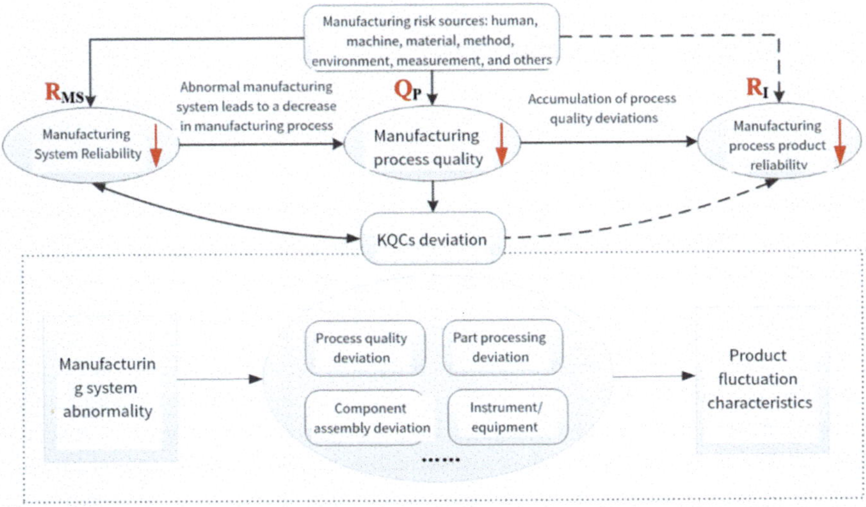

Fig. 3.7 RQR chain-based risk sources analysis for product reliability downturn in manufacturing processes

control, KQCs controlled or not, will directly determine the manufacturing process of product reliability downside risk size. Therefore, in order to prevent the risk of decline in product reliability in the manufacturing process, the key is based on the complete KQCs quality data and objective evaluation indicators for the manufacturing process deviation monitoring and control, so as to minimize the risk of decline in product reliability in the manufacturing process and to ensure that the manufacturing process quality and product reliability.

Assuming that the risk of decline in manufacturing process reliability and manufacturing process quality deviation are v_2 (Risk) and v_3 (Variation), the relationship between the two models can be expressed by the following formula:

$$R_S = F_{V \sim R_S}(V) + \varepsilon \tag{3.23}$$

where the function v_5 represents the model of the effect of manufacturing process quality deviation $A = \{A_1, A_2, A_3, A_4, A_5, A_6, A_7, A_8, A_9\}$ on the risk of manufacturing process product reliability decline $A_i = \{a_{i1}, a_{i2}, \ldots, a_{ij}\}$, $i = 1, 2, 3, \ldots, 9$, and j represents the error term. Because the monitoring of manufacturing process quality deviations is mainly based on KQCs, the quality deviation U_i in Eq. (3.23) also mainly refers to the deviation of KQCs.

II. Manufacturing Process Product Reliability Decline Risk Metrics

In order to obtain complete manufacturing process KQCs data to assess the risk of product reliability downturn in the manufacturing process, this book divides the manufacturing process quality deviation by stage into four parts covering the whole manufacturing process, namely, procurement $(u_{ij}\ (i = 1, 2, 3, \ldots 9))$, machining

$(t(t = 1, 2, \ldots, 5))$, assembling (v_t), and testing (r_{ijt}), and will calculate the procurement risk, machining risk, assembling risk, and testing risk brought about by each part of the quality deviation, as shown in Eq. (3.24) shown.

$$r_{ijt} = m_{ijt}/m \tag{3.24}$$

Quality deviation is mainly characterized by the deviation between the quality characteristics of the manufacturing process and its specified value, the size of which can be calculated by the process capability index. Based on the establishment of the reliability decline risk evaluation index system for the whole manufacturing process and the acquisition of complete quality data of KQCs, this book will use the multivariate process capability index to quantify the quality deviation of the manufacturing process and obtain the quantitative level of the reliability decline risk of the manufacturing process, which will provide a basis for the prevention and monitoring of the reliability decline risk of the manufacturing process brought about by the transmission of the RQR chain deviation.

(i) Establishment of a System of Indicators

Constructing an evaluation index system is a prerequisite for evaluating the risk of product reliability decline in the manufacturing process. In this book, from the perspective of procurement risk, processing risk, assembly risk and testing risk that lead to the risk of product reliability decline in the manufacturing process, we will establish an evaluation index system for the risk of product reliability decline in the manufacturing process covering the whole process of quality deviation as shown in Fig. 3.8, which is divided into four levels. Specifically, the procurement risk focuses on the qualified status of raw materials and the ratio of purchased parts in the product composition, i.e., the yield rate of purchased parts and the ratio of purchased parts in the product composition; the processing risk and the assembly risk mainly consider the deviation in the quality characteristics of the homemade parts and the level of technical status of the processing equipment, as well as the deviation in the quality characteristics of the assemblies and the operation errors of the personnel; the testing risk mainly considers the measurement errors and deviation in the testing ability of the equipment at the time when the product is shipped out of the factory. The risk of testing is mainly considered to be the error in the measurement of the product at the factory and the deviation of the testing capacity of the equipment. At the same time, the quantitative values of the above indicators can be calculated on the basis of the data of KQCs of the manufacturing process.

(ii) Calculation of the Risk of Decline in Reliability

The process capability index can effectively assess the magnitude of deviations in manufacturing process KQCs. When the manufacturing process KQCs data are multivariate normally distributed, the statistic $(X - \mu_0)^T \sum^{-1} (X - \mu_0)$ will obey the X^2 distribution, where μ_0 and $\sum$ are the mean and covariance matrix of the

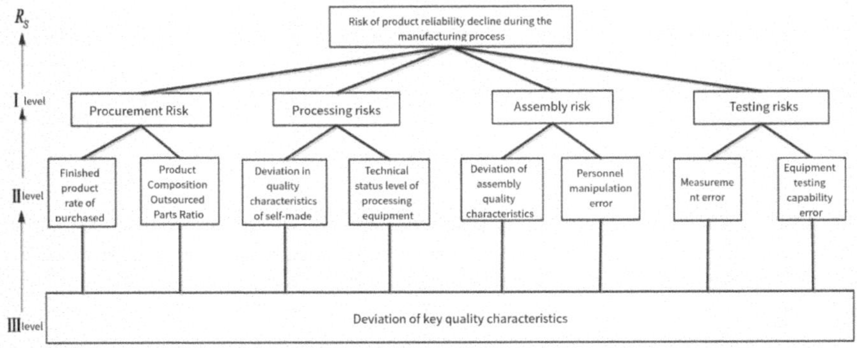

Fig. 3.8 Manufacturing process product reliability decline risk evaluation index system Calculation of the risk of decline in reliability

sample matrix X, respectively. Under 3σ quality requirement, the area formed by $(X-\mu_0)^T\sum^{-1}(X-\mu_0)\le\chi^2_{m,0.9973}$ is defined as the process area formed by m KQCs of the manufacturing process, and the following multivariate process capability index is used to monitor the fluctuation of the key quality characteristics at the bottom of Fig. 3.8, i.e., at level III of the evaluation index system, so as to obtain the corresponding manufacturing risk level R_S^{II} at level II.

$$R_S^{\text{II}} = \left[\frac{\prod_{i=1}^{m}(USL_i - LSL_i)}{\prod_{i=1}^{m}(UPL_i - LPL_i)}\right]^{\frac{1}{m}} \tag{3.25}$$

In the above formula, USL_i and LSL_i are the upper and lower limits of the quality data specification requirements, UPL_i and LPL_i are the upper and lower limits corresponding to the revised process areas in the actual production, and the volume ratio of the two is utilized to calculate R_S^{II}, m represents the number of KQCs, and $i=1,2,\ldots m$.

After that, the experts give the weights of each indicator of the evaluation index system level II, and combine with the R_S^{II} obtained by Eq. (3.25) using the method of weighted geometric mean to accumulate upward to obtain the risk of purchasing, processing, assembling and testing of the evaluation index system level I, R_S^I:

$$R_S^I = \left[\prod_{j=1}^{n}\left(R_{S(j)}^{II}\right)^{wj}\right]^{\frac{1}{\sum_{j=1}^{n} wj}} \tag{3.26}$$

Which represents the weight of the Ath indicator in the evaluation index system II level ($j=1,2,\ldots n$), w_j is an integer in the range of 1 ~ 5.

Finally, based on the hierarchical analysis method to determine the evaluation index system I level of procurement, processing, assembly and testing risk based on the final manufacturing process product reliability downturn risk Rs:

$$R_S = \sum_{k=1}^{4} R^I_{S(k)} w_k$$
$$= R^I_{S(1)} w_1 + R^I_{S(2)} w_2 + R^I_{S(3)} w_3 + R^I_{S(4)} w_4 \quad (3.27)$$

where $R^I_{S(k)}$ represents the procurement, processing, assembly and testing risks of the manufacturing process, $k = 1, 2, 3, 4$.

The greater the value of manufacturing process product reliability decline risk Rs, on behalf of the manufacturing process product reliability will face a greater decline in the probability of risk caused by a greater degree of reliability decline. Definition R_D (Degree) represents the level of manufacturing process product reliability decline risk, as shown in the following equation:

$$R_D = \begin{cases} 1, \ 1.67 < R_S \\ 2, \ 1.33 < R_S \leq 1.67 \\ 3, \ 1 < R_S \leq 1.33 \\ 4, \ 0.67 < R_S \leq 1 \\ 5, \ R_S \leq 0.67 \end{cases} \quad (3.28)$$

The R_D distribution ranges from 1 to 5, and the larger the R_D value, the higher the risk level of product reliability decline in the manufacturing process, i.e., the higher the probability of decline and the higher the degree of risk. Similarly, the probability of risk of decline and the degree of risk is also divided into 5 levels, 1 ~ 5 levels correspond to the decline of the risk probability of "low", "lower", "high", "high", "very high", and the degree of risk of decline "small", "small", "average", "high", "severe".

(iii) Evaluation of Validity Test

The accuracy of the calculation of R_S for the risk of decline in product reliability of the manufacturing process can be characterized by the error term ε, as shown in Eq. (3.1), the larger ε is, the greater the deviation of the calculation results from the actual value. The main factor affecting the size of the error term ε is the degree of adaptation of the established evaluation index system and its weights to the real state of the manufacturing process, accordingly, the higher the degree of adaptation between the two, the more accurate the results of the R_S calculation will be. At the same time, in order to make the effect of the error term ε on R_S more sensitive, this book proposes a method of using the confidence interval of R_S to monitor the risk of reliability decline in different batches of the manufacturing process.

1. Evaluation Index System Test

For the evaluation index system is determined with the real state of the manufacturing process may not be a high degree of fitness, this book will be realized through the method of structural equations to test it, as follows.

- Measurement Model Reliability and Validity Test

The structural equation is divided into two parts: measurement model and structural model, and the test of the measurement model is mainly the test of the reliability and validity of the collected raw data. Cronbach's α coefficient is the most commonly used measure in the reliability test, which characterizes the internal consistency of the raw data. A higher reliability indicates a higher degree of measurement reliability, i.e., a higher degree of consistency of the raw data under multiple measurements.

$$\alpha = \frac{k}{k-1}\left(1 - \frac{\sum S_i^2}{S_x^2}\right) \tag{3.29}$$

where k is the number of KQCs to be tested, S_x^2 is the variance of all KQCs, and S_i^2 is the variance of the AAth KQC.

The test of validity is mainly a test of the differential validity of the measurement model, the higher the differential validity indicates that the classification of the indicators in the evaluation index system is more effective, which can be realized by measuring the factor analysis of the evaluation index system.

- Structural Equation Fitness Analysis

The most important structural equation test is the structural model fitness analysis, which tests the fitness between the implied covariance matrix $\widehat{\sum}$ in the evaluation index system and the covariance matrix S of the original data, and the higher the fitness, the more consistent the assumed theoretical model is with the actual original data, i.e., the closer the $\widehat{\sum}$ is to the S. Specifically, the degree of fitness between $\widehat{\sum}$ and S can be measured by the fitness index.

This book adopts the overall fit indexes given by Wu as the fit indexes for analysis: (1) absolute fit indexes: chi-square value, GFI, RMSEA; (2) incremental fit indexes: CFI, IFI, TLI. the above fit indexes can reflect the fitness of the structural model with the samples in different aspects, and point out the direction of improvement of the model so as to correct it.

2. Calculation and Monitoring of Confidence Intervals of Evaluation Results

In the case that the R_S distribution of the risk of decline in product reliability in the manufacturing process is unknown, it is not possible to calculate the confidence intervals by precise analytical methods, but they can be obtained by simulation.

In this book, we use Monte-Carlo simulation to calculate the confidence interval of R_S. First, according to the characteristics of the actual production data of a certain batch, the simulation is used to generate n groups of normal-type simulation data that satisfy the characteristics of the KQCs of the batch (when the production of the product is stable, the quality inspection data is generally normally distributed), and the R_S value of each group of the KQCs data is calculated. After that, the R_S are sorted from smallest to largest, so that the lower R_S^L and upper R_S^U of the α confidence interval correspond to the corresponding values on the sorted position $[k_L, k_U]$ under the specified confidence level R_S:

$$[k_L, k_U] = \left[\left(\frac{1-\alpha}{2}\right)n + 1, n\left(\frac{1+\alpha}{2}\right)\right] \tag{3.30}$$

Based on the historical production data of the product in steady state and the real-time data of the normal production batch, the confidence intervals $\left[R_S^L, R_S^U\right]$ for the steady state and normal production batch R_S can be obtained respectively, and the confidence interval obtained in steady state is defined as the receiving interval. Meanwhile, in order to monitor the fluctuation of the risk of reliability degradation of each batch by using the confidence intervals, firstly, the variable ξ is introduced to monitor the overlap of the confidence intervals of normal production batch R_S with respect to the receiving interval:

$$\xi = \frac{confidence\ interval \cap The\ length\ of\ the\ receiving\ interval}{confidence\ interval \cup The\ length\ of\ the\ receiving\ interval} \tag{3.31}$$

The larger the ξ value, the more the two intervals overlap and the higher the consistency. Obviously, ξ can reflect the information of the consistency between the two intervals, but fails to reflect the relationship between the two positions. For this reason, the deviation function $f(\alpha_r)$ is then introduced to examine the information of the relative shift of the two positions:

$$f(\alpha_r) = \begin{cases} 1 - e^{-(\alpha_r - \alpha)} & \alpha_r > \alpha \\ 0 & \alpha_r = \alpha \\ -1 + e^{\alpha_r - \alpha} & \alpha_r < \alpha \end{cases} \tag{3.32}$$

where α_r denotes the midpoint of the normal production batch confidence interval, and α denotes the midpoint of the reception interval, the larger α_r deviates from α, the closer $f(\alpha_r)$ will be to -1 or 1.

Finally, by combining the overlap and location information of the confidence intervals and reception intervals of each batch and analyzing the energy clustering point of the numerical distribution of the variable $\delta = \xi \times f(\alpha_r)$, the trend and fluctuation of the confidence intervals of R_S of each batch of normal production can be reflected relative to the reception intervals in the steady-state, which can be used as a reference for further monitoring and improving the product quality and reliability level of each batch.

III. Example Verification

Take an electromechanical product batch production process as an example, the product from the initial raw material procurement to the final factory testing process, need to go through a number of complex processing, assembly and testing procedures, a large number of key quality characteristics (KQCs) of the manufacturing process of the small abnormal deviations in the mass production environment will be amplified, resulting in the batch production of the reliability of the product has a serious decline, can not meet the design requirements. In order to objectively reflect the impact of the quality deviations in the manufacturing process of this electromechanical product on the reliability of the product, the risk of decline in the reliability of the product in the manufacturing process is evaluated and analyzed according to the proposed method.

(A) Evaluation Index System Establishment and Testing

1. Establishment of Manufacturing Process Product Reliability Decline Risk Evaluation Index System

In order to accurately evaluate the risk level of reliability decline of manufacturing process products, this book starts from the KQCs that affect the quality of the product manufacturing process, and combines the methods of calculating and determining the weights of indicators at all levels in Sect. 3.3.2, and establishes the initial evaluation index system of the risk of reliability decline in the manufacturing process products, as shown in Table 3.27, where the indicators of V111, V112, V113 and so on correspond to the key quality characteristic indicators of Level III of the evaluation index system in Fig. 3.8 Indicator system level III of the key quality characteristics indicators.

2. Evaluation Index System Structure Equation Test

In order to check whether the reliability downturn risk evaluation index system established earlier can truly and effectively reflect the real state of the manufacturing process, the KQCs data of the manufacturing process in the steady state of the product

Table 3.27 Initial evaluation index system for the risk of decline in reliability of an electromechanical product manufacturing process

Manufacturing process product reliability Risk of decline RS	Procurement risk Vp	Outsourced parts yield V11	V111 V112 V113
		Ratio of purchased parts to product composition V12	V121 V122 V123
	Processing risk Vm	Deviation of quality characteristics of self-made parts V21	V211 V212
		Level of technical condition of processing equipment V22	V221 V222
	Assembly risk Va	Deviation of assembly mass characteristics V31	V311 V312
		Personnel operating error V32	V321 V322 V323
	Test risk Vt	Measurement error V41	V411 V412
		Equipment test capacity deviation V42	V421 V422 V423

are collected according to the indexes in Table 3.27, as shown in Table 3.28, and the structural equations are utilized to check the adaptability with the index system.

First of all, it is necessary to test the reliability and validity of the collected KQCs data in the steady state, after the test, the overall Cronbach's USL_i coefficient value of the KQCs data is 0.912, and the $R_S^{II} = \left[\frac{\prod_{i=1}^{m}(\text{USL}_i - \text{LSL}_i)}{\prod_{i=1}^{m}(\text{UPL}_i - \text{LPL}_i)}\right]^{\frac{1}{m}}$ coefficient values of the data of the parts of purchasing risk, processing risk, assembling risk, and testing risk are 0.91, 0.913, 0.884, and 0.92, respectively, which indicates that the data in general and the parts of the data have a very good internal consistency. Meanwhile, exploratory factor analysis was conducted to test the differential validity of the samples on the collected data of KQCs at steady state: the result of KMO (Kaiser–Meyer–Olkin) was equal to 0.896, and the approximate chi-square value of Bartlett's test of sphericity was equal to 8252.395 ($p < 0.001$), indicating that the data were suitable for factor analysis. Using factor analysis to extract the four factors obtained, corresponding to the procurement risk, processing risk, assembly risk and testing risk that lead to the risk of product reliability decline in the manufacturing process, which explains the total variance of 71.424%, indicating that the modeling setup is feasible and the samples have a good differentiated validity, i.e., the collected data samples of the KQCs in the steady state meet the requirements of both reliability and validity.

After that, AMOS 17.0 software was applied to construct the structural equation model of the evaluation index system, and the fit between the implied covariance matrix $\widehat{\sum}$ and the covariance matrix S of the KQCs data samples was examined. As shown in Fig. 3.9, the path coefficient values of procurement risk, processing risk, assembly risk and testing risk on R_S are 0.576, 0.702, 0.601 and 0.571, respectively, and all of them pass the significance test, indicating that these four indicators have a significant impact on R_S, which is in line with the previous hypothesis. Meanwhile, processing risk and assembly risk have the greatest influence on R_S, which is the key risk in the manufacturing process of this product.

The results of the AMOS 17.0 software run for the structural equation model fitness metrics are shown in Table 3.29, in which, as a key metric, the RMSEA characterizes the average difference value between the model implied covariance matrix $\widehat{\sum}$ and the data sample covariance matrix S for each degree of freedom, and it is usually assumed that the $RMSEA < 0.05$ model will have a very good degree of fitness, and the RMSEA in Table 3.27 is equal to 0.048 meets the requirements. The rest of the indicators in Table 3.27 also reach the acceptable level, indicating that the established evaluation index system is reasonable and meets the requirements, and the test is passed.

(i) Calculation and Monitoring of the Risk of Reliability Degradation

Table 3.28 Sample data for KQCs

KQCs indicators	V111	V112	V113	……	V421	V422	V423
Mean value	4.5522	4.646	4.6739	……	2.077	1.7756	1.7141
Standard deviation	0.2355	0.2094	0.157	……	0.321	0.2206	0.2499

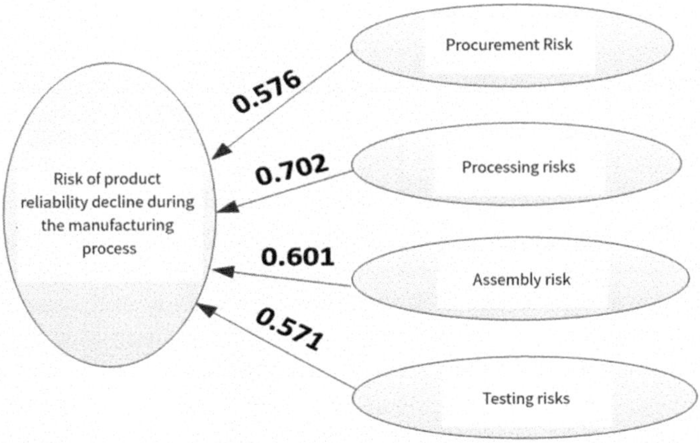

Fig. 3.9 Adaptation of the structural equations of the evaluation indicator system

Table 3.29 Results of fit tests for fitness indicators for structural equations

Model fitting index		Fitting value	Standard	Result
Absolute goodness of fit index	Chi square value	436.93	The smaller the better	Ideal
	GFI	0.931	> 0.9	Ideal
	RMSEA	0.048	< 0.05	Ideal
Incremental goodness of fit index	CFI	0.971	> 0.9	Ideal
	IFI	0.971	> 0.9	Ideal
	TLI	0.966	> 0.9	Ideal

1. Calculate the Risk of Reliability Decline and Its Confidence Interval

After the evaluation index system has been tested, it is applied to the normal batch production state to carry out the reliability decline risk evaluation. Collect the KQCs data as in Table 3.1 from batch 1 (the product batch production of 35 batches), and according to Eq. (3.25), calculate the risk quantitative value of the evaluation index system at level II, and take the calculation of the risk quantitative value of R_S^{II} for the deviation of quality characteristics of self-produced parts V21 as an example (in the following formula, $\chi^2_{(2,0.9973)} = 11.829$ and $\sum^{-1}$ represent the inverse matrix of the covariance matrix of the KQCs data):

$$R_S^{II} = \left[\frac{(3.200 - 1.700) \times (3.150 - 1.900)}{2\sqrt{\frac{\chi^2_{(2,0.9973)} \times 24.002}{\sum^{-1}}} \times 2\sqrt{\frac{\chi^2_{(2,0.9973)} \times 11.322}{\sum^{-1}}}} \right]^{1/2} = 0.808$$

wherein $\sum^{-1} = \begin{pmatrix} 11.322 & 0.498 \\ 0.498 & 24.002 \end{pmatrix}$.

Next, on the basis of determining the weights of the indicators of level II of the evaluation index system, the risks of procurement, processing, assembly and testing of level I of the evaluation index system are calculated according to Formula (3.26), and the results are (1.098, 0.908, 1.027, 1.093), respectively. Then, using hierarchical analysis to determine the weights of the manufacturing process procurement, processing, assembly and testing risk, the expert judgment matrix is shown in the following equation, which is calculated, and through the consistency test ($CR = 0.066 < 0.1$, $\lambda_{\max} = 4.117$), to get the final weights of (0.131, 0.484, 0.271, 0.114). Corresponding to the path coefficients of the previous structural equations, the processing risk (0.484) and assembly risk (0.271) are the critical risks in the manufacturing process.

$$\sum = \begin{pmatrix} 1 & 1/5 & 1/3 & 2 \\ 5 & 1 & 2 & 3 \\ 3 & 1/2 & 1 & 2 \\ 1/2 & 1/3 & 1/2 & 1 \end{pmatrix} \tag{3.33}$$

Finally, using Eq. (3.27) to get the value of R_S, as shown in Eq. (3.33), the result is equal to 0.986. According to the judgment criteria of Eq. (3.28), R_D is level 4, which means that the reliability of this batch of manufacturing process product reliability has a high probability of reliability decline and the degree of reliability decline is large.

$$\begin{aligned} R_S &= R^I_{S(1)} w_1 + R^I_{S(2)} w_2 + R^I_{S(3)} w_3 + R^I_{S(4)} w_4 \\ &= 1.098 \times 0.131 + 0.908 \times 0.484 + 1.027 \times 0.271 + 1.093 \times 0.114 \\ &= 0.986 \end{aligned} \tag{3.34}$$

To calculate the confidence interval of R_S in Batch 1. Based on the information of the mean and variance of the KQCs data in batch 1, the Matlab software is used to simulate and generate the quality characteristic data to satisfy the normal distribution, and the value of R_S is calculated using Matlab for 10,000 simulations with a confidence level of 80% (the results converge when the number of simulations is 10,000, as shown in Fig. 3.10), and the results are ranked according to Eq. (3.34), sorting the values corresponding to the 1001st and 9000th positions [0.9603, 1.0016], which are the lower and upper bounds of the confidence interval of the R_S in the desired batch 1.

2. Applying Confidence Intervals to Monitor the Fluctuation of Reliability Decline Risk in Different Batches

In order to monitor the fluctuation of reliability decline risk of manufacturing process products in different batches of the manufacturing process, the confidence interval [0.9601, 1.0014] of the historical steady-state production data is used as the receiving

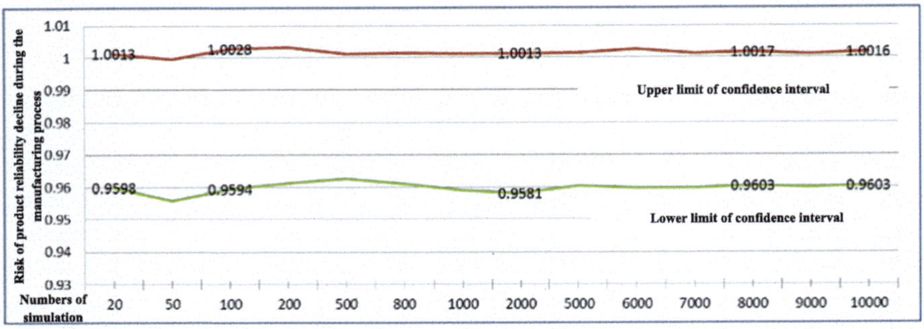

Fig. 3.10 Monte-Carlo simulation results for confidence intervals

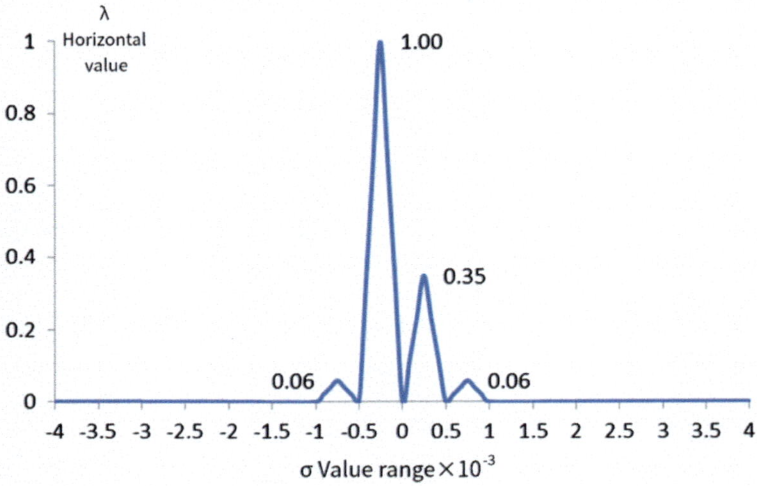

Fig. 3.11 Energy clusters for density sequences

interval, and the overlap and positional information of the confidence intervals and receiving intervals of the 35 different batches under the normal production state are compared and the energy clustering point analysis of $\delta = \xi \times f(\alpha_r)$ is performed.

Specifically, in the R_S range of $\left[-4 \times 10^{-3}, 4 \times 10^{-3}\right]$, the number of points M_n contained in each sub-interval of length $\alpha = \frac{k}{k-1}\left(1 - \frac{\sum S_i^2}{S_x^2}\right)$ is calculated to find the density sequence $\{\rho_n\}$ ($\rho_n = M_n/35,\ n = 1, 2, \ldots, 35$) of δ, and the λ level of $\{\rho_n\}$ is obtained accordingly, as shown in Fig. 3.11.

The mean value of the calculated variable ξ is 95.61% among the 35 batches, indicating a high degree of overlap and good consistency among the confidence intervals of the batches. However, it can be seen from Fig. 3.11 that the distribution of δ tends to be mostly between [0, 0.5] and the overall trend is to the left. This indicates that the manufacturing process as a whole has deviated to a certain extent and needs to be further improved in the follow-up.

References

1. The OSD manufacturing technology program in collaboration with the joint service/industry MRL working group. Manufacturing Readiness Level (MRL) Deskbook. Washington D.C.: Department of Defense, 2012.
2. M.T. Zhang. Research on the application of FMEA-based risk assessment for aerospace electronics manufacturing. University of Chinese Academy of Sciences, 2013.
3. He Yi-Hai, Chang Wen-Bing, Manufacturing Risk Assessment of Weaponry Development Projects Based on Manufacturing Maturity. Project Management Technology, 2009, No. 7: 54–58.
4. Liu Xuezong, Zhang Jian, Yu Shuyan. On the reliability and validity of scales. Journal of Capital Medical University, 2001, 22(4):314–317.
5. Xu Aiting. Application of Delphi method and its difficulties. China Statistics, 2006(9):57–59.
6. Yin, Chao. Research on the Mechanism of Product Reliability Slump and Risk Evaluation Methods in Manufacturing Processes, Master's thesis, Beijing University of Aeronautics and Astronautics, 2014.

Chapter 4
Manufacturing Process Reliability Optimization Techniques

4.1 Manufacturing System Reliability Optimization Technology

In the manufacturing process, manufacturing system reliability is a significant factor to ensure product quality and productivity, and also an important indicator to measure the performance of the manufacturing system. In principle, the manufacturing system as a special form of product existence, its reliability optimization technology can be common with the traditional reliability optimization technology, however, there are obvious defects in this approach. Firstly, the traditional reliability modeling technology focuses on the product's own function embodiment, and studies product malfunction situation. But the manufacturing system, as the medium of a certain product output, has the consistency of its function with the product's functional characteristics. And the manufacturing system also has its own specificity. Therefore, reliability optimization techniques that only consider the failure of the manufacturing system itself are obviously one-sided. Moreover, with the development of manufacturing technology, a large amount of data characterizing the manufacturing process' state can be promptly obtained. Therefore, it is very meaningful to establish a fast and practical manufacturing system reliability modeling and preventive maintenance technology based on manufacturing process quality data.

Y. He et al., *Reliability Theory and Technology in Manufacturing Process*,
https://doi.org/10.1007/978-981-96-2735-6_4

4.1.1 Manufacturing System Reliability Modeling Techniques

I. Manufacturing System Reliability Modeling Techniques Considering the Relationship between Product Quality and Manufacturing System Component Reliability

(i) Modeling Fundamentals

The mechanism and change rule of the manufacturing system component's performance degradation and failure occurrence is intricate and complex, not only from the current station manufacturing system component itself, but also from the upstream station product quality deviation [1, 2]. In the process of multi-station manufacturing system reliability analysis, it is known from the transmission of product deviation [3] that the interaction between process product quality and the reliability of manufacturing system components is also transmitted between stations, so that there is a statistical correlation between the failure of the manufacturing system components and the unqualified products.

Connotation of Manufacturing System Reliability: Manufacturing system reliability is usually defined as the ability of a manufacturing system to fulfill its expected function under operating conditions and specified time. In practice, the expected function of a manufacturing system includes not only the normal operation of components such as device and tools, but also the completion of the expected function, i.e., the output of products or semi-finished products. We hope that a manufacturing system with high reliability will not only operate normally for a long time, but also produce products with high quality and reliability. Therefore, based on the correlation between product quality and component reliability shown in Fig. 4.1, as well as the specificity of the manufacturing system function, the reliability of the manufacturing system is defined as the probability that the system components do not fail over a period of time and that the product produced is qualified.

$$\begin{aligned} R(t) = \Pr\{&\text{No system catastrophic failure before } t \\ &\cap \text{ Product quality qualified before } t\} \end{aligned}$$

Assumptions We use the following assumptions to define the operation of this manufacturing system:

(a) The deviation flow model is correct in the context of its application.
(b) The effect of critical quality characteristics of different products on the failure of manufacturing system components is independent.
(c) The manufacturing system is in series and the components are independent of each other, with their respective lifetimes following an exponential distribution.
(d) The performance degradation of multiple manufacturing system components can be described by a Gaussian Markov process for the current state and a normal distribution for the subsequent states, with the performance degradation

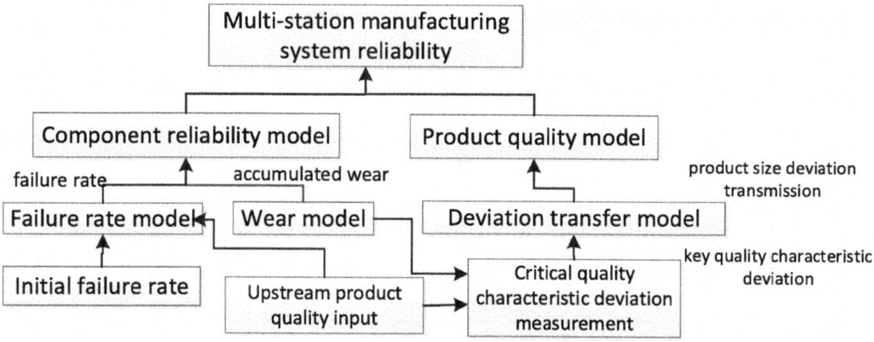

Fig. 4.1 Model of correlation between product quality and component reliability in multi-station manufacturing system

process mean and the performance decline state being correlated with the prior and current states only, and with a constant variance.

In manufacturing systems, product quality deviation is usually measured by the dimensional accuracy of key features, and product quality level is evaluated by the dimensional deviation of key product feature points.

(ii) **Modeling Process**

1. **Determine the Probability of Product Qualification**

The quality of the product at the *kth* station is denoted by $X(k)$, then $x(k) = [x_1(k), x_2(k), \ldots, x_n(k)]^T$. Where $Xi(k)$ is the *ith* quality characteristic on the part $(i = 1, 2, \ldots, n)$. From the state space equation and deviation flow theory, it can be inferred that [4]:

$$\Delta x(k) = A(k)\Delta x(k-1) + B(k)U(k) + W(k), \tag{4.1}$$

where $\Delta X(k)$ denotes the product dimensional deviation at the *kth* station and is a vector. $a(k)$, $b(k)$ and $u(k)$ denote the manufacturing system parameter matrices, and $w(k)$ denotes the residuals. And $A(k)\Delta x(k-1)$ denotes a portion of the critical product quality characteristics and $B(k)$ $U(k)$ denotes a portion of the critical control characteristics.

With $L(k)$ denoting the system parameters and $T(k)$ denoting the tolerances of the part dimensions. $f(L(k))$ denotes the quality of the product produced by the manufacturing system, $\Delta X(k)$ takes into account only the errors of the fixture parameters and does not take into account the errors of the machining features and the errors of the process control. Then:

$$f(L(k)) = \Delta X(k) - T(k), \tag{4.2}$$

where $f(L(k))$ is a nonlinear function of $x(k) = [x_1, x_2, \ldots, x_n]^T$, using Taylor expansion.

$$f(x) = a_0 + \sum_{i=1}^{n} a_i x_i, \tag{4.3}$$

where a_i is constant.

$$\beta = \frac{\mu_f}{\sigma_f} = -\frac{a_0 + \sum_{i=1}^{n} a_i \mu_{x_i}}{\sqrt{\sum_{i=1}^{n} a_i^2 \sigma_i^2 + \sum_{i=1}^{n} \sum_{\substack{j=1 \\ j \neq i}}^{n} a_i a_i Cov(x_i, x_j)}} \left(\mu_f = a_0 + \sum_{i=1}^{n} a_i \mu_{x_i} \right), \tag{4.4}$$

where $\mu_x = (\mu_{x_1}, \mu_{x_2}, \ldots, \mu_{x_n})$, is the mean point of $f(x)$. Then the mean $\mu_f = f(\mu_{x_1}, \mu_{x_2}, \ldots, \mu_{x_n})$, and the standard deviation $\sigma_f^2 = \sum_{i=1}^{n} a_i^2 \sigma_i^2 + \sum_{i=1}^{n} \sum_{\substack{j=1 \\ j \neq i}}^{n} a_i a_i Cov(x_i, x_j)$.

Then the probability that the product quality passes in the manufacturing process is:

$$\begin{aligned} R_k^Q &= \Pr\{y(k) \leq \theta_k\} = \Pr\{f(x) \leq 0\} \\ &= p\left\{ \frac{f(x) - \mu_f}{\sigma_f} < -\frac{\mu_f}{\sigma_f} \right\} \\ &= \Phi(-\beta). \end{aligned} \tag{4.5}$$

2. **Degradation of Manufacturing System Components**

The cumulative wear of a component in a manufacturing system increases with the number of cycles, denoted by $\Delta(k)$ as the amount of wear in a single cycle, then the cumulative wear of the component after k cycles $Z(k) = Z(k-1) + \Delta(k)$. If all wear degradation processes conform to an independent homogeneous distribution (most mechanical products), then:

$$Z(k) = \sum_{j=1}^{k} \Delta(j) \approx N(kE(\Delta(j)), kVarE(\Delta(j))). \tag{4.6}$$

Assuming that the failure of a single component follows an exponential distribution, the component reliability at period k can be expressed by the following equation:

$$R_k^F(t) = e^{-\lambda_k(t)t}. \tag{4.7}$$

Consider accelerated component loss from quality deviations in the input product:

$$\lambda_k(t) = \lambda_0(t) + E\big(\alpha_k (X(k) - m_k)^2\big), \tag{4.8}$$

where $\lambda_0(t)$ is the failure rate (initial failure rate) without considering the effect of the input product quality on the component, $X(k)$ is the mass of the product at the *kth* station, m_k is the standardized value of the input product quality, and α_k is the correction factor, which reflects the effect of the deviation of the input product quality on the wear of the component.

3. **Calculate the Reliability of the Manufacturing System Based on the Probability of Product Qualification**

$$\begin{aligned} R(t) &= \Pr\{\text{No system catastrophic failure before } t \\ &\quad \cap \text{Product quality qualified before } t\} \\ &= \Pr\{R^F \cap R^Q\} \\ &= \Pr\{R^F | Z(k)\}\Pr\{R^Q | Z(k)\}. \end{aligned} \tag{4.9}$$

Calculated for $\Pr\{R^F|Z(k)\}\Pr\{R^Q|Z(k)\}$, respectively.

$$\begin{aligned} \Pr\{R^F|Z(k)\} &= P\left(\prod_{i=1}^{n} R_i^F(k)|Z(k) < \eta\right) \\ &= P\left(\prod_{i=1}^{n} \exp([\lambda_0(t) + E[\alpha_i(X(i) - m_i)^2]]t_i)|Z(k) < \eta\right) \\ &= \exp\left(\sum_{i=1}^{n}(-[\lambda_0(t) + E[\alpha_i(X(i) - m_i)^2]]t_i)\right) \times \frac{1}{\sqrt{kVar(\Delta(j))}\sqrt{2\pi}} \\ &\times \left(\exp\left(-\frac{\eta - kE(\Delta(j))^2}{2kVar(\Delta(j))}\right) - \exp\left(-\frac{kE(\Delta(j))^2}{2kVar(\Delta(j))}\right)\right), \end{aligned} \tag{4.10}$$

$$\Pr\{R^Q|Z(k)\} = \Phi(-\beta). \tag{4.11}$$

When the above models are synthesized, the final expression for manufacturing system reliability based on the interaction of product quality and component reliability is as follows:

$$\begin{aligned} R(t) &= \Pr\{R^F|Z(k)\}\Pr\{R^Q|Z(k)\} \\ &= \exp\left(\sum_{i=1}^{n}(-[\lambda_0(t) + E[\alpha_i(X(i) - m_i)^2]]t_i)\right) \\ &\times \frac{1}{\sqrt{kVar(\Delta(j))}\sqrt{2\pi}} \times \left(\exp\left(-\frac{\eta - kE(\Delta(j))^2}{2kVar(\Delta(j))}\right)\right. \\ &\left. - \exp\left(-\frac{kE(\Delta(j))^2}{2kVar(\Delta(j))}\right)\right) \times \Phi(-\beta). \end{aligned} \tag{4.12}$$

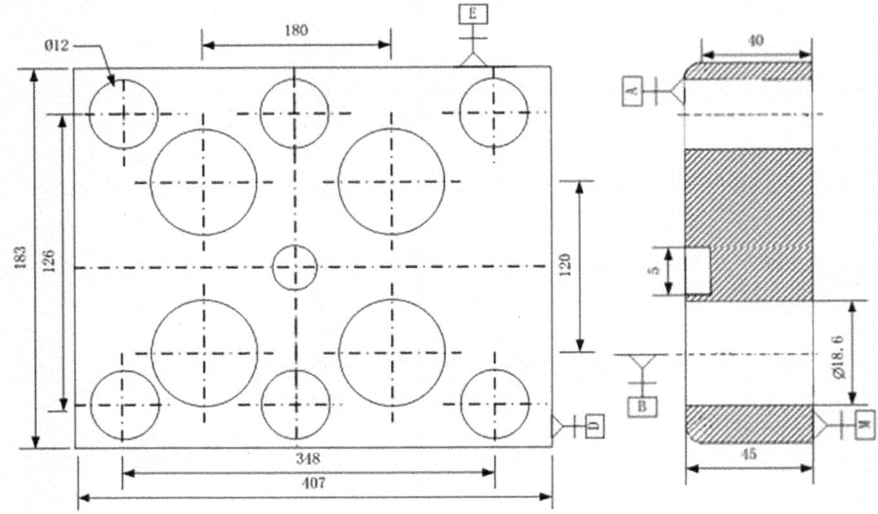

Fig. 4.2 Dimensional drawing of machined parts

(iii) **Arithmetic Analysis**

Take the machining process of a mechanical part as an example to establish a manufacturing system reliability model based on process data.

The machined dimensions of the part are shown in Fig. 4.2.

The dimensional characteristics of the part and the machining process are shown in Table 4.1.

The part parameters are as shown in Table 4.2.

Substitute the original data to get the product size output results as in Table 4.3 Column 3. The steps for solving the problem are as follows:

Step 1: First manufacturing system component decline and initial failure rate and other related parameters can be obtained from historical data:

$\Delta(j) \sim N\left(\mu_{\Delta(j)}, \sigma^2_{\Delta(j)}\right) \sim N\left(2 \times 10^{-3}, 1.6 \times 10^{-5}\right)$, initial failure rate $\lambda_0(t) = \lambda_0 = 4 \times 10^{-6}$, correction factors $\alpha_k = 0.001$, $h = 1$. Calculated from Eq. (4.8), the

Table 4.1 Parts characteristics and machining process

Workstations	Working procedures	Characteristics	Design Dimensions (mm)
Workstation 1	Milling surface M	Distance between M and A	45 ± 0.1
Workstation 2	Milling surface A Drill hole B	Hole B diameter	$18.6_{0}^{0.021}$
Workstation 3	Drill hole C	Hole C diameter	12 ± 0.1
Workstation 4	Slot milling S	Slot S radius	5 ± 0.1

Table 4.2 Part parameters

Serial number	Key features	n_x	n_y	n_z	p_x	p_y	p_z
1	Surface A	0	1	0	91.5	45	0
2	Surface M	0	1	0	91.5	0	0
3	Hole B	0	1	0	131.5	22.5	0
4	Hole C	0	1	0	31.5	22.5	0
5	Slot S	0	1	0	131.5	5	0

Table 4.3 Each station machining time and failure rate

Stations	Characteristics	Dimensions (mm)	Processing time t_i	Failure rate λ_{0i}
Station 1	Distance between M and A	44.291	295.8	8.25×10^{-6}
Station 2	Hole B diameter	18.624	141.6	4.32×10^{-6}
Station 3	Hole C diameter	11.924	99.3	1.17×10^{-5}
Station 4	Slot S radius	4.966	46.0	4.04×10^{-6}

failure rate of each station is shown in Table 4.3. The failure rate of each station is shown in the last column of Table 4.3.

Step 2: Calculate P_r, i.e., find the value of β.

The four critical dimensions of the part have tolerances of 0.2, 0.021, 0.2, and 0.2 (mm). With the same product size process, Tables 4.1 and 4.2. The data in Tables 4.1 and 4.2 are substituted into Eq. (4.1), and after Taylor expansion (Eq. 4.3), the original data are assumed to follow a normal distribution, so $\sum_{i=1}^{4}\sum_{\substack{j=1\\ j\neq i}}^{4} a_i a_i Cov(x_i, x_j) = 0$.

$$\mu_f = f\left(\mu_{x_1}, \mu_{x_2}, \mu_{x_3}, \mu_{x_4}\right) = 0.1986,$$

$$\sigma_f^2 = \sum_{i=1}^{4} a_i^2 \sigma_i^2 + \sum_{i=1}^{4}\sum_{\substack{j=1\\ j\neq i}}^{4} a_i a_i Cov(x_i, x_j) = 0.2377.$$

Step 3: Substituting the above relevant data into Eq. (4.12), we can find $R(t)$.

$$\begin{aligned} R(t) &== \Pr\left\{R^F|Z(t)\right\} \times \Pr\left\{R^Q|Z(t)\right\} \\ &= \exp\left(\sum_{i=1}^{n}\left(\left[\lambda_0(t) + E\left[\alpha_i(X(i) - m_i)^2\right]\right]t_i\right)\right) \\ &\quad \times \int_0^{\eta} \exp\left(-h\left(w_0 + e^{-\varepsilon_k t}\right)\right)d\varepsilon_k \times \Phi(-\beta) \end{aligned}$$

$$= e^{0.0044} \times \frac{1}{\sqrt{1.6 \times 2 \times 10^{-5}} \times \sqrt{2\pi}} \left[e^{-\frac{(0.019-0.008)^2}{6.4\times10^{-5}}} - e^{-0.25} \right]$$
$$\times \Phi\left(-\frac{0.1986}{-\sqrt{0.2377}}\right)$$
$$= 1.0044 \times 70.5419 \times (0.7885 - 0.7788) \times \Phi(0.407)$$
$$= 0.6873 \times 0.6808$$
$$= 0.4679.$$

II. Manufacturing System Mission Reliability Modeling Based on Quality State Task Network

(i) Modeling Fundamentals

From the production manager's point of view, for a given production task, the quality index of the product is only one aspect, and the yield index is also not to be ignored. Manufacturing system is usually combined by multiple processing device, with inherent complexity and polymorphism, the dynamic changes in the production task requirements has intensified the polymorphic nature of the manufacturing system, for the reliability assessment work has brought great challenges. Manufacturing system mission reliability refers to the ability of the manufacturing system to complete the specified production tasks under the specified conditions and within the specified time. As an effective basis for guiding production managers to carry out production scheduling, quality control and device preventive maintenance and other production activities, the accurate estimation of the mission reliability of a manufacturing system plays a pivotal role in the production process, and is a prerequisite for manufacturing enterprises to improve production efficiency and international competitiveness. How to realize the effective estimation of manufacturing system mission reliability to support the dynamic scheduling of production activities is a recognized scientific problem in the field of manufacturing and reliability engineering.

To establish a reasonable manufacturing system mission reliability model, the first step is to clarify the correlation relationship between manufacturing system mission reliability, manufacturing process quality, and produced product reliability from the perspective of system engineering. As shown in Fig. 4.3, the main mechanism is: produced product reliability requirements are mainly reflected in the use of the product performance [5], while the use of the product performance is mainly determined by the product's key quality characteristics [6]; through the decomposing mapping of the product's key quality characteristics, the key processes and related device can be identified, and then targeted mining in the manufacturing process of the accumulation of key process quality data, and the batch process reliability of the product can be utilized to use the key process quality data to [7].

In order to express the quality state of device and material and the change of task load in the manufacturing system in a visual way, the manufacturing system quality state task network model is proposed. The rectangle in the model represents the processing device and the circle represents the material state.

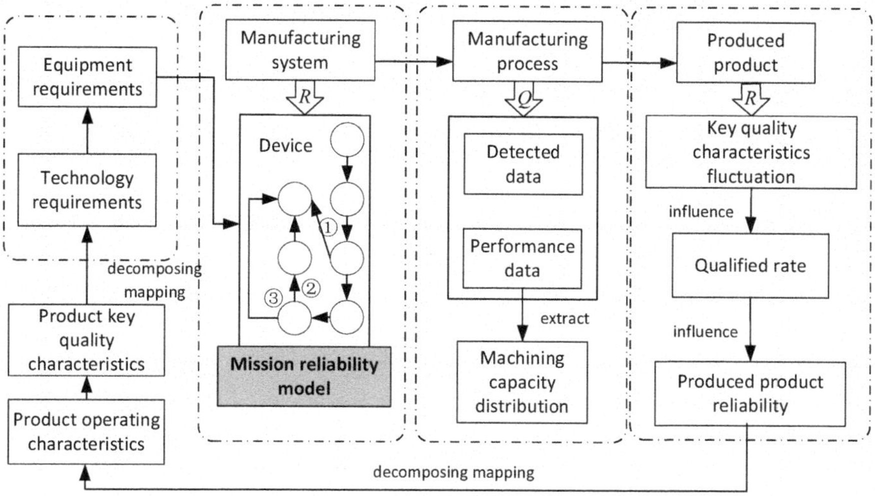

Fig. 4.3 Mission reliability and produced product reliability mapping relationship

Quality State Task Network Model: A quality state task network model for a manufacturing system consisting of three devices is shown in Fig. 4.4.

Figure 4.4a shows the local structure of the manufacturing system, and Fig. 4.4b shows the mass state task network model transformed from Fig. 4.4a. In Fig. 4.4b, the

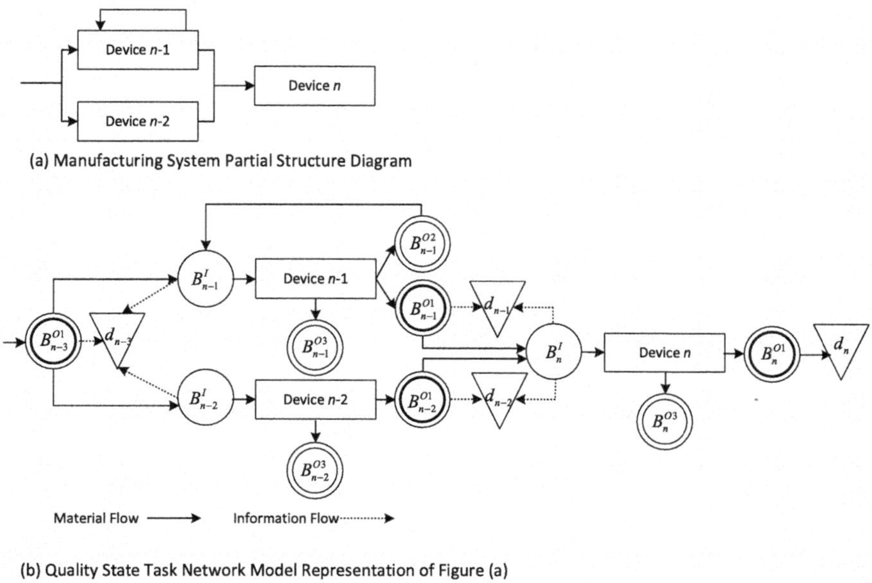

Fig. 4.4 Schematic diagram of the mass state task network model

solid line represents the material flow of the manufacturing process and the dashed line represents the information flow of the modeling process.

Material Quality State: The double line circle in the figure indicates the processed workpiece, which is macroscopically divided into three quality states: qualified state S_{i1}, defective reworkable state S_{i2}, scrapped (unqualified) state S_{i3}, because only qualified products are the effective output of the manufacturing system, so the quantification of the output product quality state in this paper is aimed at the qualified state of the product, the figure of the B_j^{O1}, B_j^{O2}, B_j^{O3}, j = 1, 2, 3, ..., n indicate the number of products in qualified state respectively; , , , j = 1, 2, 3, ..., n indicate the number of products in qualified state. ..., n denote the number of products in qualified status, defective reworkable status, and scrap (unqualified) status, respectively.

Device State: The rectangle in the figure represents the production device, whose performance state is characterized by the processing capacity state (i.e., the number of workpieces that can be processed by the device per unit time) and its probability distribution $[\boldsymbol{S_x}, \boldsymbol{P_x}]$. Where $\boldsymbol{S_x} = [S_1, S_2, S_3, \ldots, S_x, \ldots, S_M]$, S_x represents the processing capacity state of the production device unit when it is at level x, where $x = M$ is the optimal working state of the device, and x = 1 is the state of complete failure of the device, and the state probability corresponding to it, P_x, can be expressed in the form of a set as follows: $\boldsymbol{P_x} = [ep_1$, ep2, ep3, ..., ep, ..., ep]. ..., $ep_x, \ldots, ep_M]$, where p_x is a constant representing the proportionality between the probabilities of occurrence of each machining capability state, and e is a variable related to device performance. Since the device unit being in each processing capability state is an independent mutually exclusive event:

$$\sum_{x=1}^{M} ep_x = 1,\ t \geq 0. \tag{4.13}$$

Production Task: The single line circle in the diagram characterizes the state of task execution in terms of the minimum machining load required to satisfy the task requirement B_j^I. The triangular symbol in the diagram represents the production task requirement d_j, which is the hub connecting the output requirements of the upstream machine to the input requirements of the downstream machine, quantified in terms of the number of conforming products required to be output per unit of time on average.

The solid arrows in the figure indicate the material flow during processing, and each arrow contains a scale factor that is not labeled in the figure, as shown in Fig. 4.5, taking the operation process of device n − 1 as an example.

In the figure, ρ_j^{I1} indicates the percentage of qualified work-in-progress output from the upstream station input to the current station, the value of which is mainly determined by the production mode, when there is no branching structure in the manufacturing time and no reserve of the current material, $\rho_j^{I1} = 1$, ρ_j^{I2} indicates the percentage of the material with defective reworkable status input to the device j to the current device output of the material of this status. ρ_j^{O1}, ρ_j^{O2} and ρ_j^{O3} represent

Fig. 4.5 Single device quality state task network model

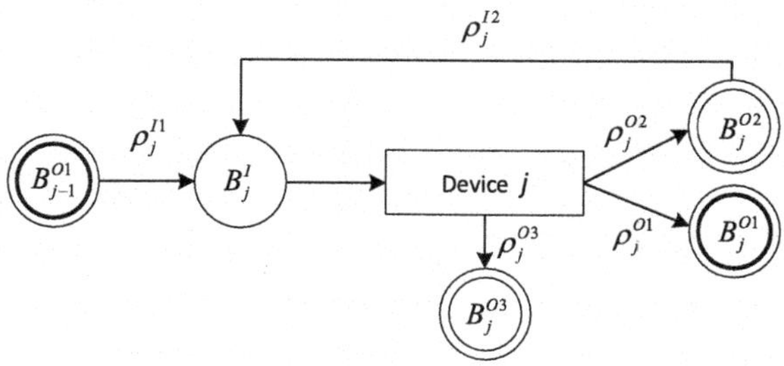

the output probability of the qualified status, the defective reworkable status, and the scrapped status respectively; without loss of generality, indicates the manufacturing qualification rate of the device; and represents the output probability of the device. output probabilities; and, without loss of generality, ρ_j^{O1} denotes the manufacturing pass rate of a device.

Quantitative Relationships

$$\begin{pmatrix} B_j^{O2} \\ B_j^{O1} \\ B_j^{O3} \end{pmatrix} = \begin{pmatrix} \rho_j^{I1}\rho_j^{O2} & 0 \\ \rho_j^{I1}\rho_j^{O1} & \rho_j^{I2}\rho_j^{O1} \\ 0 & \rho_j^{I2}\rho_j^{O3} \end{pmatrix} \begin{pmatrix} B_{j-1}^{O1} \\ B_j^{O2} \end{pmatrix}. \tag{4.14}$$

The above analyzes the input–output relationship of a single device, in a multi-station manufacturing system, there also exists the transfer and evolution of task requirements between adjacent devices. Taking a manufacturing system with n devices as an example, when the overall production task requirement of the given system is d, the intermediate device j is selected as the research object, and its task requirement is d_j, then when there is no rework process for this device, the minimum effective output to meet the task requirement should be $B_j^{O1} = d_j$, as shown in Fig. 4.6.

Quantitative Relationships

$$B_j^I = \frac{d_j}{\rho_j^{O1}}, \tag{4.15}$$

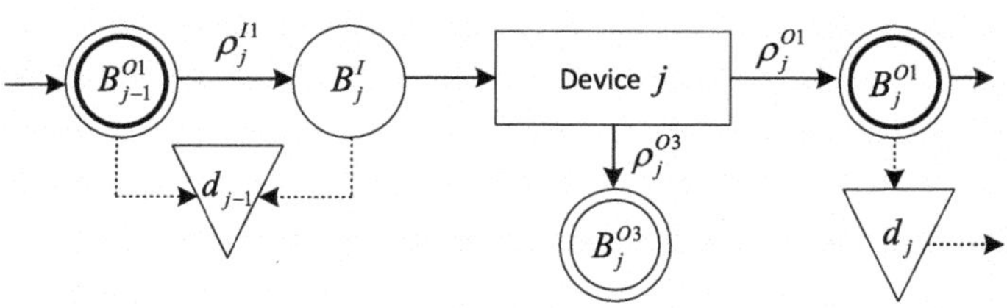

Fig. 4.6 Evolution of task requirements for general device

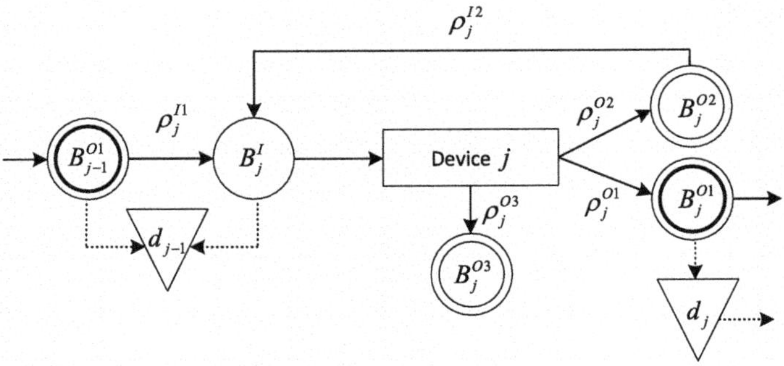

Fig. 4.7 Evolution of task requirements for device with rework processes

$$d_{j-1} = B_{j-1}^{O1} = \frac{B_j^I}{\rho_j^{I1}}. \tag{4.16}$$

The evolution of task requirements for a device with a rework process is shown in Fig. 4.7.

Quantitative Relationships

$$d_{j-1} = B_{j-1}^{O1} = \frac{d_j}{\rho_j^{I1}\rho_j^{O1}\left(1 + \rho_j^{O2}\rho_j^{I2}\right)}, \tag{4.17}$$

$$B_j^I = d_{j-1}\rho_j^{I1} + d_{j-1}\rho_j^{I1}\rho_j^{O2}\rho_j^{I2}. \tag{4.18}$$

(ii) **Modeling Process**

Hypothesis

(a) The quality status S_{i2} may only occur in processes that can be reworked and only once on the current device, i.e., if the rework still fails, it is categorized as a non-conforming scrap status S_{i3};
(b) In the quality status task network, each processing machine is followed by an inspection station, and the inspection results are absolutely reliable, so that only materials in a qualified state can move on to the next process;
(c) The manufacturing system is a tandem assembly line operation and the processing device is physically independent of each other;
(d) Only one rework process exists in the manufacturing system and it is performed only on the current device;
(e) The manufacturing pass rate follows a U-distribution.

1. **Establishment of a Manufacturing System Quality Status Task Network**

Taking the manufacturing system consisting of *n devices* shown in Fig. 4.4 as the object of study, when there is a branching structure in the mechanism of the manufacturing system, the manufacturing process can be decomposed according to the input

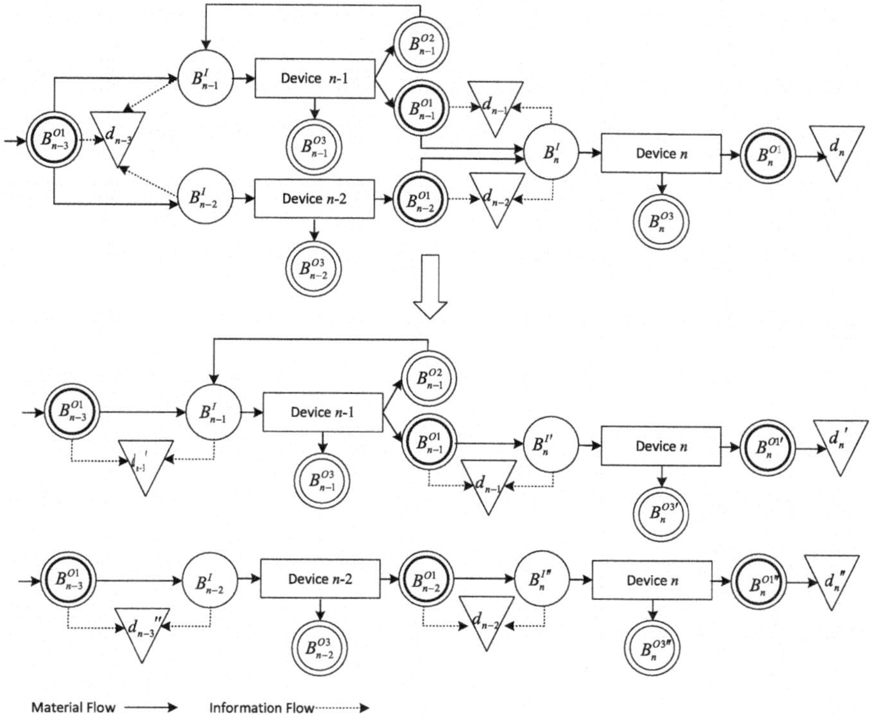

Fig. 4.8 Schematic diagram of manufacturing process decomposition

scaling factor ρ_j^{I1} in the quality state task network to characterize the manufacturing process in terms of multiple tandem manufacturing processes, as shown in Fig. 4.8.

2. **Estimation of the Manufacturing Compliance Rate**

The Bayesian method [8] is utilized to estimate the probability ρ_j^{O1} that the output material state of each device in the quality state task network is S_{i1}, and the expression $\rho_j^{O1} = \frac{a+x}{a+b+n}$ for the manufacturing pass rate is obtained. Here, a and b are the distribution parameters, n is the test sample capacity, and x is the number of test qualified samples.

3. **Analyze Device Processing Capacity State Distribution and Probability**

Based on the statistical data of the production management department over a period of time, the state and probability distribution of device processing capacity is analyzed. Device due to device failure, local failure, maintenance and other factors, the state of device processing capacity is random, so the form of intervals is selected, and the probability of the occurrence of processing capacity in the range of the intervals is counted.

Based on the concept of availability [9], the correlation between failure rate and device processing capability state and probability distribution is established

$\sum_{x=1,2...M} \frac{ep_x(S_M - S_x)}{S_M} = \frac{\tau \int_0^t \lambda(t)dt + r\tau'}{t}$, where $\int_0^t \lambda(t)dt$ represents the expectation of the number of device failures occurring in the [0, t] time period. τ' represents the length of device downtime caused by a single planned maintenance. r is the number of planned maintenance activities occurring in the [0, t] time period, and r = 0 when there are no planned maintenance activities. τ represents the expectation of device downtime caused by a single device failure. Then the expression of the parameter variable related to device performance state can be obtained. Downtime expectation, then the expression for the parameter variable e related to the device performance state can be obtained:

$$e = \frac{S_M\left(\tau \int_0^t \lambda(t)dt + r\tau'\right)}{t\sum_{x=1,2...M} p_x(S_M - S_x)}. \tag{4.19}$$

4. **Determination of Subtask Loads for Each Piece of Device**

Based on Eqs. (4.15)–(4.18) for each device, the subtask load of each device, i.e., the task execution state of each device, is determined separately.

5. **Determine the Reliability of Each Device Sub-task**

The probability that the device processing capacity meets the subtask load requirement. Set S_v as the minimum device processing capability state that satisfies $S_x \geq B_j^l$, in other words, $S_x \geq B_j^l$ holds if and only if $x \geq v$. Therefore, the mission reliability model established for device j can be expressed as follows:

$$\begin{aligned} R_{tj} &= \Pr\left\{S_x \geq B_j^l\right\} \\ &= \sum_{x \geq v} p_x \frac{S_M\left(\tau \int_0^t \lambda(t)dt + r\tau'\right)}{t\sum_{x=1,2...M} p_x(S_M - S_x)}. \end{aligned} \tag{4.20}$$

6. **Manufacturing System Mission Reliability**

$$R_d = \prod_{j=1}^{n} R_{tj}. \tag{4.21}$$

(iii) **Example Analysis**

For a certain model of automobile engine cylinder head manufacturing system, a typical multi-station manufacturing system, systematic analysis and establishment of its operating process quality state task network model to improve the state of the manufacturing system transparency, and then with reference to the above mission reliability modeling process, select three of the key device as an example of the given task requirements of the manufacturing system under the mission reliability of the modeling analysis, the process is as follows:

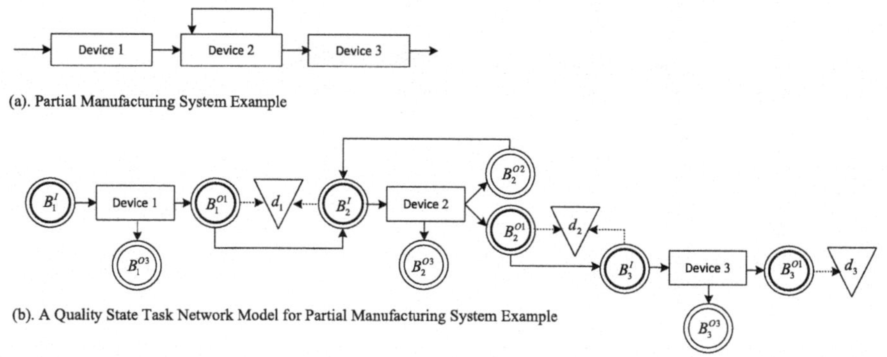

Fig. 4.9 Manufacturing system quality state task network model construction

1. Establishment of a Manufacturing System Quality Status Task Network

According to the actual process characteristics, the three key devices are represented by the device serial numbers 1, 2, and 3, respectively, and their process flow is shown in Fig. 4.9a, and the manufacturing system structure shown in Fig. 4.9a is transformed to obtain the manufacturing system quality state task network model shown in Fig. 4.9b.

2. Estimation of the Manufacturing Compliance Rate

Collecting the basic data of the manufacturing system operation in the production management department, the relevant basic parameter values can be obtained, such as the pass rate, average failure repair time, etc., and the results are shown in Table 4.4. For the change of device failure rate, its specific quantitative model can be obtained by data fitting based on reliability test data or actual historical data.

3. Analyze Device Processing Capacity State Distribution and Probability

Based on the historical maintenance data of the manufacturing device, the statistics shown in Table 4.5 were compiled.

Analyzing the proportionality between the probability of occurrence of each failure mode for each device based on the above results, a finite number of discrete device processing capability states are delineated from a state of perfect device performance to complete failure, and the state values are determined by converting the

Table 4.4 Case parameter values

Parameter	Value	Parameter	Value	Parameter	Value
τ_1	0.43	ρ_2^{O1}	0.98	ρ_3^{I1}	1
τ_2	0.3	ρ_3^{O1}	0.97	ρ_2^{I2}	1
τ_3	0.5	ρ_1^{I1}	1	ρ_2^{O2}	$1 - \rho_2^{O1}$
ρ_1^{O1}	0.96	ρ_2^{I1}	1		

Table 4.5 Device 1 failure mode related data statistical results

Device model	Failure phenomenon description	Failure exclusion time (day)	Percentage of occurrences	Total time
Machining centers x (device 1)	Machine tool failure	1	1	350
	Poor circulation	0.9	2	350
	Knotting or jamming	0.8	2	350
	Machine tool ties	0.7	3	350
	Light is not on	0.6	5	350
	False indication	0.5	6	350
	Knife holder lock not tight	0.4	7	350
	Knife holder not in place	0.3	10	350
	Knife holder failure	0.2	12	350
	Lubrication failure	0.1	18	350

unplanned downtime caused by each failure mode into depletion of the device's processing capability. For example, converting the troubleshooting time for each failure mode in Table 4.5 into a depletion of device processing capability yields the results shown in Table 4.6.

Similarly, the same method is applied to obtain the machining capacity states of device 2 and device 3 and their probability distribution as shown in Tables 4.7 and 4.8.

Based on the device failure statistics, the failure rate of the device is modeled and analyzed without considering the maintenance activities, and the following expression is fitted:

Table 4.6 Device 1 processing capacity states and their probability distributions

S_x	0	25	50	75	100	125	150	175	200	225	250
P_x	e_1	$2e_1$	$2e_1$	$3e_1$	$5e_1$	$6e_1$	$7e_1$	$10e_1$	$12e_1$	$18e_1$	$1-66e_1$

Table 4.7 Device 2 processing capacity states and their probability distributions

S_x	0	20	40	60	80	100	120	140	160	180	200
P_x	e_2	$2e_2$	$4e_2$	$5e_2$	$6e_2$	$8e_2$	$8e_2$	$10e_2$	$16e_2$	$22e_2$	$1-82e_2$

Table 4.8 Device 3 processing capacity states and their probability distributions

S_x	0	20	40	60	80	100	120	140	160	180	200
P_x	$2e_3$	$2e_3$	$3e_3$	$4e_3$	$5e_3$	$6e_3$	$7e_3$	$7e_3$	$16e_3$	$20e_3$	$1-72e_3$

$$\lambda_1(t) = 1.11^{0.00053t} \times 2.4 \times 10^{-5}t^2,$$
$$\lambda_2(t) = 1.12^{0.00079t} \times 3.91 \times 10^{-5}t^{1.5},$$
$$\lambda_3(t) = 1.10^{0.00125t} \times 2.89 \times 10^{-5}t^2.$$

Further, an expression for the parameter variable e related to the performance state of the device may be obtained as follows:

$$e_1 = \frac{250 \times 0.43 \times \int_0^t 1.11^{0.00053t} \times 2.4 \times 10^{-5}t^2 dt}{5625t},$$
$$e_2 = \frac{200 \times 0.3 \times \int_0^t 1.12^{0.00079t} \times 3.91 \times 10^{-5}t^{1.5} dt}{5740t},$$
$$e_3 = \frac{200 \times 0.5 \times \left(\int_0^t 1.10^{0.00125t} \times 2.89 \times 10^{-5}t^2 dt\right)}{5100t}.$$

4. **Determination of Subtask Loads for Each Piece of Device**

According to the production plan, for the engine head of model Q456, the overall requirement is to produce 16,200 pieces/3 months, and calculating with 30 days per month, the task requirement can be expressed as d = 180 pieces/day. That is, the manufacturing task requirement with device 3 is $d_3 = 180$ pieces/day. Then, according to the evolution flow of the manufacturing task requirement among devices described above, it can be obtained that, in order to realize the output of qualified products to meet the task requirement, the task processing load of device 3 is:

$$B_3^I = \frac{d_3}{\rho_3^{O1}} = \frac{180}{0.97} = 185.6.$$

Due to the input ratio $\rho_3^{I1} = 1$, i.e., all qualified WIP output from the upstream station is used for production and processing at the downstream station during the production process. Therefore, the input requirement of device 3 also corresponds to the output requirement of qualified WIP from device 2, i.e., $d_2 = 185.6$/day.

According to the quality state task network model shown in the case study, there is work-in-progress with defective reworkable states in device 2, so in order to accurately assess the task execution state of device 2, we need to first determine the task requirements of device 1:

$$d_1 = \frac{d_2}{\rho_2^{I1}\rho_2^{O1}\left(1 + \rho_2^{O2}\rho_2^{I2}\right)} = \frac{185.6}{0.98(1 + 0.02)} = 185.6.$$

Then, the task execution state of device 2 can be found:

$$B_2^I = d_1\rho_2^{I1} + d_1\rho_2^{I1}\rho_2^{O2}\rho_2^{I2} = 185.6 + 185.6 \times 0.02 = 189.3.$$

Finally, the task execution state of device 1 can be found:

$$B_1^I = \frac{d_1}{\rho_1^{O1}} = \frac{185.6}{0.96} = 193.3.$$

In summary, the task execution status of each device is: $B_1^I = 193.3$, $B_2^I = 189.3$, $B_3^I = 185.6$.

5. **Determine the Reliability of Each Device Sub-task**

According to the definition of manufacturing system mission reliability, the essence of mission reliability is to assess the ability of the current device performance state to meet the current production task load demand, then, the mission reliability can be expressed as the device processing capacity to meet the task state probability, the value of which is the sum of the probabilities corresponding to the processing capacity state used to meet $S_x \geq B_j^{I1}$:

$$R_1(t) = \sum_{x \geq v} p_x \frac{S_M\left(\tau_1 \int_0^t \lambda(t)dt\right)}{t\sum_{x=1,2...M} p_x(S_M - S_x)}$$
$$= 1 - 36\left(\frac{250 \times 0.43 \times \int_0^t 1.11^{0.00053t} \times 2.4 \times 10^{-5}t^2 dt}{5625t}\right),$$

$$R_2(t) = 1 - 82\left(\frac{200 \times 0.3 \times \int_0^t 1.12^{0.00079t} \times 3.91 \times 10^{-5}t^{1.5}dt}{5740t}\right),$$

$$R_3(t) = 1 - 72\left(\frac{200 \times 0.5 \times \left(\int_0^t 1.10^{0.00125t} \times 2.89 \times 10^{-5}t^2 dt\right)}{5100t}\right).$$

6. **Manufacturing System Mission Reliability**

On the basis of the above analysis, the mission reliability model of the system is established, and MATLAB is used to analyze the change of the system mission reliability with the system running time of the fetch, and the results shown in Fig. 4.10 are obtained.

As can be seen from Fig. 4.10, with the operation of the manufacturing system, the wear and tear of device burn-in and other negative impact on the execution of the task, when not taking into account the repair and maintenance of device, in the continuous operation of the device for 270 days, the system's value is close to 0. Therefore, by analyzing the trend of changes in the system's mission reliability, the operating state of the manufacturing system can be more accurate, and the state of the

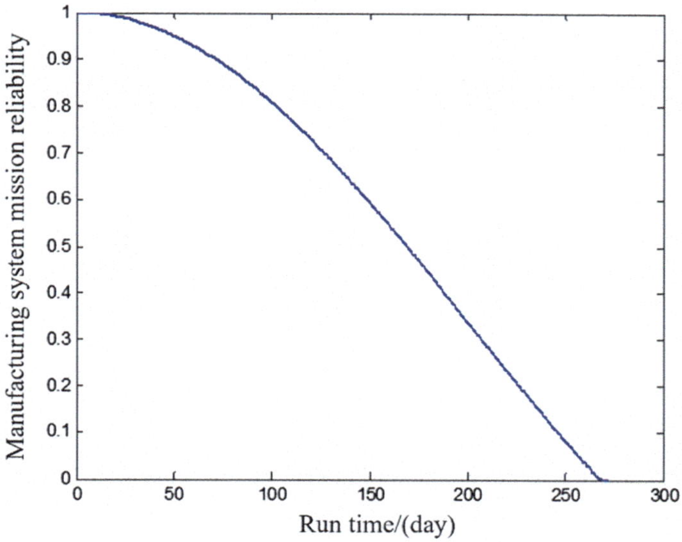

Fig. 4.10 Trend of manufacturing system mission reliability with device operating time

operation of the system can be improved transparency, which in turn guides activities such as production task scheduling and device predictive maintenance.

4.1.2 Preventive Maintenance Techniques for Manufacturing Systems

Excellent design, processing and assembly is an important guarantee of the inherent reliability of the manufacturing system, but once it enters the service phase and begins to operate, a variety of complex and changing manufacturing tasks, a variety of system operating environment uncertainty will inevitably lead to the manufacturing system and its constituent device performance status and reliability show a certain trend of degradation. Untimely maintenance activities will inevitably lead to frequent system failures, on the one hand, device failures bring unexpected downtime, directly affecting the completion of the scheduled production tasks, on the other hand, will make the system reliability cannot be guaranteed, so that the cost of maintenance has risen significantly [10]. Preventive maintenance, which has been frequent in the past, can also lead to significant economic losses [11]. In many cases, maintenance costs can range from 15 to 70% of total production costs [12]. Therefore, the scientific and rational development of maintenance strategy has become a key link to ensure the reliability and economy of manufacturing system operation.

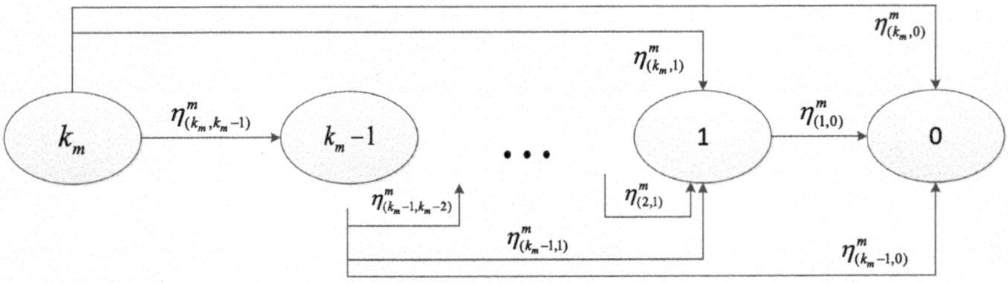

Fig. 4.11 Schematic diagram of performance state degradation

I. **Cost-Driven Preventive Maintenance Decision-Making Techniques Based on Device Reliability**

(i) **Markov Modeling of the Degradation Process of Manufacturing Device Unit Performance**

The core technologies of preventive maintenance are performance state prediction and failure isolation. We only need to realize the prediction of the performance state of the constituent device units. Assuming the device unit m in the manufacturing system, the following analysis of the performance state of the device unit m will be used to make the necessary preparations for the modeling of the performance degradation process of the system and the reliability analysis. In the process of the work of the device unit m, often due to some external influences or its own performance degradation and its performance state changes, the definition of the stochastic process for the $\{X_m(t), t \geq 0\}$.

The process of degradation of the performance state of the device unit m during the manufacturing process is shown in Fig. 4.11.

Where k_m is the optimal performance state of the device unit m, state 0 is the unacceptable performance state of the machine m and $\eta^m_{(ij)}(t)$ indicates the transfer rate of the machine m from the performance state $i \in [1, \ldots, k_m]$ to the performance state $j \in \left[0, \ldots, k_{m-1}\right]$, here $i > j$.

Instantaneous state transfer matrix for device unit m

$$A^m(t) = \begin{bmatrix} \eta^m_{(k_m,k_m-1)}(t) & \eta^m_{(k_m,k_m-2)}(t) & \cdots & \eta^m_{(k_m,0)}(t) \\ 0 & \eta^m_{(k_m-1,k_m-2)}(t) & \cdots & \eta^m_{(k_m-1,0)}(t) \\ \vdots & \vdots & & \vdots \\ 0 & 0 & \cdots & \eta^m_{(1,0)}(t) \end{bmatrix}. \tag{4.22}$$

The state probability of the device unit m is

$$p_{(m,i)}(t) = \Pr\{X_m(t) = i\}, \; i = 0, \ldots, k_m. \tag{4.23}$$

Device unit m The transfer rate between any state i and state j is

$$\eta^m_{(ij)}(t) = \lim_{\Delta t \to 0} \frac{\Pr\{X_m(t+\Delta t) = j | X_m(t) = i\}}{\Delta t}. \tag{4.24}$$

(ii) State Analysis of Manufacturing Unit Performance Degradation Processes

Define the performance state of the device unit m as $G_m(t)$, and the distribution interval of the performance state of the device unit as $[0, k_m]$, and divide the device unit into $k_m + 1$ states, and the probability of each state of the device unit m at the time of t can be obtained by the non-temporal Qi Markov theory as follows:

$$\begin{aligned} p_{(m,k_m)}(t) &= \exp\left[-\int_0^t \left(\eta^m_{(k_m,k_m-1)}(\tau) + \eta^m_{(k_m,k_m-2)}(\tau) \right.\right. \\ &\quad \left.\left. + \cdots + \eta^m_{(k_m,1)}(\tau)\eta^m_{(k_m,0)}(\tau) \right) d\tau \right] \\ &= \exp\left[-\int_0^t \sum_{i=0}^{k_m-1} \eta^m_{(k_m,i)}(\tau) d\tau \right], \end{aligned} \tag{4.25}$$

$$\begin{aligned} p_{(m,k_m-1)}(t) = \int_0^t \Bigg\{ &\exp\left[-\int_0^{\tau_1} \sum_{i=0}^{k_m-1} \eta^m_{(k_m,i)}(s) ds \right] \eta^m_{(k_m,k_m-1)}(\tau_1) \\ &\exp\left[-\int_{\tau_1}^t \sum_{i=0}^{k_m-2} \eta^m_{(k_m-1,i)}(s) ds \right] \Bigg\} d\tau_1. \end{aligned} \tag{4.26}$$

The probability of a device unit m being in the state $k_m - 2$ is more complex compared to the state $k_m - 1$. This is because there is only one way for a device unit to deteriorate from the highest performance state k_m to $k_m - 1$, namely,

$$k_m \to k_m - 1. \tag{4.27}$$

And there are two ways for the device unit to decay from the k_m state to the $k_m - 2$ state, respectively:

$$k_m \to k_m - 2, \tag{4.28}$$

$$k_m \to k_m - 1 \to k_m - 2. \tag{4.29}$$

The probability that the device unit is in the $k_m - 2$ state shall be the sum of the probabilities of the above two paths, i.e.

$$p_{(m,k_m-2)}(t) = \int_0^t \{h_1 h_2 h_3\} d\tau_2 + \int_0^t \int_0^{\tau_2} \{h_4 h_5 h_6 h_7 h_8\} d\tau_1 d\tau_2, \tag{4.30}$$

where $h_1 = \exp\left[-\int_0^{\tau_2}\sum_{i=0}^{k_m-1}\eta^m_{(k_m,i)}(s)ds\right]$, $h_2 = \eta^m_{(k_m,k_m-2)}(\tau_2)$, $h_3 = \exp\left[-\int_{\tau_2}^{t}\sum_{i=0}^{k_m-3}\eta^m_{(k_m-2,i)}(s)ds\right]$, $h_4 = \exp\left[-\int_0^{\tau_1}\sum_{i=0}^{k_m-1}\eta^m_{(k_m,i)}(s)ds\right]$, $h_5 = \eta^m_{(k_m,k_m-1)}(\tau_1)$, $h_6 = \exp\left[-\int_{\tau_1}^{\tau_2}\sum_{i=0}^{k_m-2}\eta^m_{(k_m-1,i)}(s)ds\right]$, $h_7 = \eta^m_{(k_m-1,k_m-2)}(\tau_2)$, $h_8 = \exp\left[-\int_{\tau_2}^{t}\sum_{i=0}^{k_m-3}\eta^m_{(k_m-2,i)}(s)ds\right]$.

And so on, the probability that the device unit m is in any state i

$$p_{(m,i)} = \int_0^t \cdots \int_0^{\tau_2} \exp\left[-\int_0^{\tau_1}\sum_{i=0}^{k_m-1}\eta^m_{(k_m,i)}(s)ds\right] \eta^m_{(k_m,k_m-2)}(\tau_2)\prod_r^{n-1}\left\{\exp\left[-\int_{\tau_i}^{\tau_{r+1}}\sum_{i=0}^{x-1}\eta^m_{(k_x,i)}(s)ds\right]\eta^m_{(k_x,k_r+1)}(\tau_{r+1})\right\} \exp\left[-\int_{\tau_n}^{\tau_i}\sum_{i=0}^{k_n-1}\eta^m_{(k_n,i)}(s)ds\right]\eta^m_{(k_n,i)}\exp\left[-\int_{\tau_i}^{t}\sum_{j=0}^{i=1}\eta^m_{(i,j)}(s)ds\right]d\tau_1\cdots d\tau_{n+1}, \tag{4.31}$$

where n is the number of intermediate states to pass from state k_m to i. k_n is $k_m - 1$, ..., $i + 1$. r is 1, 2, ..., n.

Assuming that the device unit m is in a brand new state k_m, $p_{(m,k_m)}(0) = 1$, $p_{(m,k_{m-1})}(0) = 0$, $p_{(m,k_{m-2})}(0) = 0$, ..., $p_{(m,0)}(0) = 0$ at the time of $t = 0$ as the initial condition, and that the probability of each performance state satisfies the full probability condition $\sum_{i=0}^{k_m} p_{(m,i)}(t) = 1$, the steady state probability of the device unit in each state can be found (Table 4.9).

(iii) Predictive Maintenance Decision-Making Based on Maintenance Costs

Maintenance costs are a key factor in conducting maintenance strategy studies, and a more comprehensive and accurate calculation of maintenance costs will result in more targeted maintenance activities.

The maintenance cost of manufacturing system C is mainly composed of the following aspects: material cost C_m, labor cost C_h, downtime cost C_s, and product quality loss cost C_p. Namely:

$$C = C_m + C_h + C_s - C_p. \tag{4.32}$$

Table 4.9 Steady state probabilities for different performance state of device unit m

Performance values	$g_{(m,0)}$	$g_{(m,1)}$	...	$g_{(m,i)}$	$g_{(m,k_m)}$
Steady-state probability	$p_{(m,0)}$	$p_{(m,1)}$	...	$p_{(m,i)}$	$p_{(m,k_m)}$

When the reliability of the manufacturing system falls below the minimum acceptable value R_L, it is necessary to calculate the group of device units that need to be repaired $U\{u_1, u_2, u_3, u_4, \ldots\}$ and make it possible to reach the target value of the reliability of the manufacturing system R_U while minimizing the total cost. Considering that in the context of advanced manufacturing systems, the cost of downtime C_s will be a relatively high part of the total maintenance cost, the manufacturing system reliability target value R_U should be increased as much as possible within the acceptable range to minimize the total cost when downtime maintenance is performed.

For each maintenance activity, the number and type of device units do not have a significant impact on the total maintenance labor costs and downtime costs, so when selecting the group of device units to be repaired, only the material cost of the device units C_m, the cost of lost product quality C_p, and the impact on the overall mission reliability of the manufacturing system need to be considered.

For any device unit i, the cost of raw materials is C_{m_i}, the coefficient of influence on the reliability of the task is δ_i, and δ_i is calculated as follows:

$$\delta_i = \frac{R'_{m_i}}{R_{m_i}}, \tag{4.33}$$

where R_{m_i} is the reliability of the device unit i before repair, R'_{m_i} is the reliability of the device unit i after repair, and R'_{m_i} is calculated as:

$$R'_{m_i} == \Pr\{k_i > g'\} = \sum_{q=g'}^{k_m} p_{(m,q)}, \tag{4.34}$$

where g' is the performance status of the device unit after repair.

There are some device units with high importance in the manufacturing system, and the reduction of the reliability of these device units will bring higher risks for the operation of the manufacturing system, so the importance of the device units should be taken into account when comparing the efficiency ratio of the device units. The importance of the score by experts in the manufacturing system before the operation of the assessment, no special impact of the importance of the device unit that is 1, the impact of the higher take 1 above the appropriate real number can be.

In addition, since the repair method determines the performance status of the device unit after the repair activity, and the performance status determines the product quality situation in the production process afterwards, it is necessary to incorporate the changes in the cost caused by the changes in the rate of product conformity under different repair methods into the repair cost model. The cost of product quality loss C_p is determined by the change in the rate of conformity of the product being manufactured, assuming that the change in the rate of conformity of the product before and after the repair activity is Δp, the cost of product quality loss C_p is:

$$C_p = \Delta p \times A \times C_0. \tag{4.35}$$

The cost-effectiveness ratio of the device unit i ROI_i is therefore

$$ROI_i = I_i \times \frac{\delta_i}{C_m - C_p}. \tag{4.36}$$

The group of device units to be repaired $U\{u_1, u_2, u_3, u_4, \ldots\}$ is selected in descending order of the device unit cost ratio until the expected reliability of the repaired task reaches the target value.

(iv) **Arithmetic Analysis**

Here we analyze for an individual device unit 1 and divide the degradation process into seven performance states, the first state being the optimal performance state (state 6) and the last state being the unacceptable performance state (state 0). In this case, we assume that the system is repaired when it degrades to the last state and is repaired as new. From the formulas and assumptions derived above, we utilize MATLAB software and program the system's transfer probability as:

$$A^1 = \begin{bmatrix} 0.2174 & 0.5948 & 0.1625 & 0.0225 & 0.0025 & 0.0002 & 0 \\ 0 & 0.3647 & 0.5026 & 0.1156 & 0.0153 & 0.0016 & 0.0002 \\ 0 & 0 & 0.3876 & 0.4879 & 0.1086 & 0.0142 & 0.0017 \\ 0 & 0 & 0 & 0.3965 & 0.4822 & 0.1059 & 0.0155 \\ 0 & 0 & 0 & 0 & 0.4013 & 0.4791 & 0.1196 \\ 0 & 0 & 0 & 0 & 0 & 0.4043 & 0.5957 \\ 0 & 0 & 0 & 0 & 0 & 0 & 0 \end{bmatrix}$$

From the derivation process of (ii), we can get the probability that the device 1 is in each state under the smooth state, and the calculation can be realized with the help of MATLAB software programming, and the results obtained are shown in Table 4.10.

Here we assume that unit device 1 has no special impact on the manufacturing system, i.e., the importance of the device is 1. In accordance with Eqs. (4.29)–(4.31), we obtain the coefficients for the impact of material cost and product loss cost of device 1 based on the reliability of the task as shown in Table 4.11.

Therefore, according to Eq. (4.32), the efficiency ratio of device unit 1 ROI_1 is

Table 4.10 Steady state probabilities for different performance state of device unit 1

Performance values	$g_{(1,6)}$	$g_{(1,5)}$	$g_{(1,4)}$	$g_{(1,3)}$	$g_{(1,2)}$	$g_{(1,1)}$	$g_{(1,0)}$
Steady-state probability	0.1441	0.1350	0.1490	0.1517	0.1532	0.1542	0.1128

Table 4.11 Parameters values in the calculated cost-effectiveness ratio of device unit 1

Parameters	C_{m1}	C_p	δ_1
Specific values	5.93 (million dollars)	3.96 (million dollars)	1.2775

$$ROI_1 = I_1 \times \frac{\delta_1}{C_m - C_p} = 0.648.$$

By analogy, we can obtain the efficiency ratio of all device units in the manufacturing system that need to be repaired, in descending order, to determine the order of repairs, knowing that the expected reliability of the task after repair reaches the target value can be.

II. **Cost-Driven Preventive Maintenance Techniques Based on Mission Reliability Status**

The previous section introduced the modeling methods for manufacturing system mission reliability and their superiority, and accordingly, the study of dynamic preventive maintenance strategies based on dynamic mission reliability prediction results has equally important academic and engineering implications.

(i) **Preventive Maintenance Mechanism Based on Mission Reliability State**

Based on the connotation of manufacturing system mission reliability, it is indicated that the manufacturing system mission reliability state is determined by parameters such as device processing capability and manufacturing pass rate in the case of a given task requirement. Therefore, preventive maintenance decision-making based on the mission reliability state investigates the predictive maintenance problem for a manufacturing system within a given task cycle. In general, a single manufacturing mission's time span is much smaller than the device's life cycle, so the case of device replacement is not considered. According to the connotation of manufacturing system mission reliability and the manufacturing system's operating mechanism, this paper firstly provides the single device integrated predictive maintenance decision-making mechanism, as shown in Fig. 4.16.

As shown in Fig. 4.12, the solid line represents the material flow of the actual manufacturing process and the dashed line represents the modeling information flow. In the figure, B_j^O indicates the number of qualified products outputs per unit time, and B_j^I indicates the processing load required by the device to meet the task requirements, i.e., the task execution status. From a macroscopic point of view, the probability of unplanned downtime and the length of time caused by device failure reflects the state of device performance, which can be characterized by the processing capacity of the device and the distribution probability, and is affected by the failure rate of the device, the failure of the maintenance time and other parameters. According to the definition of mission reliability, the comprehensive device performance state and task execution state of the device mission reliability modeling, analysis of the device's operating state, to carry out preventive maintenance decision-making based on the mission reliability state. Specifically, the mission reliability index that comprehensively characterizes the operating state of the device is used as the threshold value for the device to perform planned maintenance activities, and the optimal integrated predictive maintenance threshold value of the device is determined by analyzing the total production cost under different mission reliability thresholds, with the criterion

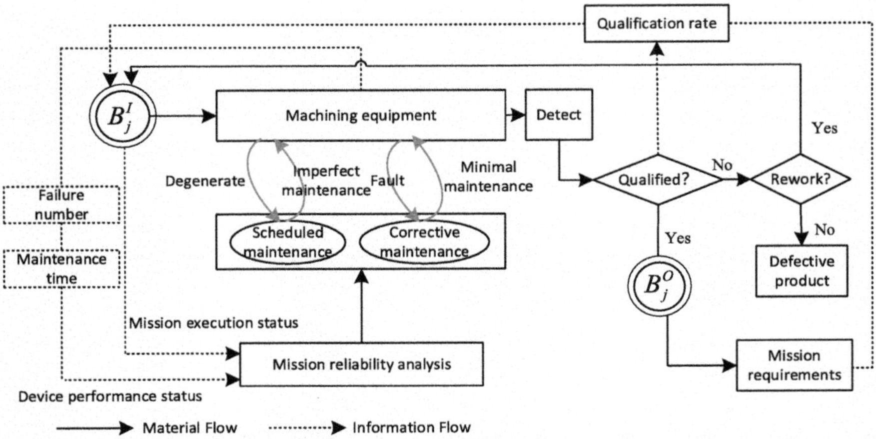

Fig. 4.12 Degraded state relationship diagram for mission reliability device in a maintenance environment

of minimizing the total cost, and once the threshold value is reached, the device is subjected to planned maintenance activities.

Maintenance activities are primarily considered to be corrective maintenance versus preventive maintenance. Corrective maintenance is minimal maintenance and is used only for restoration of functional operation of the device in case of occasional failures, with no impact on the performance of the device. Preventive maintenance is proactive maintenance based on the mission reliability state and for device performance, according to the service age decreasing factor a_k and risk rate increasing factor b_k defined in the literature [13], the failure rate after preventive maintenance can be expressed as Eq. (4.37); in order to simplify the study, it is assumed that the change of manufacturing qualification rate during the manufacturing process is very small and negligible.

$$\lambda_{k+1}(t) = b_k \lambda_k (t + a_k t_k), \tag{4.37}$$

where $0 < a_k < 1$, $b_k > 1$. k is the number of preventive maintenance cycles in time period [0, T] and t_k denotes the *kth* preventive maintenance cycle time.

Assuming that the manufacturing system starts operating from a brand new state and the device failure rate obeys a two-parameter Weibull distribution, that is, $\lambda_{k0}(t) = \left(\frac{m_k}{\eta_k}\right)\left(\frac{t}{\eta\, k}\right)^{m_k - 1}$, where m_k and η_k represent the shape parameter and the scale parameter, respectively, and combining with Eq. (4.37), the failure rate of the device after the *kth* repair can be expressed as follows:

$$\lambda_{l+1}(t) = \varsigma_l \prod_{i=1}^{h} \left(\frac{\theta_i}{\theta_i - \beta_i}\right)^{v_i t} \frac{m}{\eta} \left[\frac{\left(t + t_V + \sum_{f=1}^{l} a_f t_f\right)}{\eta}\right]^{m-1}, \tag{4.38}$$

where t_V indicates the virtual service age of the device, when the device is brand new, $t_V = 0$.

(ii) Integrated Cost-Oriented Calculations

Based on the previously defined manufacturing system mission reliability's connotation, since this paper assumes that the change in the probability of qualified product manufacturing is sufficiently small to be negligible, the key parameters affecting the mission reliability can be determined as the state of the device processing capability and the distribution probability. And then, based on the established relationship between the device degradation state and the mission reliability, the unavailability of different device under different mission reliability states is analyzed, and then the determine the failure probability. Next, based on the device unavailability and cumulative failure probability, a comprehensive cost model including capacity loss and maintenance cost is established. Finally, the optimal predictive maintenance strategy based on mission reliability is determined based on the principle of minimizing the comprehensive cost (Fig. 4.13).

Four specific types of costs incurred in the finite interval [0, T] are considered here, including the minimum repair cost in case of failure c_1, the cost of preventive maintenance c_2, the loss of device capacity due to the production time taken up by carrying out the maintenance activities c_3, and indirect losses due to delays in the tasks c_4.

The minimum repair cost in the event of a failure c_1 can be calculated using the following formula:

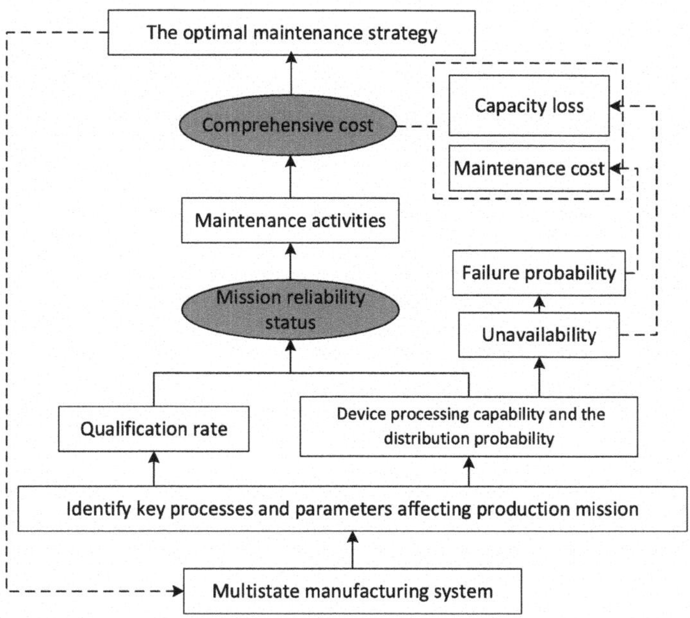

Fig. 4.13 Preventive maintenance strategy optimization problem framework

$$c_1 = c_r\left(\sum_{k=1}^{E}\int_0^{t_k}\lambda_k dt + \int_0^{\varepsilon}\lambda_{E+1}dt\right), \tag{4.39}$$

$$\varepsilon = T - \sum_{k=1}^{E} t_k - E\tau', \tag{4.40}$$

where c_r denotes the cost expectation of a single corrective maintenance activity, E + 1 denotes the number of planned maintenance cycles in time period T, E denotes the number of planned maintenances, and ε denotes the time remaining after the last planned maintenance activity to the end of the planning time period, which is the length of the E + 1st planned maintenance cycle.

The cost of preventive maintenance c_2 can be calculated using the following formula:

$$c_2 = Ec_p, \tag{4.41}$$

where c_p is the expected cost of a single preventive maintenance visit.

Loss of production due to production time taken up by carrying out maintenance activities c_3 can be calculated using the following formula:

$$c_3 = \kappa S_M\left(\sum_{k=1}^{E+1}\tau\int_0^{t_k}\lambda_k dt + E\tau'\right). \tag{4.42}$$

The formula κ denotes the loss cost per unit of processing capacity loss, and τ denotes the expected device downtime due to a single device failure.

The cost of indirect losses due to delays in missions c_4 can be calculated as follows

$$c_4 = \sigma\left(\frac{\sum_{k=1}^{E}(t_k + \tau')}{T}(1 - R_T) + \frac{\varepsilon}{T}(1 - R_\varepsilon)\right). \tag{4.43}$$

The expected cost of indirect losses (σ) is related to the importance of the production task and its value can be derived from expert evaluation. Then the implicit quality loss in time period T can be expressed as follows:

Then the combined cost in the interval [0, T] can be expressed as:

$$c = c_1 + c_2 + c_3 + c_4. \tag{4.44}$$

This preventive maintenance strategy is to find the manufacturing system mission reliability threshold that minimizes the overall cost. The iterative numerical optimization process can be expressed as follows:

(1) Collect basic operational data on the device, such as failure, maintenance and economic data. Determine several constant parameters in the relevant model.

Create a table of state distributions of device processing capacity and analyze the proportionality between the probabilities of each state distribution based on the probability of occurrence of each failure mode $[S_x, ep_x]$;

(2) Set the initial threshold $R_T = R_T^{min}$;
(3) Based on the given task requirements, the parameter value of e is determined, which in turn yields the scheduled maintenance time t_l based on the failure rate model;
(4) Determine the remaining time based on Eq. (4.40) and determine the device failure rate function for the remaining time period, and calculate the device unavailability, the cumulative number of failures, and the probability of distribution of the device processing capacity state $[S_x, e'p_x]$, and the mission reliability for the remaining interval;
(5) Based on Eqs. (4.39)–(4.44), calculate the total cost of production at this maintenance threshold;
(6) Set the search step R_T, $R_T = R_T + \Delta R_T$;
(7) Check to determine if the optimization process should be terminated, and if $R_T > 1$, terminate the optimization process and perform step (8). Otherwise, perform step (3);
(8) Output the optimal mission reliability threshold, and the corresponding planned maintenance strategy.

(iii) Example Analysis

On the basis of the arithmetic example of the part of mission reliability modeling of manufacturing system based on quality state task network in Sect. 4.2.1, the preventive maintenance decision analysis based on mission reliability is carried out, in order to realize the case validation work of this part, in addition to the data collected and processed in the previous section, other types of data mainly cost data are needed to realize the maintenance, the specific parameters and the values of which are shown in Table 4.12.

The validation is carried out using device 3 as an example. The performance evolution of this device considering the impact of maintenance activities is first modeled:

Table 4.12 Parameter values

Parameters	Value	Parameters	Value	Parameters	Value	Parameters	Value
a_{1l}	0.1	τ_3'(day)	0.6	c_{2r}(\$)	24.3	γ	0.03
a_{2l}	0.085	c_{1m}(\$)	120	c_{3r}(\$)	20	t_V (day)	14.24
a_{3l}	0.0712	c_{2m}(\$)	110	κ_1(\$)	0.17	T (day)	100
τ_1' (day)	0.55	c_{3m}(\$)	100	κ_2(\$)	0.2	σ(\$)	3000
τ_2' (day)	0.47	c_{1r}(\$)	22.4	κ_3(\$)	0.2		

$$\lambda_{l+1}(t) = 1.10^{0.00125t} \times 2.89 \times 10^{-5}\left(t + 14.24 + 0.712\sum_{f=1}^{l} t_f\right)^2.$$

On this basis, the trends of corrective maintenance cost, planned maintenance cost, processing capacity loss, and mission delay loss with mission reliability are analyzed separately. The minimum mission reliability threshold is set as 0.01, the numerical search interval is $R_T \in (0.001, 1.000)$ and the step size is $\Delta R_T = 0.001$.

MATLAB is used to analyze the trend of each type of cost under different mission reliability thresholds and the results are shown in Fig. 4.14.

According to Fig. 4.14, it can be seen that when the mission reliability threshold is greater than 0.79, with the increase of the mission reliability threshold, the corrective maintenance cost of the device gradually decreases, the corrective maintenance cost increases, the loss of machining capacity shows a decreasing and then an increasing trend, while the task delay loss shows a decreasing trend in general, which is due to the fact that, under the conditions of this case, by the effect of the length of the planning time period, when the predictive maintenance threshold of the device (mission reliability threshold) is less than 0.79, in the planning time period without

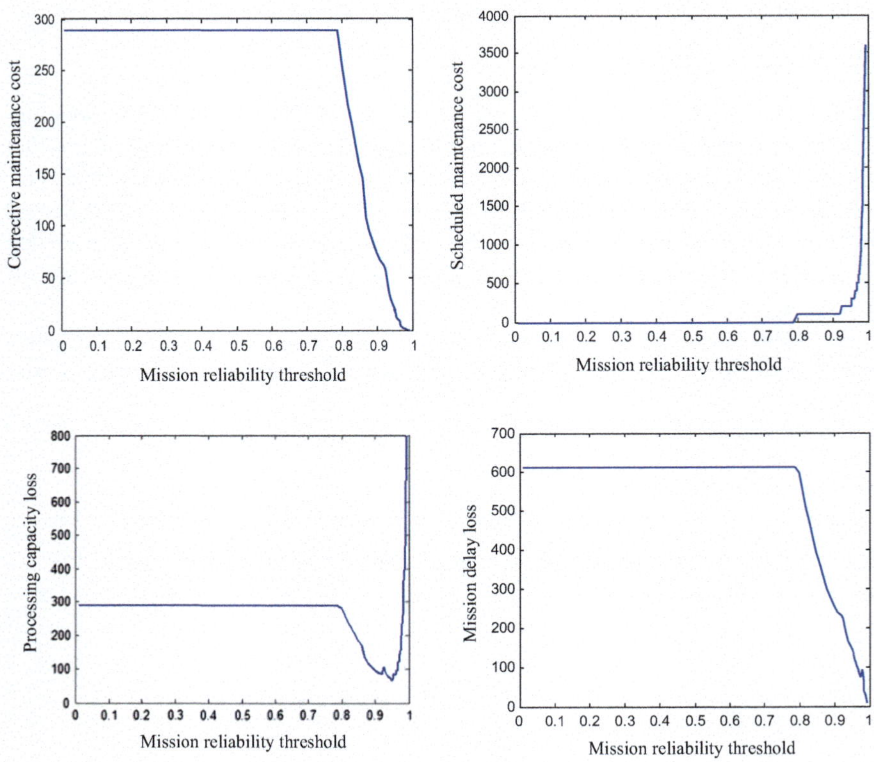

Fig. 4.14 Corrective maintenance cost trends at different maintenance thresholds

the need to carry out planned maintenance activities, so the various maintenance costs in the form of a straight line; when the predictive maintenance threshold is greater than 0.79, as the predictive maintenance threshold is raised, the requirements for device performance are gradually raised, which leads to an increase in the number of planned maintenance, the number of corrective maintenance down, the combination of the two effects, the device total The downtime of the device first shows a decreasing trend.

Finally, the four production costs mentioned above were combined to analyze the changes in the combined costs for different predictive maintenance thresholds.

As shown in Fig. 4.15, since there is no implementation of planned maintenance activities when the predictive maintenance threshold is less than 0.79, the trend of the total cost change within the range of mission reliability threshold taken as (0.8, 1.0) is analyzed, and the overall trend is first decreasing and then increasing, and the minimum value occurs when the mission reliability threshold is taken as 0.951, so the optimal integrated predictive maintenance threshold for the device in this case is 0.951, and its corresponding integrated predictive maintenance strategy is shown in Table 4.13.

In addition, in order to further analyze the scientific and advanced nature of this method, further comparative analyses are carried out, which are based on the device performance or reliability of the depending maintenance decision-making methods, respectively.

Under the conditions of this case, the optimal maintenance strategy of the traditional device performance or reliability-based decision-making approach for the visual maintenance is analyzed; as the name suggests, under this maintenance model, when the performance state of the device reaches the predetermined thresholds of the model, the planned maintenance activities will be executed. In order to achieve the scientific nature of the comparison, the decision modeling of the deeming maintenance is still based on the four cost types mentioned above, and the MATLAB

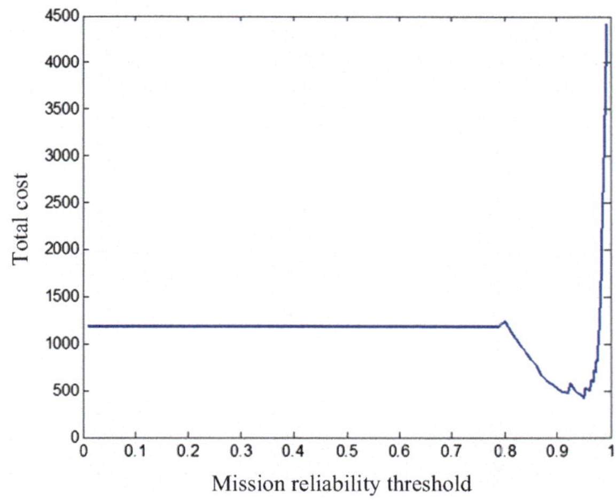

Fig. 4.15 Trends in total costs at different maintenance thresholds

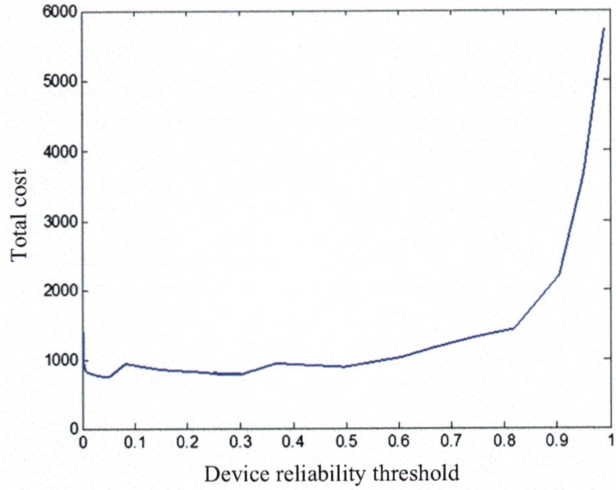

Fig. 4.16 Trends in total costs for different device reliability thresholds

Table 4.13 Optimally integrated predictive maintenance strategies

Scheduled maintenance interval (t_l)		Optimal predictive maintenance threshold	C_T
1	2		
37.8	33.8	0.951	431.33

software is also used to analyze the trend of the total cost under different maintenance thresholds (device reliability indicators), and the results are shown in Fig. 4.16.

Analyzing Fig. 4.16, it can be seen that with the gradual increase of the device reliability threshold, the total cost shows an overall trend of decreasing and then increasing, which is due to the fact that the system is in the state of under-maintenance when the device reliability threshold is set too low, while the system is in the state of over-maintenance when the device reliability threshold is set too high. Analyzing the optimal maintenance threshold of the device, it can be obtained that when the reliability of the device is 0.05, its total cost is the lowest, which is 743.93, and the optimal maintenance strategy in this mode is compared with the optimal maintenance strategy obtained from the method based on the reliability of the task, which is added over as shown in Table 4.14.

Table 4.14 Comparison results of the proposed method with traditional maintenance method

Maintenance mode	Scheduled interval (t_l)		C_T	Cost savings rate (%)
	1	2		
The proposed method	37.8	33.8	431.33	42.02
Traditional dependent care maintenance	53.65		743.93	–

It can be found from the comparative study that this method can save 42.02% of the cost compared with the traditional visual maintenance method, so this method can guide the maintenance activities more scientifically and reduce the production cost when guiding to make predictive maintenance decisions.

4.2 Product Early Failure Rate Optimization Technology

4.2.1 Early Root Cause of Failure Identification Techniques

I. **Early Root Cause of Failure Interrelated Tree**

Engineering practice and experience has shown that the direct or indirect factors that induce or lead to early product failures are an important part of failure analysis. The identification of failure factors is the most direct reference for preventing the next similar failure. Even so, the fact that a large number of products are still subject to similar early failures in subsequent use shows that avoiding similar failures based on previously descriptive failure factors such as improper operation, lack of management, poor environment, etc. is not effective. How to break through the direct influence of the product's early failure, the launch of the potential root cause to failure in-depth analysis is the only way to realize the fundamental prevention of early failure and effective.

Root Cause Analysis [14] theory is an incident-oriented process of backward and forward projection of the accident analysis tool, which originally originated from the Swiss cheese theory, through the light through the layers of stacked cheese on the holes in the incident shows that the occurrence of errors is a series of potential vulnerabilities in the joint action of the results, and root cause analysis is based on the idea of the potential failures and factors and the root cause of the relevant backward traceability, in order to avoid the occurrence of similar failures or accidents. The root cause analysis is based on the idea of reverse tracing of potential failures and factors and related root causes to avoid similar failures or accidents.

The root cause itself is different from the direct cause, and has the characteristic of potentiality. At the same time, it can be used to fundamentally explain the cause of a failure or accident through data collection, listing of causal factors, identification of the root cause, and development and implementation of an improvement plan, and to realize the prevention of the same kind of failures in advance. The purpose of root cause analysis is to actively ask the right questions at the right time for a system incident or product failure, to break down the sub-causes or sub-results through the five-stage questioning of the cause or effect, and to ultimately determine the underlying root cause of the sub-causes [15].

Similarly, in order to realize the orderly decomposition of early product failures by finding and counting the surface or direct causes of early product failures, and then further finding the hidden root causes, while avoiding the huge and complex decomposition workload of directly related product failures and their structural trees to ensure the concentration of information and maximize the convenience of in-depth analysis of the deep-seated reliability problems of the products, this book especially forms a hierarchical structure that includes the design function problems, physical structure defects and process parameter variations to establish the mapping relationship between the product early failure characteristics and the influencing factors at all levels to quantitatively trace back the root of the problems. The book is specially designed to form a hierarchical structure that includes design function problems, physical structure defects, and process parameter variations, and establishes a mapping relationship between the early failure characteristics of the product and the influencing factors at various levels to quantitatively trace back the root of the problem. Drawing on the logical thinking of "goal-how to achieve" among the four domains in the theory of Axiomatic design (AD) [16]: customer attributes (CA), functional requirements (FR), design parameters (DP) and process variables (PV) neighboring domains, we introduce a tree structure of association between different domains, which helps to establish a hierarchical structure that includes design function problems, physical structure defects and process parameter variations. The tree structure of different domains is introduced to help realize the mutual mapping decomposition process among "Functional requirements", "Design parameters" and "Process variables" related to the early failure of the product, as shown in Fig. 4.17.

As shown in Fig. 4.17, the Symptoms of product infant failures (SIFs) are mapped to the functional requirements FRs, which derive the functional tree structure, and then, the functional requirements FRs give each functional requirement related to the failure characteristics, which is independently satisfied by the corresponding design parameters in DPs without affecting the other functional requirements, design tree structure is derived. Subsequently, the design parameters DPs satisfy the independent axioms and are finally expanded to the process variables PVs, which correspond to the specific process execution parameters. In turn, the design parameters DPs, which satisfy the independence axiom, are finally unfolded and mapped to the process variables PVs, which correspond to specific process execution parameters, to derive the process tree structure.

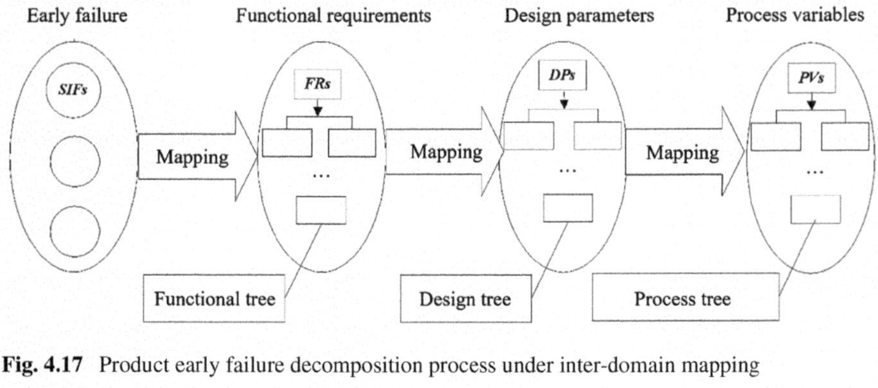

Fig. 4.17 Product early failure decomposition process under inter-domain mapping

Further, based on the top-down interrelated tree decomposition direction, the interrelated tree conceptual structure of early product failure is established as shown in Fig. 4.18.

As shown in Fig. 4.18, the early failure of a product is located at the top of the constructed interrelated tree. After top-down vertical functional decomposition, design decomposition, and process decomposition, supplemented by horizontal multilevel decomposition for the first-level functional elements, the interrelated tree structure system oriented to the characteristics of early failures realizes the coverage of the factors affecting the occurrence of early failures at all levels. The quantification and assignment of key nodes can help to identify the root cause of the failure.

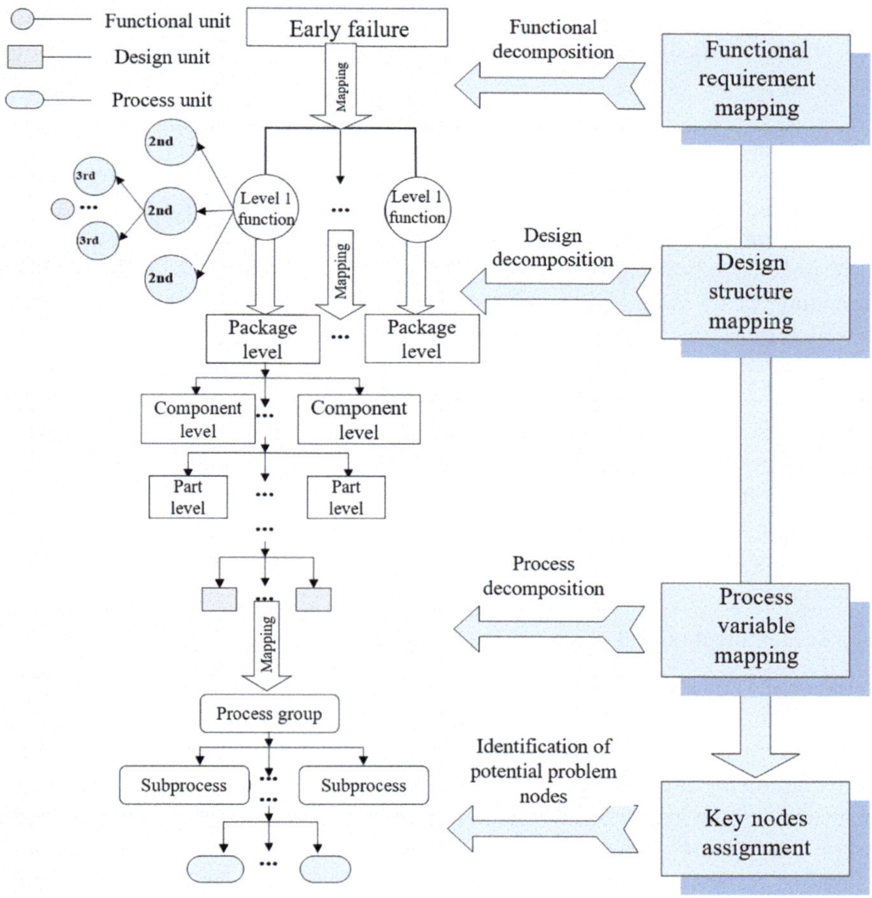

Fig. 4.18 Conceptual structure of interrelated tree for early product failure

II. Interrelated Tree Based Early Product Failure Root Cause Analysis

(i) Establishment of a Top-Down Product Early Failure Interrelated Tree

The interrelated tree building for early product failures is mainly based on the following four steps of mapping and decomposition, specifically.

Step 1: Functional Requirements Decomposition and Mapping

As shown in Fig. 4.18, the main task of functional requirements decomposition and mapping is to establish the mapping relationship between early product failures and their related first-level functional models. The Design Object Analysis (DOA) method of Function-Means Tree (FME) in product design can realize the systematic decomposition of the first-level functional nodes of the failures, forming a multi-level FRs-type tree structure containing the first, second, and third levels of functionality and so on.

Step 2: Design Structure Decomposition and Mapping

Based on the results of the functional decomposition and the mature design planning of the product, the mapping from the first level functional model to the corresponding physical structure component model can be carried out by the designer, and then the corresponding sub-assembly process information can be determined by referring to the coding of each assembly process in the Product Data Management (PDM) system, as well as the corresponding component diagrams and technical documents.

Step 3: Process Variable Mapping and Process Decomposition

Based on the results of the design structure decomposition and the existing manufacturing process planning, production engineers or specific process personnel can expand the first level of physical structure model to the first level of the sub-process nodes of the mapping, reference to the Enterprise Resource Planning (ERP) system, to search for the corresponding sub-process information and the related process key variables etc.

Step 4: Identification and Assignment of Process Key Nodes

Based on the constructed early failure interrelated tree structure, the key node identification ideas affecting early failures are as follows: first, the node design matrix is oriented to start the identification of potential problem nodes in accordance with the requirements of the two basic axioms of Axiomatic Design, "the axiom of independence and the axiom of information minimization"; then the noise factors and failure information of the nodes are collected from the aspects of robustness and reliability, respectively; finally the failure related weight of the potential problem nodes of concern are calculated from the sub-process nodes in the process variables in a bottom-up manner. Then, the noise factor and failure information of the nodes are collected from the aspects of robustness and reliability; finally, the failure related weight of the potential problem nodes of concern are calculated from the sub-process nodes in the process variables in a bottom-up order.

(ii) Measurement of the Weights of Nodes in a Bottom-Up Failure Interrelated Tree

The simplicity and practicability of the underlying data information collection determines the bottom-up logic to carry out the calculation of the weights of the synthetic nodes in the failure interrelated tree, which in turn needs to measure the size of the failure related weight of the nodes between the three domains in order to further assess the priority of the contribution of the failures of each node. Here, the rough set knowledge is mainly applied to mine the decision rules of node failure related weight, and then the decision rules of node failure related weight can be calculated according to the multi-attribute decision making method Fuzzy Approximation to Ideal Solution (TOPSIS) for Order Preference by Similarity to Ideal Solution (Fuzzy). (TOPSIS) to give systematic and comprehensive predictive evaluation and ranking of node failure related weight to node quality association weights [17]. In practical analysis, sufficient quality and reliability data information is the basis for evaluating the failure related weight (W_{FR}). For this reason, for the execution of the underlying process, the failure information related to the process nodes in the process variables, such as failure modes, failure frequency, and failure impact analysis, as well as the noise information that drives the occurrence of failures, such as batch and manufacturing variations, noise from the external environment, material consumption and wear, etc., need to be analyzed and counted in order to dig out the scientifically effective Failure Correlation Weights, and to accurately locate the root causes of the early failures.

The determination of the failure related weight W_{FR} of the lowest process node of the failure interrelated tree is the basis for the analysis of the node weights of the entire interrelated tree. The top-down tree structure determines the bottom-up analysis process of the failure related weight of each node, i.e., when the failure related weight of all process nodes is calculated, the failure related weight of the relevant structural nodes and functional nodes are determined with reference to the method of the maximum value of the weights of the child nodes. The specific process is as follows:

(1) Selected Process Node Objects

Considering the requirements of decision-making efficiency and accuracy, when the number of process tree nodes is too large, the calculation can be carried out in batches according to the similarity of quality attributes. Of course, if there is a large difference in the similarity of quality attributes, different batches still need to be considered.

(2) Determine the Evaluation Index System of Each Process Node Influence on Failure Related Weight W_{FR}

Starting from the influence of process nodes' own attributes on the two characteristics of process quality and reliability, the probability of variations in the attributes of process nodes and the effect of variations on process quality are analyzed, while the probability of process node failures and the detectability and severity of its failure effect are analyzed for the role of process reliability. At the same time, in order to effectively express the technical meaning of each evaluation index, it is necessary to determine the corresponding characteristic attribute set of each evaluation index based on the statistical quality and reliability information. The specific indicator system is shown in Fig. 4.19, and the related feature attribute sets are shown in Table 4.15.

(3) Use Rough Sets to Determine Association Degree of Node Indicators and Decision Rules

Obviously, the weight of influential factor (W_{IF}) of the five indicators in the proposed evaluation index system reflects the degree of contribution of the sub-indicators to the weight of the process node failure association, and its value is different and satisfies $\sum_{j=1}^{5} W_{IFj} = 1$. In order to avoid the inaccurate determination of the weight of Influential Factor by individual subjective tendency, based on the historical quality data,

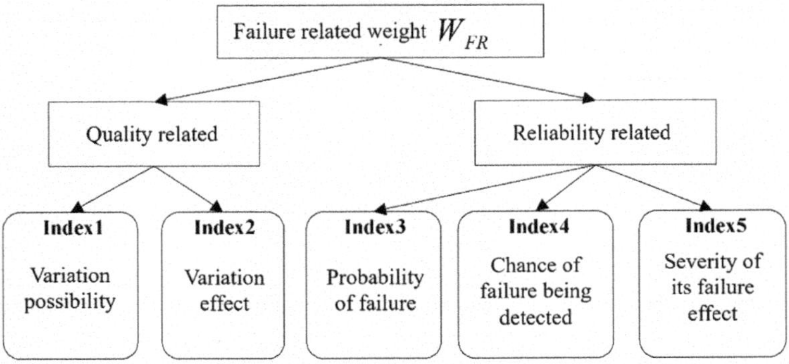

Fig. 4.19 Evaluation index system for failure related weight

Table 4.15 Characteristic attribute sets for evaluation indicators

Indexes	No.				
	Characteristic attribute C1	Characteristic attribute C2	Characteristic attribute C3	Characteristic attribute C4	Decision degree D
Variation possibility	Noise factors' number	Variation parameters' number			Influence degree
Variation effect	Quality loss	Cost loss	Time loss		Influence degree
Probability of failure	Failures' number	Failure mode type			Influence degree
Chance of failure being detected	Design and testing capabilities	Manufacturing and testing capabilities	Detection capabilities	Maintenance capability	Influence degree
Severity of its failure effect	Maximum impact level	Remediation costs			Influence degree

this book adopts the rough set theory to calculate the value of weight of influential factor of nodes. Firstly, collect the conditional attributes in the set of characteristic attributes as the sample data. Secondly, discretely normalize all the values of characteristic attributes in the sample data and form a decision table by this way. After that, determine the basic equivalence classes corresponding to decision attribute set D, and solve the positive domain set associated with each conditional attribute value. Finally, the indicator influence factor weights are calculated by the base ratio of the respective positive domain set: $M_{C-c_i(D)} = 1 - \frac{\text{card}(\text{POS}_{C-c_i}(D))}{\text{card}(POS_C(D))}$. The determination of decision rules for association degree of node indicators can be extracted by applying the rough set knowledge mining method. Specifically, based on the constructed decision table, after maintaining the coordination of the decision table, the same primary principle and attribute columns in the table can be removed to realize the simplification of the decision table, and then, through the computation of the corresponding kernel values of conditional attributes of the decision table, the minimum set of decision rules can be outputted to meet the decision-making requirements.

(4) **Determine the Evaluation Fuzzy Number of the Node's Corresponding Indicator Based on the Existing Decision Rule**

The triangular fuzzy number is used to quantify the subjective evaluation of fuzzy. The triangular fuzzy number is generally expressed in the form of $M = (l, m, u)$ and the characteristic affiliation function $\mu_M(x) : R \to [0, 1]$ is equivalent to the following equation:

$$\mu_M(x) = \begin{cases} 0, & x < l \text{ or } x > u \\ (x-l)/(m-l), & l \le x \le m \\ (x-u)/(m-u), & m \le x \le u \end{cases}. \tag{4.45}$$

Among them, $l \le m \le u$, l denote the left value level of the fuzzy number M, u denotes the right value level of the fuzzy number M, and m denotes the median level of the fuzzy number M. Generally speaking, the importance of subjective evaluation weights is divided into 8 levels in increasing order, as follows.

$$W = \begin{cases} w_1 \to \text{Absolutely Less Important, ALI} \\ w_2 \to \text{Very Strongly Less Important, VSLI} \\ w_3 \to \text{Strongly Less Important, SLI} \\ w_4 \to \text{Weakly Less Important, WLI} \\ w_5 \to \text{Equally Important, EI} \\ w_6 \to \text{Weakly More Important, WMI} \\ w_7 \to \text{Very Strongly More Important, VSMI} \\ w_8 \to \text{Absolutely More Important, AMI} \end{cases}. \tag{4.46}$$

Correspondingly, the ratio of triangular fuzzy number transformations is shown in Fig. 4.20, and the comparison table is shown in Table 4.16.

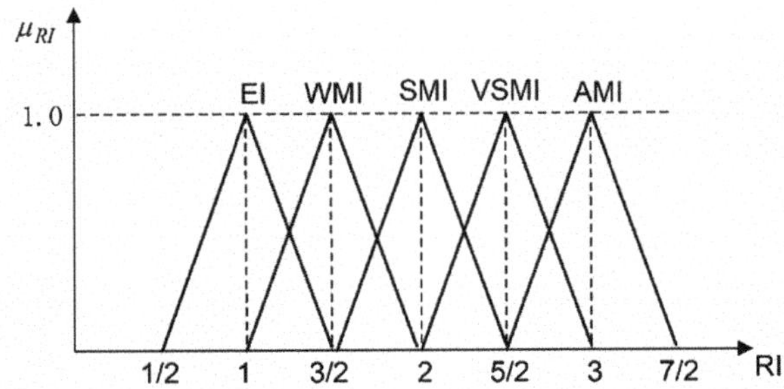

Fig. 4.20 Transformation ratio of indicator weights

Table 4.16 Weighted triangular fuzzy number transformation comparison table

Weight name	Triangular fuzzy number	The reciprocal of triangular fuzzy number
EI	(1/2, 1, 3/2)	(2/3, 1, 2)
WMI	(1, 3/2)	(1/2, 2/3, 1)
SMI	(3/2, 2, 5/2)	(2/5, 1/2, 2/3)
VSMI	(2, 5/2, 3)	(1/3, 2/5, 1/2)
AMI	(5/2, 3, 7/2)	(2/7, 1/3, 2/5)

(5) **Prioritize the Failure Related Weight of the Ordered Nodes Based on Fuzzy TOPSIS Decision Theory**

The main process is as follows. Firstly, take $r_{ij} = \frac{x_{ij}}{\sqrt{\sum_{i=1}^{n} x_{ij}^2}}$ as the statistic to form the normalized decision matrix, and take w_j as the weight to process the statistic r_{ij} in the decision matrix to get the weighted value $v_{ij} = r_{ij}w_j$; then, determine the positive ideal node $A^+ = \{(\max_i v_{ij} | \mathrm{j} \in \mathrm{J}), (\min_i v_{ij} | \mathrm{j} \in \mathrm{J}')\} = \{v_1^+, v_2^+, \ldots, v_n^+\}$ and the negative ideal node $A^- = \{(\min_i v_{ij} | \mathrm{j} \in \mathrm{J}), (\max_i v_{ij} | \mathrm{j} \in \mathrm{J}')\} = \{v_1^-, v_2^-, \ldots, v_n^-\}$, of which J indicates the benefit index, which belongs to the hope of large characteristics, and J indicates the cost index, which is as small as possible; furthermore, the proximity between each node and the ideal node is measured with the distance formula $S_i^+ = \sqrt{\sum\left(v_{ij} - v_j^+\right)^2}, i = 1, 2, \ldots, n$ or $S_i^- = \sqrt{\sum\left(v_{ij} - v_j^-\right)^2}, i = 1, 2, \ldots, n$. Finally, the expression for the relative proximity of each node is given as $C_i^+ = \frac{S_i^-}{S_i^+ + S_i^-}$.

(6) **Calculate the Ranking Weight Vector of a Node as Its Failure Association Weight According to the Node Proximity**

Specifically, based on the neutral decision, i.e., an optimism–pessimism coefficient of 0.5, the expected value of the node closeness is determined as $I\left(C_i^+\right) = (l_i + 2m_i + u_i)/4$ and consequently, the relative weights of the ranking vectors are determined as $w_i = I\left(C_i^+\right) / \sum_{i=1}^{n} I\left(C_i^+\right)$.

Similarly, the W_{FR} determination for any of the nodes may determine the failure related weight corresponding to the FRs Functional Tree Structure nodes and the DPs Physical Tree Structure nodes in the overall product early failure interrelated tree in accordance with the method of calculating and ranking the process node failure related weight W_{FR}.

Based on the aforementioned mechanism analysis and modeling methods, this book takes the series of early failures of a certain model of washing machine as a case study object, and through carrying out the early failure mechanism analysis of the vibration and noise of the washing machine cabinet, modeling the early failure rate of the washing machine PCB board module, and controlling the truncation characteristics of the manufacturing process and optimization of the burn-in time, etc., the book expects to be able to complete the case study of the preceding theoretical methods and to propose relatively perfect techniques and methodological bases for recognizing and controlling the reliability problems of the relevant modules of the washing machine. propose a relatively perfect technology and methodological basis for recognizing and controlling the reliability problems of washing machine related modules.

III. **Algorithm Analysis**

For the typical early failure of washing machine cabinet vibration and noise, the systematic establishment of the interrelated tree structure of this typical early failure is conducive to the effective identification and localization of the root cause of the failure, so as to take timely and appropriate improvement and preventive measures to reduce the possibility of the occurrence of cabinet vibration and noise failure, and to enhance customer satisfaction. Referring to the above process of establishing the early failure tree and calculating the weights of the nodes, the process of analyzing the box vibration and noise failures based on the interrelated tree is as follows.

(i) **Establishment of Top-Down Interrelated Tree Between Box Vibration and Noise Failures**

In accordance with the axiomatic design of inter-domain mapping, the decomposition process of early failure is carried out, which is composed of "Early Failure → Functional Tree Structure → Design Tree Structure → Process Tree Structure", and firstly, the first-level functional set affecting the acoustic vibration characteristics is formed, and the mapping relationship between the box vibration and noise failures to the functional requirements is established. And then, based on the functional tree method, product structure coding and ERP system, the functional nodes, structural components and process engineering are sequentially decomposed, which covers the bottom-level functional elements, structural feature elements and sub-process units, finally the establishment of the interrelated tree of the box vibration and noise failures is realized, as shown in Fig. 4.21.

(ii) **Bottom-Up Solution of Node Weights for Failure Interrelated Trees**

In view of the large size of the constructed tree structure, and considering that the aforementioned method has focused on the calculation of the weights of the process tree nodes in the process variables, the algorithmic example selects a demonstration of the process, which calculates the weights of the structural component nodes' failure associations in the design domain.

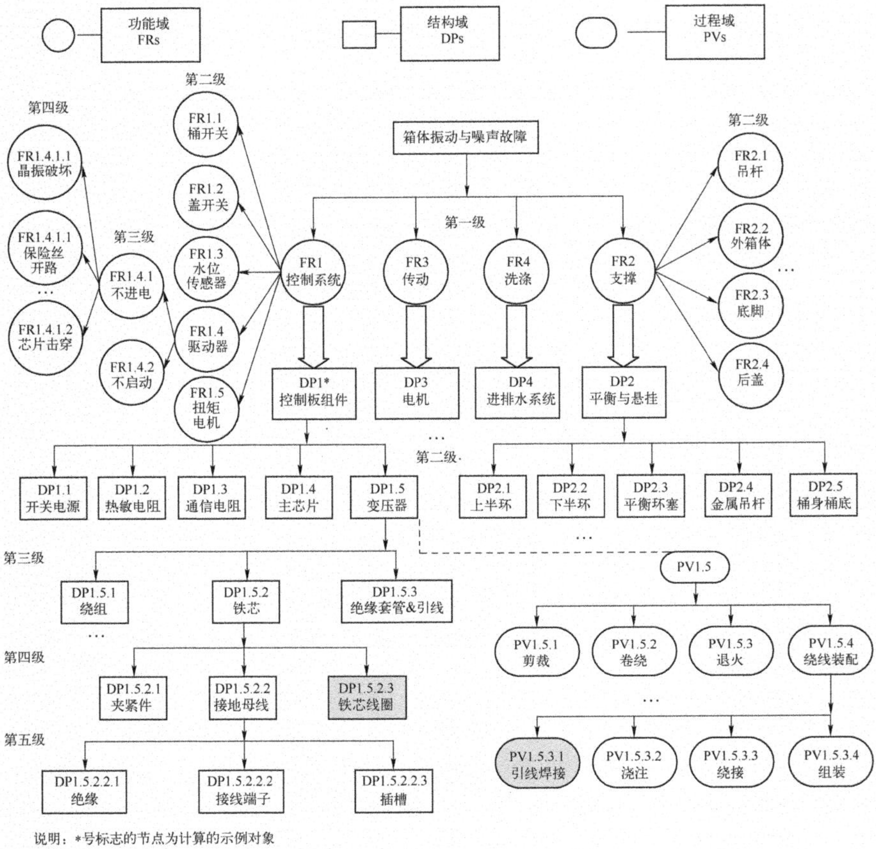

Fig. 4.21 Box vibration and noise failure interrelated tree

① Select the Structural Component Node Object

The structural component node DP1.5 (transformer assembly in the PCB board) at the second level is located and DP1.5.1 (winding), DP1.5.2 (iron core), DP1.5.3 (insulating sleeve and lead wire) at the third level is selected as its node weight solving object.

② Construct the Structural Component Nodes' Evaluation Index System Affecting the Failure Related Weight and Determine Its Influence Factor

Where, the evaluation index system still consists of five major indicators of variation probability, variation effect, probability of failure, Chance of failure being detected, and Severity of its failure effect. Based on the historical quality data, the Rough Set model is used to quantify the influence factors related to each evaluation index, and the specific, decision-oriented attribute information of the sample data is shown in Table 4.17.

Table 4.17 Failure related weight evaluation index system influence factor decision table

Sample number	Evaluation indicators					
	Indicator properties					Decision attributes
	Index 1	Index 2	Index 3	Index 4	Index 5	D
X1	2	1	3	2	2	1
X2	3	4	3	1	3	3
X3	5	4	2	3	5	4
X4	4	3	5	4	5	4
X5	3	2	3	1	3	2
X6	5	4	5	5	5	5
X7	1	3	2	2	1	1
X8	4	3	2	4	3	3
...						

Firstly, the sample data in Table 4.17 are discretely normalized to effectively construct a relevant decision table. Specifically, the mean A_D and the standard deviation V_D of the final score D are calculated based on the sample data given in the table, and the division of D into categories 1–5 is performed according to the following rules.

$$D \xrightarrow{Classification} \begin{cases} D \geq A\_D + V\_D \rightarrow & D_1 \\ A\_D \leq D < A\_D + V\_D \rightarrow & D_2 \\ A\_D - 1/2V\_D \leq D < A\_D \rightarrow & D_3 \\ A\_D - V\_D \leq D < A\_D - 1/2V\_D \rightarrow & D_4 \\ D < A\_D - V\_D \rightarrow & D_5 \end{cases} \tag{4.47}$$

Following this, the equivalence classes based on the decision attributes are derived as follows:

$$\begin{aligned} &X1 = U/\mathrm{IND}(D1) = \{x1, x7\}, X2 = U/\mathrm{IND}(D2) = \{x5\}, \\ &X3 = U/\mathrm{IND}(D3) = \{x2, x8\}, X4 = U/\mathrm{IND}(D4) = \{x3, x4\}, \\ &X5 = U/\mathrm{IND}(D5) = \{x6\}. \end{aligned}$$

Finally, the following positive domains are derived about the samples X with different sets of conditional attributes:

$$\begin{aligned} &\mathrm{POS}_C(D) = \{\mathrm{x1}, \mathrm{x3}, x5, x7\}, POS_{C-c1}(D) = \{x3, x5\}, POS_{C-c2}(D) = \{x1\}, \\ &POS_{C-c3}(D) = \{x1, x7\}, POS_{C-c4}(D) = \{x1, x5\}, POS_{C-c5}(D) = \{x3\}. \end{aligned}$$

On the basis of the positive domain values regarding the different sets of conditional attributes, the influence factor of each conditional attribute can be derived according to the formula $M_{C-c_i(D)} = 1 - \frac{\text{card}(POS_{C-c_i}(D))}{\text{card}(POS_C(D))}$, $(i = 1 \sim 5)$ and the final results are shown in Table 4.18.

③ Use Rough Sets to Determine Association Degree of Node Indicators and Decision Rules

Indicator 4 (Chance of failure being detected) of node DP1.5.2 is selected for analysis, and a collection of sample data is collected to support the validity of the decision. When measuring the detection ability by the size of the value (3–2–1), the rule can be set as follows: the larger the value, the fewer detectable problems instead; and the classification of the decision attributes is divided into 8 levels by the influence of their final score D (see the following equation), the

$$D \xrightarrow{influence} \begin{cases} 1 \to ALI \\ 2 \to VSLI \\ 3 \to SLI \\ 4 \to WLI \\ 5 \to EI \\ 6 \to WMI \\ 7 \to VSMI \\ 8 \to AMI \end{cases} \quad (4.48)$$

Further, the decision table for node DP1.5.2 for metric 4 (chance of failure being detected) can be given as shown in Table 4.19.

Expanding the simplification process on the rows and columns of Table 4.19 yields the simplified failure related weight decision table (Table 4.20).

When the decision table is in a coordinated state, based on the support of the theory that the computational scope of each decision rule can be realized with relative simplicity, the kernel value table for the failure related weight decision rule shown in Table 4.21 can be given.

On the basis of Table 4.21, its simplest rule is solved to give Table 4.22.

④ Determine the Evaluation Fuzzy Number of the Node's Corresponding Indicator Based on the Existing Decision Rule

Table 4.18 Failure related weight evaluation index system influence factor value table

Indicator number	Indicator 1	Indicator 2	Indicator 3	Indicator 4	Indicator 5
Indicator name	Variation possibility	Variation effect	Probability of failure	Chance of failure being detected	Severity of its failure effect
Impact factor	0.15	0.25	0.15	0.15	0.3

Table 4.19 Failure related weighting decision table

Sample number	Evaluation indicators				
	Conditional properties				Decision attributes
	Index 4-C1	Index 4-C2	Index 4-C3	Index 4-C4	D
X1	2	2	3	3	5
X2	2	2	2	2	6
X3	1	2	1	1	2
X4	1	2	2	3	2
X5	1	1	2	1	1
X6	3	3	2	3	8
X7	2	3	3	1	6
X8	3	1	3	2	6
X9	1	2	2	2	2
X10	2	3	3	3	6
X11	2	2	2	2	6
X12	2	1	2	2	4

Table 4.20 Simplified failure related weight decision table

Sample number		Evaluation indicators			
		Conditional properties			Decision attributes
		C1	C2	C3	D
X1′	X5	1	1	2	1
X2′	X3	1	2	1	2
X3′	X4, X9	1	2	2	2
X4′	X12	2	1	2	4
X5′	X1	2	2	3	5
X6′	X2, X11	2	2	2	6
X7′	X7, X10	2	3	3	6
X8′	X8	3	1	3	6
X9′	X6	3	3	2	8

Based on the simplest decision rule given in Table 4.22, the evaluation fuzzy number of indicators corresponding to nodes is determined based on their weights as shown in Table 4.23.

⑤ Unfold the Multi-factor Prioritization Based on the Decision Theory of Fuzzy TOPSIS

First, the normalized decision matrix shown in Table 4.24 is constructed according to the normalization algorithm applicable to the triangular fuzzy number model.

Table 4.21 Failure related weight decision rule kernel table

Sample number	Evaluation indicators			
	Conditional properties			Decision attributes
	C1	C2	C3	D
X1′	1	1	2	1
X2′	1	2	1	2
X3′	1	2	2	2
X4′	2	1	2	4
X5′	2	2	3	5
X6′	2	2	2	6
X7′	2	3	3	6
X8′	3	1	3	6
X9′	3	3	2	8

Table 4.22 Failure related weight decision simplest rule

Sample number	Evaluation indicators			
	Conditional properties			Decision attributes
	C1	C2	C3	D
X1′	1	1	–	1
X21′	1	2	–	2
X22′	1	–	2	2
X31′	–	2	1	2
X32′	2	2		2
X4′	2	1	–	4
X5′	–	2	2	5
X6′	2	2	3	6
X71′	–	3	3	6
X72′		3	2	6
X81′	3	2	–	6
X82′	2	3		6
X91′	3	–	3	8
X92′		3	3	8

Table 4.23 Evaluated values of failure related weight evaluation metrics

Node	Evaluation indicators				
	Index 1	Index 2	Index 3	Index 4	Index 5
DP1.5.1	VSLI	WMI	SLI	WMI	EI
DP1.5.2	WMI	VSLI	EI	EI	VSLI
DP1.5.3	SLI	EI	WLI	EI	WMI

Table 4.24 Failure related weight normalization decision matrix

Node	Evaluation indicators				
	Index 1	Index 2	Index 3	Index 4	Index 5
DP1.5.1	(0.154, 0.254, 0.443)	(0.256, 0.487, 0.872)	(0.208, 0.218, 0.821)	(0.343, 0.728, 1.5)	(0.128, 0.324, 0.655)
DP1.5.2	(0.462, 0.920, 1.2)	(0.512, 0.811, 1.15)	(0.260, 0.436, 1.2)	(0.172, 0.485, 1.2)	(0.512, 0.811, 1.3)
DP1.5.3	(0.185, 0.307, 0.590)	(0.128, 0.324, 0.655)	(0.260, 0.291, 1.1)	(0.172, 0.485, 1.2)	(0.256, 0.608, 0.873)

Next, the metric weights determined in Table 4.18 are multiplied by the normalization matrix demonstrated in Table 4.24 to determine the failure related weight-weighted normalization matrix shown in Table 4.25.

Further, based on the weighted normalization matrix in Table 4.25, the ideal and negative ideal nodes are solved, corresponding to the results in Table 4.26.

Specifically, Eq. (4.47) gives the process of calculating the distance of each node from the ideal node to the negative ideal node, and the exhaustive results are shown in Table 4.27.

$$
\begin{aligned}
S_1^+ = \{ & [(0.023, 0.037, 0.066) - (0.069, 0.138, 0.18)]^2 \\
& + [(0.064, 0.122, 0.218) - (0.128, 0.203, 0.288)]^2 \\
& + [(0.031, 0.033, 0.821) - (0.039, 0.065, 0.821)]^2
\end{aligned}
$$

Table 4.25 Failure related weight-weighted normalized decision matrix

Node	Evaluation indicators				
	Index 1	Index 2	Index 3	Index 4	Index 5
DP1.5.1	(0.023, 0.037, 0.066)	(0.064, 0.122, 0.218)	(0.031, 0.033, 0.821)	(0.051, 0.109, 0.225)	(0.038, 0.097, 0.196)
DP1.5.2	(0.069, 0.138, 0.18)	(0.128, 0.203, 0.288)	(0.039, 0.065, 0.18)	(0.026, 0.073, 0.18)	(0.153, 0.243, 0.39)
DP1.5.3	(0.028, 0.046, 0.089)	(0.032, 0.081, 0.164)	(0.039, 0.044, 0.165)	(0.026, 0.073, 0.18)	(0.182, 0.183, 0.262)

Table 4.26 Failure related weight ideal and negative ideal nodes

Node	Evaluation indicators				
	Index 1	Index 2	Index 3	Index 4	Index 5
A^+	(0.069, 0.138, 0.18)	(0.128, 0.203, 0.288)	(0.039, 0.065, 0.821)	(0.051, 0.109, 0.225)	(0.182, 0.243, 0.39)
A^-	(0.023, 0.037, 0.066)	(0.032, 0.081, 0.164)	(0.031, 0.033, 0.165)	(0.026, 0.073, 0.18)	(0.038, 0.097, 0.196)

$$+ [(0.051, 0.109, 0.225) - (0.051, 0.109, 0.225)]^2$$
$$+ [(0.038, 0.097, 0.196) - (0.182, 0.243, 0.39)]^2\}^{1/2}$$
$$= (0.174, 0.198, 0.683)$$

Further, the calculation of the distance between each node and the ideal node can be given as shown in the following equations respectively.

$$C_1^+ = \frac{S_1^-}{S_1^+ + S_1^-} = (0.053, 0.217, 0.327),$$
$$C_2^+ = \frac{S_2^-}{S_2^+ + S_2^-} = (0.525, 0.83, 1.25),$$
$$C_3^+ = \frac{S_3^-}{S_3^+ + S_3^-} = (0.103, 0.293, 1.007).$$

The ranked weight vector of the association is solved according to the node proximity and this is used as the failure association weight value for this node. First of all, the measure of the expected value of closeness is determined mainly based on the expression of the expected value in the triangular fuzzy number model as: $I(C_1^+) = \frac{0.053+2*0.217+0.327}{4} = 0.2035$, $I(C_2^+) = 0.8588$, $I(C_3^+) = 0.424$. Therefore, the relative weights of the nodes under study are given as: $w_1 = \frac{0.2035}{0.2035+0.8588+0.424} = 0.137$, $w_2 = 0.578$, $w_3 = 0.285$. According to the results of the calculations, it is concluded that the three nodes mentioned above, according to the size of their weights, have been determined as follows in the order of priority: DP1.5.2, DP1.5.3, DP1.5.1.

Repeating the above process, the calculation of the weights of the nodes of the whole interrelated tree can be realized, and the size of the failure related weight of the functional nodes, the physical structure nodes and the process production nodes can be obtained. The final calculation structure shows that **FR1.4.1.2, DP1.5.2.3** and **PV1.5.3.1** have the highest priority, and they will be the most important links for the prevention and rectification of the box vibration and noise, which is an early failure.

Table 4.27 Distance between failure related weight and ideal nodes

Distance from ideal node	Weighting	Distance from the negative ideal node	Weighting
S_1^+	(0.174, 0.198, 0.683)	S_1^-	(0.04, 0.055, 0.07)
S_2^+	(0.036, 0.038, 0.045)	S_2^-	(0.138, 0.181, 0.218)
S_3^+	(0.107, 0.169, 0.206)	S_3^-	(0.036, 0.07, 0.144)

4.2.2 *Optimization Technique for Burn-in Time Considering Quality Deviation Loss*

The early failure rate of a product is an important indicator for top management to understand the reliability of a product, and is also an important basis for quality management decisions [18]. In order to reduce the early failure rate, reliability screening, such as refining tests, before the actual field use and early failure of parts eliminated in advance, but often used to screen and eliminate products with quality defects but bring considerable refining costs. It is different from the conventional quality inspection and assumes that the performance of the product before the screening is good, the purpose is not to check out the test that is before the screening is a bad product. For a certain batch of products, the non-destructive reliability screening test, its failure mechanism and failure distribution should not be affected, but played an "aging" or "burn-in" role, so also known as "burn-in test". For electronic components, to carry out reliability screening has more prominent practical significance. Obviously, reliability screening needs to pay a certain price, such as the increase in cost, the extension of the production cycle, etc., but relative to the installation, commissioning and use of the site to find and replace the cost of early failure caused by the cost is insignificant. Especially for high-reliability engineering or aviation, aerospace and other sophisticated systems is more significant.

I. **Analysis of Quality Costs Associated with Manufacturing Defects**

(i) **Definition of Quality Bias Effects Towards Manufacturing Defects**

The product manufacturing process affected by variations in material purity and process precision often results in unavoidable manufacturing defects. Simultaneously, along with the manufacturing process quality deviation, the hidden deviation loss follows. The processing of electronic devices is often accomplished by multi-station manufacturing systems, and in mass production, defects inevitably become the key quality feature of electronic device processing due to the inherent variations in the manufacturing process such as changes in process conditions, device conditions, and raw materials, leading to varying degrees of quality and reliability problems. When describing the quality characteristics of manufacturing defects in terms of dimensional characteristics, given the critical dimensional threshold of manufacturing defects x° that can lead to product failure and non-conformity, manufacturing defects of different sizes have different deviation effects. This is shown in Fig. 4.22.

In Fig. 4.22, let x be the observed value of the dimensional characteristics of the manufacturing defect ($x > 0$), and given the critical dimensions of the manufacturing defect x°, the following determination can be made:

(1) If there is a defect size $x > x^\circ$, regard such defects as Type-I large-size characteristic manufacturing defects, because it can directly lead to the failure of the device fatal defects, also known as fatal defects, can be detected in the factory test to be rejected, so the corresponding defects effect is: affect the level of yield of the device manufacturing process; (Note: Manufacturing yields are often used

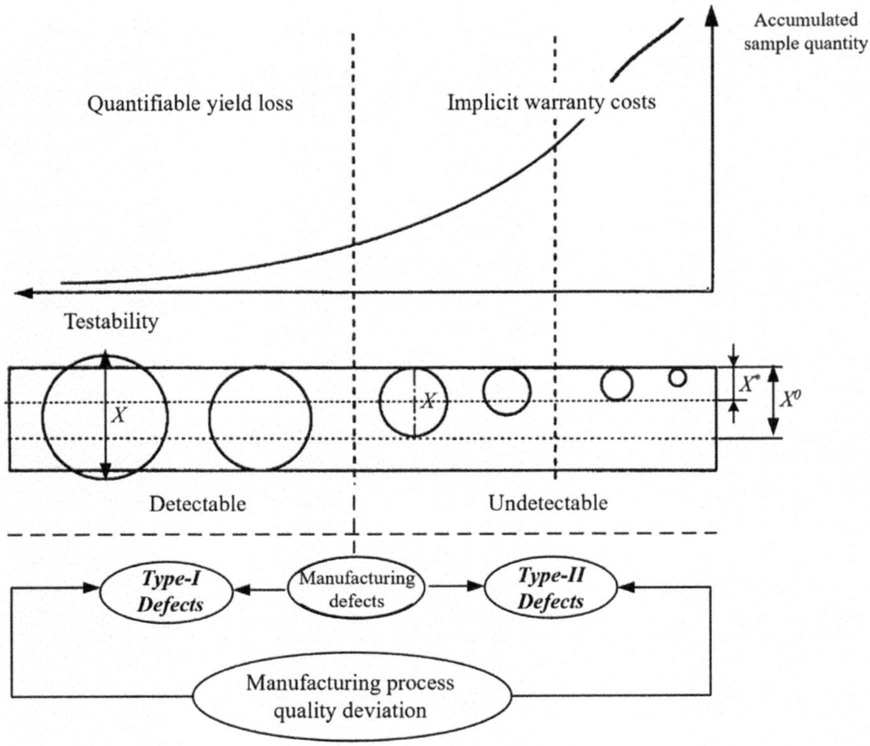

Fig. 4.22 Bias effects for different types of manufacturing defects

to measure the level of quality of the manufacturing process of the electronic products that contain manufacturing defects.)

(2) If there is a defect size $x \leq x^{\circ}$, regard this type of defect as Type-II small size characteristics of manufacturing defects, that is, non-fatal defects, different from Type-I can be detected and eliminated defects, Type-II manufacturing defects Missed factory inspection, more with the role of time stress in the use of the process leading to the failure of some components, due to the use of components closely related to the reliability of the device, also for the "Reliability defects", corresponding to the defect effect is: affect the level of device reliability;

From the cost perspective of quality deviation, given the factory inspection after electrical performance testing and functional testing, electronic devices with fatal defects of large size characteristics are easily eliminated, while non-fatal defects of small size characteristics, i.e., reliability defects, as hidden defects exist as a technological bottleneck that is difficult to detect, and are more likely to cause stress- and time-related early failures during the initial period of use and lead to the early failure rate of electronic components, etc. The early failure rate of electronic components remains high. In turn, units with large-size fatal defects are rejected, resulting in yield loss costs associated with factory testing, while units with small-size non-fatal

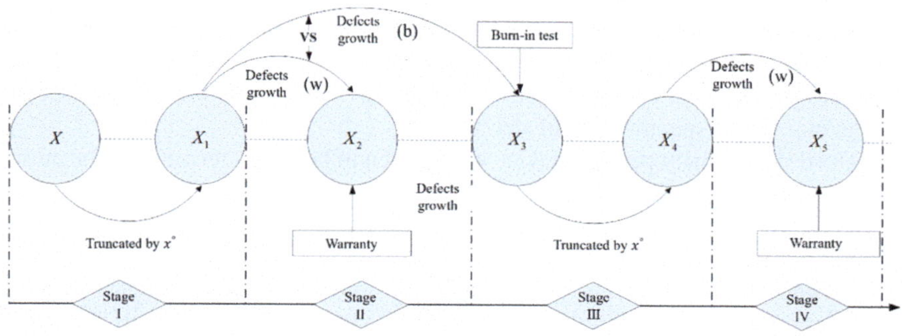

Fig. 4.23 Schematic diagram of the four-stage defect size evolution process

defects also incur hidden costs, which grow in size and are concentrated in the early failure stage, resulting in warranty costs under a specific product liability policy.

Among the manufacturing defects, Type-II defects are time-related defects, whose corresponding product parameters and performance are generally regarded as qualified until such defects evolve into Type-I fatal defects, which are characterized as hidden. Burn-in test is mainly to deal with hidden manufacturing defects of Type-II, which is an effective inspection program to find out the hidden problems of early failure products and ensure the reliability of products in the field. It can effectively eliminate early failure parts by accelerating the size growth of Type-II non-fatal defects with small size characteristics.

Along with the evolution of manufacturing defect sizes, different stages of defect sizes exhibit different behavioral characteristics that require special attention and consideration. Based on this perspective, this book proposes a four-stage defect size evolution framework as shown in Fig. 4.23.

In Fig. 4.23, X is the initial defect size characteristic value, X_1 is the defect size characteristic value truncated after the defects of $X \geq x^{\circ}$ size are rejected by the factory inspection test, X_2 is the defect size characteristic value after the defects growth in the warranty time after putting into use w, X_3 is the defect size characteristic value after the defects growth in the burn-in period b, X_4 is the defect size characteristic value truncated after the burn-in test size meets the size of $X_4 < x^{\circ}$, and X_5 is the defect size characteristic value after the defects growth in the warranty time after putting into use w. The value of the dimensional characteristics of the defects.

For the purpose of the study, the basic assumptions are as follows:

(1) Electronics are not repairable;
(2) The warranty policy is a free replacement warranty during the warranty period;
(3) The burn-in environment of the device is approximately the same as the outfield use environment;
(4) The old refinement test does not induce any other new failure modes;
(5) Brand loss due to degradation of device performance is not considered;

(6) The product quality characteristic of electronic components is defined as the size of the defects, which is an expected small size characteristic;
(7) A device is considered to have failed if the size of the manufacturing defect is greater than a specific critical size x°;
(8) The defect density per unit device was chosen to be characterized by a negative binomial distribution.

(ii) **Yield Loss Modeling Associated with Type-I Fatal Manufacturing Defects**

Along with the device manufacturing process, the defects present different size characteristics, which makes the determination of Type-I fatal and Type-II non-fatal manufacturing defects and the mining of more correlations rely on the more statistically significant defect size distribution to unfold. Here, the widely used defect size feature distribution $s_0(x)$ is used as follows:

$$S_0(x) = \begin{cases} \frac{x}{x^{*2}}, & 0 < x \le x^* \\ \frac{x^{*2}}{x^3}, & x^* < x < \infty \end{cases}. \tag{4.49}$$

In particular, x^* refers to the value of the point in the distribution of defect sizes where there is a clear tendency to concentrate, representing the general level of defect sizes, which can be referred to as mass sizes, as can be seen in Fig. 4.23. It is easy to determine that the mass size x^* is smaller than the critical size x° compared to the critical size x° of manufacturing defects, which is an important reference for determining the type of defects and the degree of effect. Therefore, based on the established characteristic distribution of defect size $s_0(x)$, the probability of device failure due to defect size exceeding the critical size x° after factory test inspection can be determined by the following formula p_1:

$$\begin{aligned} p_1 &= \Pr(x > x^\circ | \mathrm{s_0(x)}) \\ &= \int_{x^0}^{\infty} \frac{x^{*2}}{x^3} dx = \frac{1}{2}\left(\frac{x^*}{x^0}\right)^2. \end{aligned} \tag{4.50}$$

Let c_0 be the selling price per unit of the device, and the factory yield loss cost of the electronic device judged to be unqualified by the effect of fatal defects after the factory inspection of the electronic device by the electrical performance test and the functionality test is as follows Y_0:

$$Y_0 = c_0 * \left[1 - (1 - p_1)^{Np_1}\right], \tag{4.51}$$

where N refers to the number of manufacturing defects contained per unit of electronics, which can be portrayed by a defect density distribution defined by the classical negative binomial distribution incorporating defect aggregation effects in the literature, namely,

$$\Pr(N = n) = \int_0^\infty \frac{e^{-\Lambda}\Lambda^n}{n!}\frac{1}{\Gamma(\alpha)\gamma^\alpha}\Lambda^{\alpha-1}e^{-\Lambda/\gamma}d\Lambda$$
$$= \frac{(n+\alpha-1)!}{n!(\alpha-1)!}\left(\frac{\lambda/\alpha}{1+\lambda/\alpha}\right)^n(1+\lambda/\alpha)^\alpha, \quad (4.52)$$

where $\lambda = \alpha\gamma$. α is the defect aggregation effect factor, which ranges from 0.5 to 5, and the smaller the value of α, the larger the corresponding aggregation effect. Accordingly, the expected value of N, $E(N) = \sum_0 n * \Pr(N = n)$.

As the tests of the factory inspection rejected fatal defects beyond the critical size x°, the initial characteristic distribution of defect sizes $s_0(x)$ was changed and refined to a truncated size distribution $s_1(x)$ with defect sizes conforming to $x_1 \le x^\circ$ as determined by the following distribution:

$$S_1(x) = \begin{cases} \frac{1}{1-p_1}\frac{x}{x^{*2}}, & 0 < x \le x^* \\ \frac{1}{1-p_1}\frac{x^*}{x^3}, & x^* < x \le x^0 \end{cases}, \quad (4.53)$$

where, p_1 is an estimate of the probability of device failure due to a FAT-tested defect size exceeding the critical size x° as described in the previous section.

(iii) **Modeling of Warranty Costs Associated with Type-II Non-detectable Reliability Defects**

The classical bathtub-type failure rate curve is often used to describe and quantify the failure distribution of electronic devices. Failure rates start out high in the "early failure" phase of a product's life, but decrease rapidly. The intrinsic mechanism of the decreasing failure rate function is often traced back to the manufacturing process due to potentially undesirable factors in raw materials, device, processes, etc. Non-fatal defects seriously affect the reliability of the device, and non-fatal defects can be stimulated in the actual stress environment, resulting in the early failure of low-life products.

As produced product reliability represents the quality level of the product on the timeline, warranty has rightly become a very important factor in the product marketing process, the enterprise's perfect product liability system, that is, the comprehensive warranty policy, brings excellent reliability assurance at the same time, does also bring the production side of the operating costs.

Under normal temporal stress conditions, it is assumed that the defect size growth law conforms to RULE-1: the rate of growth of the defect size is proportional to the current defect size by the factor $k_1x_1(x_1 \sim s_1(x))$, i.e., the following relationship exists:

$$\frac{dx}{dt} = k_1x_1 \quad (4.54)$$

For a given warranty length w, the corresponding growth of the defect size x_2 is $x_2 = x_1 e^{k_1 w}$. After the defect size growth process at $s_1(x) \overrightarrow{_{RULE1}} s_2(x)$, the factory-tested truncated size distribution $s_1(x)$ evolves into a new size characterization distribution $s_2(x)$ with the following form:

$$s_2(x) = \begin{cases} \frac{1}{1-p_1} \frac{x}{e^{k_1 w}(x^*)^2}, & 0 < x \le x^* e^{k_1 w} \\ \frac{1}{1-p_1} \frac{(x^*)^2 \left(e^{k_1 w}\right)^3}{x^3}, & x^* e^{k_1 w} < x \le x^\circ e^{k_1 w} \end{cases}. \tag{4.55}$$

At this point, the probability of device failure due to an increase in the size x° of a non-fatal defect or reliability defect beyond the critical size p_2 The derivation process is as follows:

For $w < \frac{\ln x^0 - \ln x^*}{k_1}$, $p_2 = \Pr(x_2 > x^\circ | s_2(x))$ can be expressed as,

$$\begin{aligned} p_2 &= \Pr\left(x_2 > x^0 | \mathrm{s}_2(\mathrm{x})\right) \\ &= \int\limits_{x^0}^{x^0 e^{h_1 w}} \frac{1}{1-p_1} \frac{(x^*)^2 \left(e^{k_1 w}\right)^3}{x^3} dx. \end{aligned} \tag{4.56}$$

In turn, for $\ge \frac{\ln x^\circ - \ln x^*}{k_1}$, $p_2 = \Pr(x_2 > x^\circ | s_2(x))$ can be expressed as,

$$\begin{aligned} p_2 &= \Pr(x_2 > x^\circ | \mathrm{s}_2(\mathrm{x})) \\ &= \int\limits_{x^*}^{x^* e^{h\eta^{**}}} \frac{1}{1-p_1} \frac{x}{e^{k_1 w}(x^*)^2} dx + \int\limits_{x^* e^{\frac{1}{2}}}^{x^0 e^{\xi 1/w}} \frac{1}{1-p_1} \frac{(x^*)^2 \left(e^{k_1 w}\right)^3}{x^3} dx. \end{aligned} \tag{4.57}$$

Let c_1 be the unit failure cost of the device over the warranty period w. Warranty costs over the use of the warranty period W_0 are as follows, subject to dimensional growth of non-fatal defects:

$$W_0 = c_1 * \left[1 - (1 - p_2)^{(1-p_1)Np_2}\right], \tag{4.58}$$

where, p_2 is an estimate of the probability of device failure due to the RULE-1 growth defect size exceeding the critical size x° as described in the previous section, and $(1 - p_1)Np_2$ corresponds to the number of fatal defects remaining after the Stage I truncation and functioning within the warranty period.

In summary, the deviation loss of manufacturing process quality can be considered and analyzed by integrating the deviation loss of both Type-I and Type-II manufacturing defects.

Let L_0 be the quality deviation loss per unit device, as this book starts from the hidden loss of quality variation in the manufacturing process, the manufacturing quality deviation loss will be defined as the additional cost of production cost caused by manufacturing defects due to quality variation, while considering the effect of defects of different sizes, L_0 should include the loss of yield cost for the devices failing the factory test Y_0 and the warranty cost for the qualified devices under a specific warranty strategy W_0, i.e., there are

$$L_0 = Y_0 + W_0 = c_0 * \left[1 - (1 - p_1)^{Np_1}\right] + c_1 * \left[1 - (1 - p_2)^{(1-p_1)NN_2}\right]. \quad (4.59)$$

II. **Dynamic Characterization Simulation of the Implicit Quality Deviation Loss Model under the Burn-in Effect**

As shown in Fig. 4.23, the size evolution patterns in Stage II are divided into two different categories, corresponding to the growing sizes X_2 and X_3. Here, due to the unusual excitation effect of the burn-in effect on the defect size, the size X_3 will be significantly different from X_2, which is more prone to cause early device failure. Given the direct modulation of implicit mass deviation loss by the deviation effect of size-dependent Type-I and Type-II defects in Sect. 4.3.1, the evolutionary pattern of defect size greatly influences the magnitude of actual mass deviation loss. Therefore, it is of great significance to analyze the dynamic characteristics of the implicit quality deviation loss model under the burn-in effect based on the defect evolution process.

The introduction and rollout of burn-in tests will inevitably result in fixed costs for preparing the burn-in environment c_2. At the same time, the burn-in test is mainly aimed at the excitation of small-size characteristic non-fatal defects. Under a certain defect density distribution, a certain defect size distribution and a certain defect size growth law, small-size featured non-fatal defects carry out the growth evolution of b, when the defect size grows to more than the critical size of $x°$, by the defect effect of fatal defects, the device will be induced to fail due to the growth of the defect size, which leads to an increase in the cost of refining test, and at the same time, must bear the failure costs associated with the refining time. Failure costs associated with the burn-in time W_b.

(i) **Modeling of Increased Costs of Ageing**

It is assumed that the pattern of growth of the defect size during the burn-in time b corresponds to RULE-2: the rate of growth of the defect size is proportional to the current size of the defect x_1 ($x_1 \sim s_1(x)$) by the factor k_2, i.e. the following relation exists:

$$\frac{dx}{dt} = k_2 x_1. \quad (4.60)$$

Given the ageing time b, the corresponding growth of the defect size x_3 is $x_3 = x_1 e^{k_2 b}$. After the defect size growth process at $s_1(x)\overrightarrow{_{RULE2}}s_3(x)$, the factory-tested truncated size distribution $s_1(x)$ evolves into a new size characterization distribution $s_3(x)$ of the following form:

$$s_3(x) = \begin{cases} \frac{1}{1-p_1}\frac{x}{e^{k_2 b}(x^*)^2}, & 0 < x \le x^* e^{k_2 b} \\ \frac{1}{1-p_1}\frac{(x^*)^2\left(e^{k_2 b}\right)^3}{x^3}, & x^* e^{k_2 b} < x \le x^0 e^{k_2 b} \end{cases}. \tag{4.61}$$

The probability p_3 of device failure due to an increase in the size of a non-fatal or reliability defect beyond the critical size x° in the burn-in test environment is derived as follows:

For the threshold size x° satisfying the relation $x^\circ > x^* e^{k_2 b}$, i.e., $b < \frac{\ln x^0 - \ln x^*}{k_2}$, the expression for the second segment of the distribution function corresponding to the size $s_3(x)$, p_3 is as follows.

$$\begin{aligned} p_3 &= \Pr\left(x_3 > x^0 | \mathrm{s}_3(\mathrm{x})\right) \\ &= \int_{x^0}^{x^0 e^{k_2 b}} \frac{1}{1-p_1}\frac{(x^*)^2\left(e^{k_2 b}\right)^3}{x^3} dx. \end{aligned} \tag{4.62}$$

Further, for the threshold size x° satisfying the relation $x^\circ \le x^* e^{k_2 b}$, i.e., $b \ge \frac{\ln x^0 - \ln x^*}{k_2}$, the expression of the first segment distribution function p_3, corresponding to the size $s_3(x)$ is as follows.

$$\begin{aligned} p_3 &= \Pr\left(x_3 > x^0 | \mathrm{s}_3(\mathrm{x})\right) \\ &= \int_{x^*}^{*e^* e^{k_2 b}} \frac{1}{1-p_1}\frac{x}{e^{k_2 b}(x^*)^2} dx + \int_{x^* e^{k_2}}^{x_0^0 k^{k_b b}} \frac{1}{1-p_1}\frac{(x^*)^2\left(e^{k_2 b}\right)^3}{x^3} dx. \end{aligned} \tag{4.63}$$

where c_3 is the time-dependent burn-in cost per unit of time, and c_4 be the failure cost per unit of device within the burn-in time duration b. The burn-in related costs x° accumulated in the burn-in time are as follows:

$$C_b = c_2 + c_3^* b. \tag{4.64}$$

Additionally, the failure costs associated with the burn-in hours W_b are as follows:

$$W_b = c_4 * \left[1 - (1 - p_3)^{(1-p)Np_3}\right], \tag{4.65}$$

where, p_3 represents the probability that a growing non-fatal or reliability defect size exceeds the critical size w and causes device failure, and $(1 - p_1)Np_3$ represents the number of remaining fatal defects truncated by Stage I and acting within the ageing timeframe.

(ii) **Modeling Changes in Warranty Costs**

After the screening of electronic devices by rational and effective burn-in test, under the environment of temperature stress, electrical stress, mechanical stress or some combination of stresses, low-life products containing manufacturing defects and prone to premature failure can be eliminated before delivery to the user to effectively ensure the user's experience. In turn, electronic devices that pass the burn-in test are put into service with a warranty period of w. Since the burn-in test eliminates some defect samples whose size increases beyond the critical size x° among non-fatal defects, the characteristic distribution of burn-in defect sizes $s_3(x)$ changes, and is refined to a truncated size distribution $s_4(x)$, whose defect sizes conform to the size of $x_4 \leq x^{\circ}$ as determined by the following distribution:

$$S_4(x) = \begin{cases} \frac{1}{1-p_3}\frac{1}{1-p_1}\frac{x}{e^{k_2 b}(x^*)^2}, & 0 < x \leq x^* e^{k_2 b} \\ \frac{1}{1-p_3}\frac{1}{1-p_1}\frac{(x^*)^2\left(e^{k_2 b}\right)^3}{x^3}, & x^* e^{k_2 b} < x \leq x^0 \end{cases}. \tag{4.66}$$

That is, after the burn-in test, the characteristic distribution of defect sizes of electronic devices is updated to a truncated distribution $s_4(x)$, at which time the number of small-sized non-fatal defects is $(1-p_1)N(1-p_3)$. Where p_1 and p_3 are the parameters derived in the previous section, and k_2 is the scaling factor under the RULE-2 type growth model.

Under normal use stress conditions, it is known that the defect size growth law conforms to RULE-1, i.e., the growth rate of the defect size is proportional to the current defect size by the coefficient $k_1 x_4 (x_4 \sim s_4(x))$, i.e., the following relationship exists:

$$\frac{dx}{dt} = k_1 x_4. \tag{4.67}$$

Given the length of the warranty w, the size of the growing defect x_5 is $x_5 = x_4 e^{k_1 w}$. However, based on the distribution of segment sizes $s_4(x)$, the relationship between the size of $x^* e^{k_2 b + k_1 w}$ corresponding to $x^* e^{k_2 b}$ at the decomposition point and $x^0 e^{k_1 w}$ corresponding to x^0 at the upper limit, along with the growth of defect size x_5, is not clear. Therefore, after the defect size growth process of $s_4(x)\overrightarrow{_{RULE1}}s_5(x)$, the truncated size distribution $s_4(x)$ after the old refinement test grows and evolves into a new size characterization distribution $s_5(x)$ with the following expression:

$$S_5(x) = \begin{cases} \frac{1}{1-p_3}\frac{1}{1-p_1}\frac{x}{e^{k_2 b + k_1 w}(x^*)^2}, & 0 < x \leq \min\left(x^* e^{k_2 b + k_1 w}, x^0 e^{k_1 w}\right) \\ \frac{1}{1-p_3}\frac{1}{1-p_1}\frac{(x^*)^2\left(e^{k_2 b + k_1 w}\right)^3}{x^3}, & \min\left(x^* e^{k_2 b + k_1 w}, x^0 e^{k_1 w}\right) < x \leq x^0 e^{k_1 w} \end{cases}. \tag{4.68}$$

Considering the actual use of the device in the field, the final defect size distribution $s_5(x)$ is often used to determine the role of manufacturing defects in influencing the occurrence of early device failures. That is, the derivation of the probability of device failure due to non-fatal or reliability defects increasing in size beyond the critical size p_4 can only be carried out based on the current complete and well-defined size distribution $s_5(x)$. The details are as follows:

When there is a size relationship between $x^0 e^{k_1 w}$ and $x^* e^{k_2 b + k_1 w}$, i.e. $x^0 e^{k_1 w} \leq x^* e^{k_2 b + k_1 w}$, the length of the old practice $b \geq \frac{\ln x^0 - \ln x^*}{k_2}$, the size distribution $s_5(x)$

corresponds to $\min\left(x^{*}e^{k_2b+k_1w}, x^0e^{k_1w}\right)x^0e^{k_1w}$, in which case there is a contradiction in $x_5 = x^0e^{k_1w}$ and $s_5(x)$ does not have the generalized statistical distribution connotation for estimating the probability $p_4 = \Pr(x_5 > x^\circ|s_5(x))$. Therefore, this case is not considered.

On the other hand, when there is a size relationship $x^0e^{k_1w}$ and $x^{*}e^{k_2b+k_1w}$, i.e. $x^0e^{k_1w} > x^{*}e^{k_2b+k_1w}$, the length of the old refinement satisfies $b < \frac{\ln x^0 - \ln x^{*}}{k_2}$, the size distribution $s_5(x)$ in $\min\left(x^{*}e^{k_2b+k_1w}, x^0e^{k_1w}\right) = x^{*}e^{k_2b+k_1w}$ and has the following expression, the

$$S_5(x) = \begin{cases} \frac{1}{1-p_3}\frac{1}{1-p_1}\frac{x}{e^{k_2b+k_1w}(x^{*})^2}, & 0 < x \le x^{*}e^{k_2b+k_1w} \\ \frac{1}{1-p_3}\frac{1}{1-p_1}\frac{(x^{*})^2\left(e^{k_2b+k_1w}\right)^3}{x^3}, & x^{*}e^{k_2b+k_1w} < x \le x^0e^{k_1w} \end{cases}. \tag{4.69}$$

In this way, the distribution of the dimension $s_5(x)$ is finalized and the derivation of the probability $p_4 = \Pr(x_5 > x^\circ|s_5(x))$ can thus be developed. However, the same ambiguous relationship exists for the specific location of the threshold dimension x° in the interval $\left(0, x^{*}e^{k_2b+k_1w}\right)$ or $\left(x^{*}e^{k_2b+k_1w}, x^0e^{k_1w}\right)$ and needs to be further developed for categorization as discussed below:

When x° lies in the interval $x^{*}e^{k_2b+k_1w}, x^0e^{k_1w}$, i.e., $x^{*}e^{k_2b+k_1w} < x^\circ \le x^\circ e^{k_1w}$, the discussion of $\Pr(x_5 > x^\circ|s_5(x))$ is oriented to the second part of the function of the size distribution $s_5(x)$. At this point, the warranty duration satisfies the size relationship $w < \frac{\ln x^0 - \ln x^{*} - k_2b}{k_1}$ and corresponds to the probability p_4 of the expression that

$$p_4 = \Pr\left(x_5 > x^0|s_5(x)\right) = \int\limits_{x}^{x} \frac{1}{1-p_3}\frac{1}{1-p_1}\frac{(x^{*})^2\left(e^{k_2b+k_1w}\right)^3}{x^3}dx. \tag{4.70}$$

Conversely, when x° lies in the interval $0 < x^\circ < x^{*}e^{k_2b+k_1w}$, the discussion of $\Pr(x_5 > x^\circ|s_5(x))$ is oriented towards the first part of the size distribution $s_5(x)$ function. At this point, the warranty length satisfies the size relation $w \ge \frac{\ln x^0 - \ln x^{*} - k_2b}{k_1}$ and corresponds to the probability p_4 with the expression, the

$$p_4 = \Pr\left(x_5 > x^0|s_5(x)\right) = \int\limits_{x^{*}}^{x^{*}e^{k_2b+k_1w}} \frac{1}{1-p_3}\frac{1}{1-p_1}\frac{x}{e^{k_2b+k_1w}(x^{*})^2}dx$$

$$+ \int\limits_{x^{*}e^{k_2++kk^2w}}^{x^0e^{k_1w}} \frac{1}{1-p_3}\frac{1}{1-p_1}\frac{(x^{*})^2\left(e^{k_2b+k_1w}\right)^3}{x^3}dx. \tag{4.71}$$

Accordingly, use the warranty-type fee W_1 below for failures due to burn-in test incentives during the warranty period:

$$W_1 = c_1 * \left[1 - (1-p_4)^{(1-p_1)N(1-p_3)p_4}\right], \tag{4.72}$$

where, p_4 represents the probability of device failure due to non-fatal defects or reliability defects exceeding the critical dimensions x° as they increase in size during the warranty period, and $(1 - p_1)N(1 - p_3)p_4$ represents the number of fatal defects that are still active during the warranty period after the secondary truncation of Stage I and Stage III.

Based on the above analysis and discussion, let L_1 be the mass deviation loss per unit device under the burn-in test plan, then L_1 has the following expression:

$$\begin{aligned} L_1 = Y_0 + W_1 + W_b = c_0 * \left[1 - (1 - p_1)^{Np_1}\right] + c_1 * \left[1 - (1 - p_4)^{(1-p_1)N(1-p_3)p_4}\right] \\ + c_4 * \left[1 - (1 - p_3)^{(1-p_1)Np_3}\right]. \end{aligned} \tag{4.73}$$

III. Determination of Optimal Burn-in Time Considering Hidden Quality Deviation Loss

The analysis of optimal burn-in time is often based on specific burn-in criteria, such as the expectation of a certain level of reliability of the aged product from the user's point of view, including the maximization of the average residual life, the minimization of the task failure rate, and the predetermined reliability, etc., and from the producer's point of view, the control of the process deviation loss to achieve the minimization of the cost of quality has become the primary criterion for the implementation of optimal burn-in. This book is based on the production side of the device and aims to determine the optimal burn-in time from an economic point of view by weighing the increased burn-in cost Δ_1 against the reduced quality deviation loss Δ_2.

Specifically, when considering the burn-in screening of electronic devices, the increased burn-in cost Δ_1 consists of the following two components: a fixed cost for the preparation of the burn-in test environment c_2, and a time-dependent burn-in test cost $c_3 * b$, which is the burn-in cost within the burn-in time duration C_b, which is known as

$$\Delta_1 = C_b = c_2 + c_3^* b. \tag{4.74}$$

At the same time, based on the loss of mass deviation per unit device under the burn-in test program L_1 and the loss of mass difference per unit device under the normal environment L_0 the reduced loss of mass deviation can be measured as $\Delta_2 = L_0 - L_1$.

Then, the objective function $g(b)$ of the cost control oriented optimal burn-in analysis is established with burn-in time b as the independent variable as follows:

$$\begin{aligned} g(b) &= \Delta_2 - \Delta_1 = (L_0 - L_1) - C_b = W_0 - W_1 - W_b - \left(c_2 + c_3^* b\right) \\ &= c_1 * \left[1 - (1 - p_2)^{(1-p_1)Np_2}\right] - c_1^*\left[1 - (1 - p_4)^{(1-p_1)N(1-p_3)p_4}\right] \\ &\quad - c_4 * \left[1 - (1 - p_3)^{(1-p_1)Np_3}\right] - \left(c_2 + c_3^* b\right), \end{aligned} \tag{4.75}$$

where, c_1, c_2, c_3 and c_4 are the four basic cost factors described in the previous section: failure cost per unit of device in field use, fixed burn-in preparation cost, burn-in cost per unit of time per unit of device, and burn-in failure cost per unit of device; p_1, p_2, p_3 and p_4 correspond to the probability of device failure due to oversizing beyond the threshold size in the four-phase evolution of the defect size in the preceding sections, respectively; and furthermore, the expressions $(1 - p_1)Np_2$, $(1 - p_1)Np_3$ and $(1 - p_1)N(1 - p_3)p_4$ denote the number of remaining lethal defects after truncation in the four-phase evolution of the size, respectively. number of fatal defects remaining after truncation during the four-stage evolution process.

Theoretically, $g(b)$ has to be greater than or equal to 0 for conducting burn-in tests to be meaningful. Further, from an economical point of view, the optimal burn-in time can be determined by solving the following equation.

$$\frac{\partial g(b)}{\partial b} = \frac{\partial[(L_0 - L_1) - C_b]}{\partial b} = 0. \tag{4.76}$$

In particular, by quadratic fitting of the discrete type $g(b)$, $g(b)$ can be approximately transformed into a certain continuous distribution, and then the optimal burn-in duration b^* and its corresponding burn-in gain $g(b^*)$ can be solved, which provides practical guidance for more objective, economical and reasonable setting of burn-in test duration.

IV. Analysis of Examples

According to the after-sales maintenance department of the actual collection of failure data can be statistically a model of PCB board main early failure mode examples of charts, as shown in Fig. 4.24.

From the failure performance mode, in Fig. 4.24, the breakdown of the PCB board chip or transistor insertion misalignment can directly lead to power supply disorders or transistor itself burned, triggering the early failure of the PCB board;

Fig. 4.24 Diagram of the main early failure examples of a model of PCB board

and photocoupler components, such as soldering, although in the obvious failure alarm can still maintain the system before the alarm, there is always the hidden danger of not working. The burn-in test provides technical support for reducing the high early failure rate.

At the present stage, a large number of research ideas on the optimization of burn-in test time are based on various constraints, the analysis of failure modes and effects after the failure, and the modeling and analysis of optimal burn-in for feedback maintenance guidance. Tracing the root causes of high-frequency early failures, it can be found that the inevitable quality variations from the manufacturing process have an irreversible effect on the formation of conformity-type key characteristics, and the deviation of the key characteristics from the target value ultimately determines the existence of manufacturing defects of varying degrees. In Fig. 4.25, the multi-station manufacturing process for this model of PCB board is shown, while the nodes marked in red are considered to be the key steps to ensure product stability and quality.

In actual production, even if the PCB board process as shown in Fig. 4.25 has set up a number of inspection stations and test points, from the manufacturing process of personnel errors, machine degradation, material defects, improper methods, environmental discomfort and measurement error of the many uncertainties will still lead to key characteristics of the variations and deviations, and ultimately bring about explicit or implicit manufacturing defects, which greatly affects the quality of the manufacturing process and the use of early produced product reliability. Reliability. It can be seen that the quality deviation information that can be collected in the manufacturing process of PCB boards will be valuable a priori data for carrying out the analysis of optimal refining. Therefore, this book will pay full attention to the hidden losses related to quality deviations in the manufacturing process from the cost perspective, and develop the following applications as new constraints in the optimization process of refining.

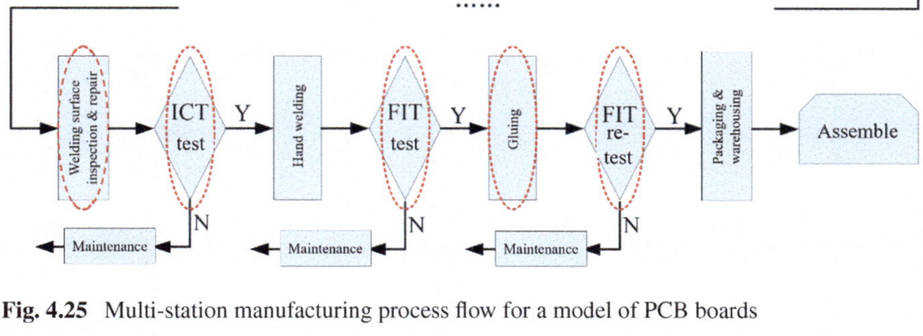

Fig. 4.25 Multi-station manufacturing process flow for a model of PCB boards

Table 4.28 gives the basic cost and cost of each stage of the product life cycle and the key parameters related to manufacturing defects to further support the development of modeling and optimization of the burn-in test.

The variation of manufacturing defect dimensions x under the burn-in effect over the product life cycle, given the burn-in duration b and warranty duration w, can be refined into a typical four-stage evolution as shown in Fig. 4.26.

In Fig. 4.26, Phase I and Phase II mainly monitor and characterize the changes in the state of x_1 and x_3 after the initial defect size x is truncated by the threshold value of x° in the factory testing session, and after the accelerated growth of the defect size in the burn-in test with the length of b. Meanwhile, based on the preset data listed in Table 4.28 and programmed by MATLAB, the growing size of x_3 at the end of Phase II has different morphological characteristics under different burn-in durations such as $b = 48$ and $b = 60$, as shown in Fig. 4.27.

In Fig. 4.27, the initial defect size x labeled by the green line is truncated to the size $x_1 \left(\leq x^0 = 450\right)$ labeled by the blue line due to the size exceeding the threshold x°, which corresponds to a higher and more concentrated defect density. Further, the

Table 4.28 List of basic data preset for a model of PCB board

Notation	Descriptive	Reference point
c_0	Failure cost for factory tested devices	850
c_1	Failure cost per unit of device use	2000
c_2	Fixed environmental preparation costs for burn-in tests	18
c_3	Time-dependent cost per unit time of burn-in	5
c_4	Failure cost per unit of device within burn-in time $x^0 e^{k_1 w} > x^* e^{k_2 b + k_1 w}$	60
x°	Critical defect threshold size (nanometer)	450
x^*	Median defect size (nanometer)	220
k_1	Proportionality coefficient (related to the growth rate of defect size at normal time stress)/h	2.5E−05
k_2	Proportionality coefficient (related to the growth rate of defect size at burn-in time stress)/h	0.01

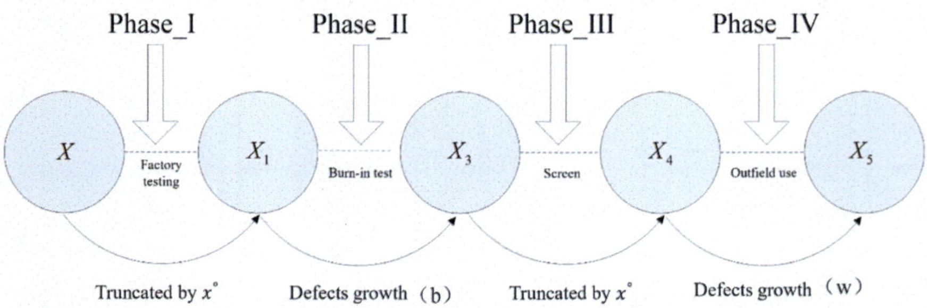

Fig. 4.26 Four-stage evolution of defect size under burn-in effect

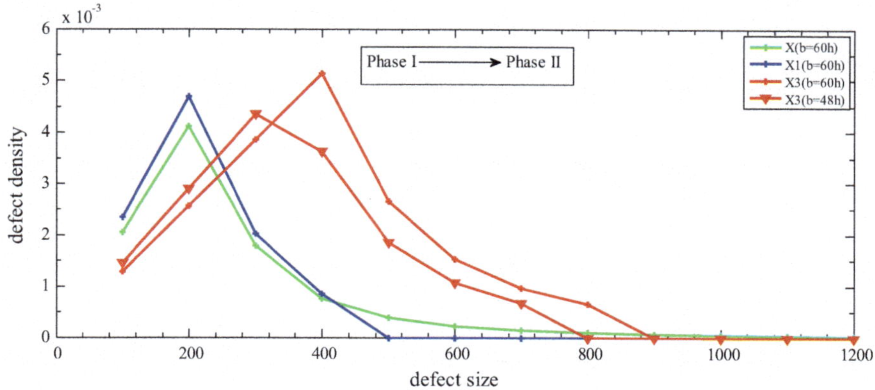

Fig. 4.27 Characterization of the evolution of the size x_3 for different burn-in durations of the Phase I–Phase II process

increased size of x_3 labeled by the red line corresponds to a larger size with higher defect density in a longer burn-in time $b = 60$ environment, and a larger defect size would be more likely to bring fatal defects leading to early failure.

Similarly, the evolution of the screened truncated dimensions x_4 in Phase III and Phase IV with different burn-in lengths and of the grown dimensions x_5 under usage conditions is characterized in Fig. 4.28.

In Fig. 4.28, the size distribution of the defects at the longer burn-in time $b = 60$ marked by the solid line is located farther to the right than that of the dashed line $b = 48$, where the density of defects takes a higher value, thus corresponding to a greater degree of early failure risk. The increase in the size of defects during burn-in at x_3, marked in red, is significantly higher than the increase in the size of defects during use at x_5, which further suggests that the burn-in environment has a stronger incentive to increase the size of defects than the normal use stress environment.

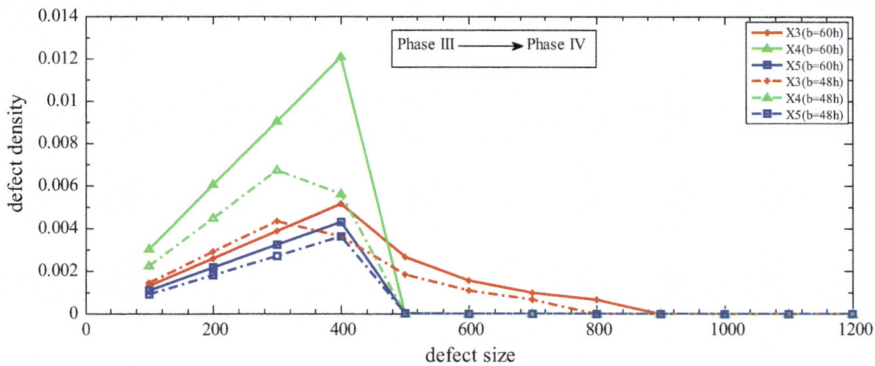

Fig. 4.28 Evolutionary characteristics of the dimensions x_4 and x_5 for different burn-in durations of the Phase III–Phase IV process

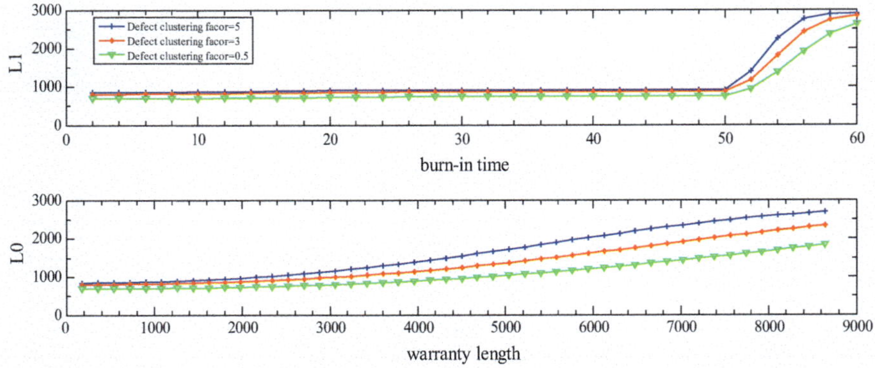

Fig. 4.29 Characterization of the dynamics of implicit quality loss L_1 and L_0 under different defect aggregation effects

Since the aggregation effect of defects acts on the yield loss related to Type-I fatal defects and the warranty cost related to Type-II non-fatal defects at the same time, the implicit quality deviation loss models L_1 and L_0 defined in this book take the burn-in time b and warranty time w as independent variables respectively, and then, by considering the aggregation effect of defects of different strengths and weaknesses at α, the dynamic simulation and analysis of the characteristics of the implicit quality loss model and with different lengths of warranty time b can be carried out respectively. The dynamic simulation and analysis of the change characteristics of L_0 under different levels of burn-in time L_1 and different lengths of warranty time w can be carried out respectively, as shown in Fig. 4.29.

In Fig. 4.29, the quality losses L_1 and L_0 are affected by different levels of defect aggregation effect α, and show significant differences in the longer burn-in time b and longer warranty time w parts, respectively. It is also shown that the smaller the defect aggregation effect, i.e., the larger the value of the defect aggregation factor α, the quality losses L_1 and L_0 are higher than the quality losses corresponding to the smaller value of the defect aggregation factor α, which further demonstrates the necessity of defect aggregation monitoring in quantifying the actual hidden quality deviation losses of the manufacturing process.

Based on the previous analysis and conclusion, and based on the objective function $g(b) = (L_0 - L_1) - C_b$ proposed in this book, we can finally carry out an in-depth analysis of the optimal refining time considering the role of different warranty lengths w. Here, the expressions of L_1, L_0 and C_b are as follows.

$$
\begin{aligned}
L_0 &= 850 * \left[1-(1-0.1195)^{N*0.1195}\right] + 2000 * \left[1-(1-p_2)^{(1-0.1195)Np_2}\right] \\
L_1 &= 850 * \left[1-(1-p_1)^{N*p_1}\right] + 2000 * \left[1-(1-p_4)^{(1-p_1)N(1-p_3)p_4}\right] \\
&\quad + 60 * \left[1-(1-p_3)^{(1-p_1)Np_3}\right] \\
C_b &= 18 + 5 * b
\end{aligned}
$$

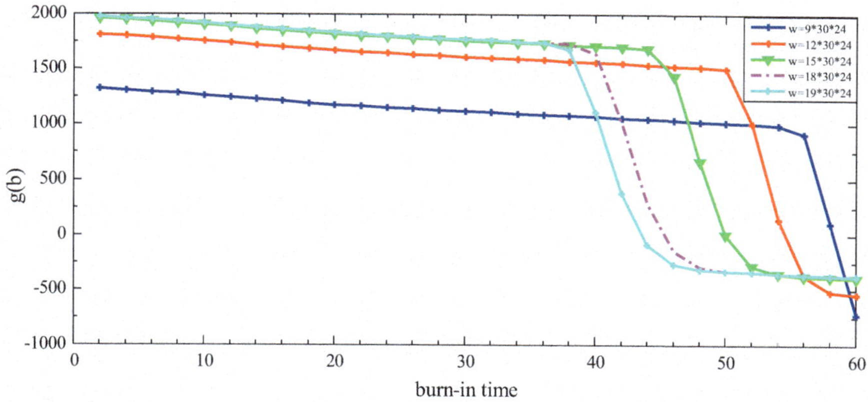

Fig. 4.30 Characteristics of changes in burn-in returns with different warranty lengths

where $p_1 = 0.1195$, and other parameters such as N, p_2, p_3 and p_4 can be estimated one by one according to the formulas given in the previous sections and have a more complex structure of expression.

With the help of MATLAB programming, the simulation of the dynamic characteristics of the burn-in gain $g(b)$ under different warranty lengths w can be developed as shown in Fig. 4.30. In the figure, the longer warranty length setting corresponds to the higher burn-in yield $g(b)$, and this phenomenon only lasts for the first 15 months. In the following 4 months, i.e., from month 15 to month 19, the difference in the refining yield $g(b)$ between different warranty lengths $g(b)$ is not significant. Meanwhile, overall, it seems that the $g(b)$ curve of the refining gain decreases rapidly after the refining duration exceeds 35 h, with the b curve corresponding to the longest warranty duration being the first one.

Based on Fig. 4.30, a continuous quadratic curve in the form of x° is fitted to the discrete burn-in gains x^* to further determine the optimal burn-in time k_1 and the corresponding burn-in gains k_2 for different warranty durations, as shown in Table 4.29.

Table 4.29 Quadratic fitting and optimal solution for burn-in gains $g(b)$

w	x	w				
3	5	w = 9*30*24	w = 12*30*24	w = 15*30*24	w = 18*30*24	w = 19*30*24
A		− 0.56972	− 1.3157	− 1.539	− 1.3408	− 1.2017
B		21.115	53.482	56.057	35.399	24.63
C		1101.3	1363.9	1541.4	1772.1	1872.8
Residual		1418.2	1772.2	1626.1	1758.9	1860.2
b^*		18.53	20.32	18.21	13.2	10.25
$g(b^*)$		1296.94	1907.4	2051.86	2005.75	1999

Comparing the value of the burn-in gain $g(b^*)$ in Table 4.29, it can be seen that fixing the length of the warranty at x has the largest burn-in gain, and corresponds to a burn-in time of the right length. For the boards in this book, the optimal burn-in time is given in the table x^0, and the associated burn-in gain is given in the table b.

Mathematically, the burn-in gain x_1 in the table shows different characteristics around the optimal length x_3. In particular, even if the burn-in gain $b = 48$ on the left side of x_3 becomes larger in spite of the increase in burn-in time, $b = 60$ has a higher value for a shorter burn-in time x_3.

Overall, the dynamic characteristics of the implicit quality deviation loss model in the burn-in environment are simulated in the arithmetic part by incorporating the size growth factor of defects, taking into account the deviation and cost effects associated with Type-I and Type-II manufacturing defects. After giving the four-stage evolution of defect size under different burn-in refining durations x, the dynamic monitoring of quality loss under different defect aggregation effects x^0, x_1 $\left(\le x^0 = 450\right)$ and x_3, respectively, is carried out, and it is concluded that the defect aggregation factor is an indispensable key parameter to accurately assess the implicit quality deviation loss. Then, after comparing the burn-in gains x_4 under different warranty lengths $b = 60$ and performing the quadratic fitting of the burn-in gains x_5, the optimal solution related to the burn-in test of the proposed computerized board is determined. At the same time, the paper gives suggestions on the actual warranty policy setting and product planning for practical manufacturing and reliability management decisions.

References

1. Yong Chen, Jionghua Jin, Jianjun Shi. Reliability modeling and analysis of multi-station manufacturing processes considering the quality and reliability interaction. IEEE, 2001, vol.4, 2093–2098.
2. He Y, He Z, Wang L, Gu C. Reliability Modeling and Optimization Strategy for Manufacturing System Based on RQR Chain. Mathematical Problems in Engineering, 2015, 2015(15):1–13.
3. Yu Ding, Jianjun Shi, Ceglarek Dariusz, Diagnosability Analysis of Multi-Station Manufacturing Processes. Journal of Dynamic Systems, Measurement and Control, 2002. vol. 124, no. 1, pp. 1–13.
4. Faping Zhang, Jiping Lu, Yan Yan, et al. Dimensional Quality Oriented Reliability Modeling for Complex Manufacturing Processes. International Journal of Computational Intelligence Systems, 2011, 4(6):1262–1268.
5. Guigui Fu, Yun Shangguan, Xingkuan Shi, et al. A process system reliability model based on produced product reliability. Journal of Beijing University of Aeronautics and Astronautics, 2009, 35(1):9–12.
6. Yihai He, Xiaoqing Tang. Product quality assurance design based on critical quality characteristics. Journal of Aeronautics, 2007, 28(6):1468–1481.
7. Yihai He, Changchao Gu, Zhenzhen He, Jiaming Cui. Reliability-oriented quality control approach for production process based on RQR chain. Total Quality Management & Business Excellence. 2018, 29(5–6):652–672.
8. Jun Lian, Zhongqin Lin, Xinmin Lai, et al. A Bayes method for dynamic estimation of pass rate in small samples. Journal of Shanghai Jiao Tong University, 2002, 36(8):1064–1067.

9. Minzhi Ruan, Qingmin Li, Yingwu Peng, et al. Availability assessment of device systems for multilevel maintenance supply under string-piece repair countermeasures. Journal of Aviation, 2012, 33(4):658–665.
10. Dehayem Nodem F I, Kenné J P, Gharbi A. Simultaneous control of production, repair/replacement and preventive maintenance of deteriorating manufacturing systems. International Journal of Production Economics, 2011, 134(1): 271–282.
11. Sarkar A, Panja S C, Sarkar B. Survey of maintenance policies for the Last 50 Years. International Journal of Software Engineering & Applications, 2011, 2(3): 130–149.
12. Madu C N. Competing through maintenance strategies. International Journal of Quality & Reliability Management, 2000, 17(9):937–949.
13. Zhou XJ, Xi LF, Lee J. Reliability-centered predictive maintenance scheduling for a continuously monitored system subject to degradation. Reliability Engineering & System Safety. 2007; 92(4):530–534.
14. Rooney, J. J., L. N. V., Heuvel. Root Cause Analysis for Beginners. Quality Progress, 2004, 29: 45–53.
15. Vidyasagar, A. The Art of Root Cause Analysis. Quality Progress, 2003, rube.asq.org.
16. Suh, N. P. Axiomatic design: advances and applications. New York: Oxford University Press, 2001.
17. Yihai He, Linbo Wang, Zhenzhen He, Min Xie, 2016. A fuzzy TOPSIS and rough set based approach for mechanism analysis of product infant failure. Engineering Applications of Artificial Intelligence. 47:25–37.
18. Linbo Wang. Research on analysis and optimization of product early failure rate considering manufacturing quality deviation. Master's thesis, Beijing University of Aeronautics and Astronautics, 2015.

Chapter 5
Reliability Control Approach of Manufacturing Process

5.1 Key Process Characteristics Identification Approach

5.1.1 Product Reliability Manufacturing Process Decomposition Technology

1. **Basic Connotation of Product Reliability in Manufacturing Process**

A product, from demand to delivery, must go through two stages: design and manufacturing. At present, the more mature reliability ideas in the academic circle are all aimed at the design stage. However, there is not enough research on the reliability of manufacturing process. Manufacturing process reliability can be defined as the ability of equipment (production line, process system, etc.) to maintain the indicators that guarantee product quality under specified time and environment. Its basic connotation is to control and improve some process indicators in the equipment (production line, process system, etc.), so that the product can meet the specified reliability requirements.

However, there are many process indicators involved in a manufacturing process, which are limited by the time and cost of product production, and it is impossible to comprehensively control and improve them. Therefore, this book proposes the model and method of product reliability index decomposition for manufacturing process. This method quickly determines the process improvement measures that have the greatest impact on product reliability through the five-step decomposition from system reliability to product structure, to manufacturing characteristics, to fault causes, and to process improvement measures, and then through the corresponding optimization methods. The optimal process improvement scheme is obtained under the specified time and capital constraints. Figure 5.1 shows the principle of this method.

Y. He et al., *Reliability Theory and Technology in Manufacturing Process*,
https://doi.org/10.1007/978-981-96-2735-6_5

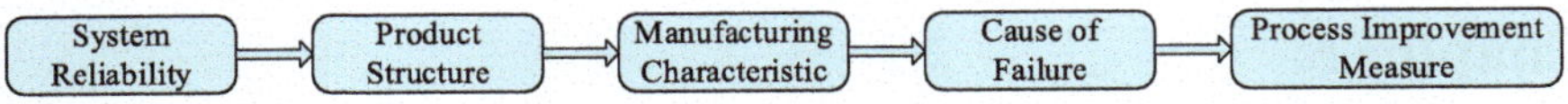

Fig. 5.1 Flowchart of product reliability breakdown in the manufacturing stage

2. Manufacturing Reliability Decomposition Model

In order to translate the product reliability index (MTBF in this book) into process improvement measures, we first apply the traditional reliability assignment to the subcomponents. Then, we will use the structure-manufacturing feature matrix to investigate the influence of manufacturing characteristics on the reliability of these subcomponents. Then, the manufacturing feature-fault cause matrix is used to determine the importance of each fault cause. Finally, the influence of process improvement measures on manufacturing characteristics is investigated by using manufacturing feature-process improvement measures matrix. Figure 5.2 shows the idea of the model. Through the transmission of the above four steps, we can get the impact of processing technology improvement measures on the overall reliability index of the product.

3. Decomposition Method and Result Optimization

(1) Decomposition Method

The decomposition method mentioned in this book uses the House of Mass waterfall decomposition model in QFD [1]. According to the four transformation steps

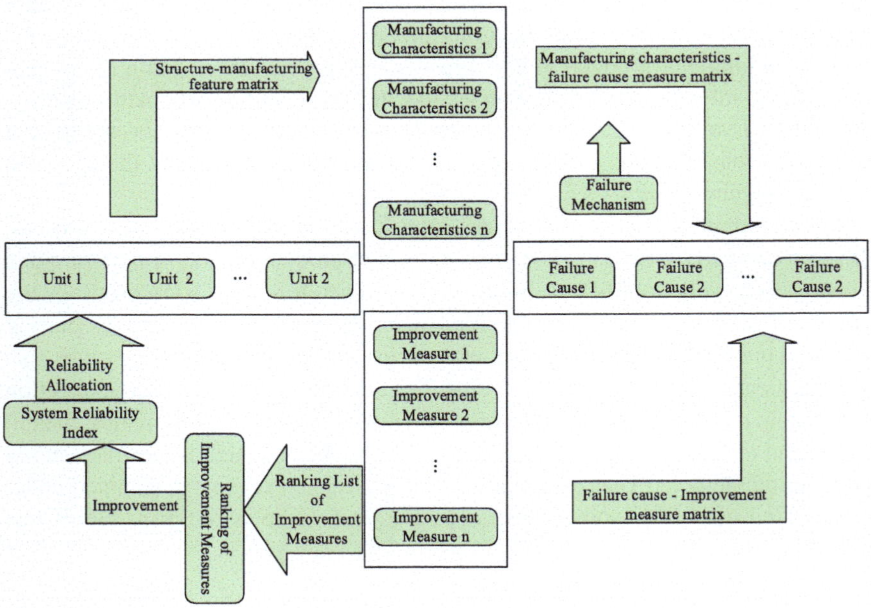

Fig. 5.2 Conversion diagram of reliability index and process improvement measures

of decomposition, four transformation matrices are established, and the reliability index is transferred to the process improvement measures by matrix multiplication. Different products will have different manufacturing characteristics and process improvement measures. Here the book takes the general machining process as an example to elaborate the specific decomposition method.

(i) With the total reliability index of the product as input and the product structure as output, the reliability index of the product (MTBF in this book) is assigned to the appropriate sub-components. The method can be referred to the relevant content of reliability allocation in product reliability design. After that, according to the specific distribution results and corresponding engineering experience, the importance of reliability of each component is scored by experts (the score range is 1–10), and the scoring results are filled into the first matrix.

(ii) Take the structure of the product as the input and the manufacturing characteristics of the product as the output to complete the transmission from the product structure to the manufacturing characteristics. According to the classification of machining manufacturing characteristics in the book "Technical Specifications for Reliability, Maintainability and Support of Model Equipment" edited by Kang Rui, I divided the main manufacturing feature requirements into four parts: size, shape, position and surface characteristics, and the corresponding failure modes are respectively size error, shape error, position error, and surface quality unqualified. For different parts, the requirements for these four aspects are often different. We complete the second matrix according to the influence of different manufacturing characteristics on the reliability of different components.

(iii) Take the fault mode of the product manufacturing characteristics as the input, and the fault cause leading to this fault mode as the output, and transfer the reliability from the manufacturing characteristics to the processing technology, all the process defects here are still based on the "Model Equipment Reliability, maintainability and support Technical specifications" edited by Kang Rui (a total of 18 major fault modes). In this matrix, let i = 1, 2, 3, 4; j = 1, 2, … 18 is the degree of influence of the JTH fault cause on the fault mode of the i manufacturing feature. The size can also be 0, 1, 3, 5, 7, 9. Where 0 indicates that the fault cause has no influence on the manufacturing feature fault mode, and 1–9 indicates that the fault cause has very light, light, average, heavy, and extremely heavy influence on the manufacturing feature fault mode. The rating for this part is given by experts or based on engineering experience.

(iv) With manufacturing feature fault mode as input and improvement measures as output, the manufacturing feature fault model-improvement measures matrix is established, the purpose of which is to determine which improvement measures can best improve product reliability. Let i = 1, 2, …, 18; j = 1, 2, … 18 is the improvement degree of the jth improvement measure to the ith fault mode. The score can be 0, 1, 3, 5, 7, 9. Where 0 indicates that the improvement measure has no impact on the fault mode, and 1–9 indicates that the improvement measure has very light, light, average, heavy, and extremely heavy impact on the fault mode. The rating for this part is given by experts or based on engineering experience.

(v) Calculate the four matrices obtained. We use matrix multiplication: $\left[P_1\ P_2\ \cdots\ P_{11}\ P_{12} \right] = A \times W \times G \times F$. Finally, the P-matrix obtained is the improvement measure matrix. And P_j is the weight of improvement measures for item j. Next, we require the relative improvement measure weight P'_j.

Then, we included the relative improvement measures weights in Table 5.1 to get the ranking of the importance of improvement measures to improving product reliability.

(2) Optimization Method

The above ranking of improvement measures is only the result of the ideal situation, and does not take into account the financial and time constraints. This book will use multi-layer fuzzy comprehensive evaluation and combined with planning methods to find several process improvement measures that contribute the most to the improvement of product reliability in practical engineering. Method is as follows:

(i) Establish factor set. For the evaluation factors of improvement measures, there are mainly three aspects: the degree of improvement of product reliability, the funds required for improvement measures and the time required for improvement measures. Accordingly, the factor set is set as $U = \{u_1, u_2, u_3\}$, where u_1 is the degree of improvement of product reliability, u_2 is the funds required for improvement measures and u_3 is the time required for improvement measures.
(ii) Given weight vector. According to different engineering tasks, the weight vector is given as $\underline{A} = (a_1, a_2, a_3)$.
(iii) Determine the evaluation set.$V = \left\{v_1, v_2, \ldots, v_j\right\}$ is established for different process improvement measures.

Table 5.1 Weight table of process improvement measures to improve product reliability

Improvement measure	Relative improvement measure weight
First roughing, then finishing	0.057
Select the right cutting amount	0.066
Select the right tool	0.053
Add coolant	0.051
Choose a reliable way of clamping	0.066
Align or align a part or fixture to an acceptable level	0.019
Arrange aging treatment between processes	0.025
Thick toolbar	0.044
Eliminate machine errors	0.019
Set the correct coordinates	0.014
Install vertically or parallel	0.050
Reduce machining speed	0.035

(iv) Establish evaluation matrix $\underset{\sim}{R}$.
(v) Comprehensive evaluation calculation.

Through the above five processes, the optimal process improvement measures can be obtained, taking into account the impact of money and time.

4. **Case Study**

The component "shell assembly" of a missile rocket engine is welded by five parts, including shell cylinder, front-end ring, back-end ring, wing and fixed piece. Based on the reliability decomposition model and method presented in this book, we will investigate which process improvements can most efficiently improve the reliability of this "shell assembly".

Using the method mentioned above, the product improvement measure matrix is obtained through the decomposition model of "product reliability → product structure → manufacturing characteristics → fault cause → improvement measure".

$$\begin{aligned}\left[P_1\ P_2\ \cdots\ P_{11}\ P_{12}\right] &= A \times W \times G \times F\\ &= [\,159{,}828\ 186{,}008\ 148{,}408\ 144{,}764\\ &\quad 185{,}452\ 53{,}460\ 70{,}440\ 124{,}452\\ &\quad 53{,}388\ 41{,}544\ 140{,}436\ 98{,}056\,]\end{aligned}$$

According to the improvement measures matrix, the weight table of process improvement measures aimed at improving product reliability can be obtained (Table 5.1).

Next, consider the time and cost factors, and evaluate the results comprehensively.

The time and cost of each process improvement measure are scored (1–10), and the time–cost table of each process improvement measure is shown in Table 5.2

As mentioned above, the factors of fuzzy comprehensive judgment are determined as the degree of reliability improvement, time and cost. Since the missile is a high-tech product of national defense and has relatively high requirements for reliability, the weight vector of the three factors is determined as $\underset{\sim}{A} = (0.8, 0.1, 0.1)$. The evaluation sets of these three factors correspond to Tables 5.1 and 5.2.

The relation between factor set and evaluation set can be represented by fuzzy relation matrix through membership function. Among them, the membership function of the degree of reliability improvement is:

$$\mu(u_1) = \begin{cases} 0 & u_1 \le 0.01 \\ \frac{u_1 - 0.01}{0.07 - 0.01} & 0.01 \le u_1 \le 0.07 \\ 1 & u_1 \ge 0.07 \end{cases}$$

Similarly, the membership functions of time and cost are:

Table 5.2 Time–cost table of improvement measures

Improvement measure	Time increment L_0	Cost increment L_1
First roughing, then finishing	8	5
Select the right cutting amount	2	1
Select the right tool	1	4
Add coolant	1	3
Choose a reliable way of clamping	4	2
Align or align a part or fixture to an acceptable level	5	1
Arrange aging treatment between processes	5	6
Thick toolbar	1	4
Eliminate machine errors	3	1
Set the correct coordinates	3	1
Install vertically or parallel	5	1
Reduce machining speed	8	7

$$\mu(u_2) = \begin{cases} 0 & u_2 \le 0.01 \\ \frac{u_2-1}{10-1} & 1 \le u_2 \le 10 \\ 1 & u_2 \ge 10 \end{cases}, \quad \mu(u_3) = \begin{cases} 0 & u_3 \le 0.01 \\ \frac{u_3-1}{10-1} & 1 \le u_3 \le 10 \\ 1 & u_3 \ge 10 \end{cases}.$$

The membership degree of improvement scheme 1 "roughing first, finishing later" to improve product reliability is as follows:

$$r_{11} = \mu(u_1) = \frac{0.057 - 0.01}{0.07 - 0.01} = 0.783.$$

Similarly, $r_{21} = \mu(u_2) = \frac{8-1}{10-1} = 0.778$, $r_{31} = \mu(u_3) = \frac{5-1}{10-1} = 0.444$.

According to the similar method, the membership degree of other schemes can be obtained, and the following evaluation matrix can be obtained after combining.

$$\underline{R} = \begin{bmatrix} 0.783 & 0.993 & 0.717 & 0.683 & 0.933 & 0.15 & 0.25 & 0.567 & 0.15 & 0.067 & 0.067 & 0.417 \\ 0.778 & 0.111 & 0 & 0 & 0.333 & 0.444 & 0.444 & 0 & 0.222 & 0.222 & 0.444 & 0.778 \\ 0.444 & 0 & 0.333 & 0.222 & 0.111 & 0 & 0.556 & 0.333 & 0 & 0 & 0 & 0.667 \end{bmatrix}$$

The method of fuzzy transformation is used to calculate the comprehensive evaluation.

$$b_j = \overset{3}{\underset{i=1}{\vee}} \left(a \wedge r_{ij}\right) \; (j = 1, 2, \ldots, 12)$$

$$\underline{B} = \underline{A} \times \underline{R} = (0.783, 0.8, 0.717, 0.683, 0.8, 0.15, \\ 0.25, 0.567, 0.15, 0.1, 0.667, 0.417).$$

From the calculation results, it can be seen that the effects of the third and sixth process improvement measures are optimal.

5.1.2 Product Reliability and Process Dimension Correlation Model

In the theory of deviation flow analysis, it is generally believed that the structure D of the rigid part, the size X of the product, the material M of the product and some other design factors O greatly affect the reliability of the part, and the reliability R of the product is a function of D, X, M, O. As follows:

$$R = F_{D,X,M\sim R}(D, X, M, O) \tag{5.1}$$

In the product design concept stage, when collecting customers' demand for product reliability, after corresponding reasonable analysis, the reliability design requirements of the product established in the design finalize can be marked as R_{d}. According to the CAD model of the product given in the detailed design, the geometric size data of the product can be obtained, which is recorded as X_{d}. At the same time, the material required by the design is, the product structure of the design is, and the other factors are recorded as O_{d}. According to Eq. (5.1), the formula is as follows:

$$R_d = F_{X\sim R}(D_d, X_d, M_d, O_d) \tag{5.2}$$

In the process of batch production, the processing is completed after m stations. The size of the product is X_e, the product structure is D_e, the material used is M_e. Other design factors are the reliability R_{e} of the O_{e} product.

$$R_e = F_{X\sim R}(D_e, X_e, M_e, O_e) \tag{5.3}$$

Comparing Formula (5.3) with Formula (5.2), ΔR is the deviation between the reliability of the manufactured product and the reliability required by the design.

$$\Delta R = R_e - R_d \tag{5.4}$$

$$R_e = R_d + \Delta R \tag{5.5}$$

$$\Delta R = F_{D,X,M\sim R}(D_e, X_e, M_e, O_e) - F_{D,X,M\sim R}(D_d, X_d, M_d, O_d) \tag{5.6}$$

Formula (5.3) indicates that there is a deviation between the reliability of the product and the reliability required by the design. Formula (5.4) indicates that the reliability deviation ΔR is caused by the difference in the size of the product and the design size, the subtle changes in the product structure during the manufacturing process, and the defects of the material.

Use d to represent the products were produced in the deviation between the structure and design of structure, x represents the size of the products were produced in the deviation between the size and design, m represents the products were produced in the deviation between the material and the design requirements of material, o represents the design of the products were produced in the other factors and the deviation of the design type 5.4 can be written as:

$$\begin{aligned} \Delta R &= F_{d,x,m\sim\Delta R}(d, x, m, o) \\ &= F_{D,X,M\sim R}(D_d + d, X_d + x, M_d + m, O_d + o) \\ &\quad - F_{D,X,M\sim R}(D_d, X_d, M_d, O_d) \end{aligned} \tag{5.7}$$

In the batch production process, because the product has been in the design finalize state, the structure of the product basically does not change, that is, d equals 0. The other design factor o is basically unchanged, that is, o equals 0. In the case of the selected supplier, the source of the material does not change much, and the change of the material is also small, that is, m equals 0. The deviation x between the manufactured product size and the size required by the design contributes greatly to the deviation between the reliability of the product and the reliability required by the design. In this context, Eq. (5.7) can be written as:

$$\Delta R = F_{x\sim\Delta R}(x) + C + \nu \tag{5.8}$$

Formula (5.8) considers that the reliability of the product in the batch manufacturing stage is determined by the size of the product. Here, v is the system noise, C is the constant, the same below. As mentioned above, the material can not fully meet the design requirements, the process design defects, resulting in the batch production process may not fully complete the design requirements, the reliability of the manufactured product and the design requirements there is a deviation between the C constant to compensate. Usually C is negative. On the other hand, the randomness of the size deviation x and the noise of the system cause the reliability of the product to fluctuate. According to Eq. (5.9), the deviation flow model can be written as:

$$x_k = F_{x_k\sim x_{k-1}}\left(x_{k-1}, e_k^s, e_k^m\right) + \nu \tag{5.9}$$

where e_k^s is the positioning deviation of the k processing station, e_k^m is the processing deviation of the k processing station.

Product fluctuations are caused by the accumulation of fluctuations in the various processes:

$$x_j = F_{x_i \sim x_j}\left(x_i, e^s_{i+1}, \ldots, e^s_j, e^m_{i+1}, \ldots, e^m_j\right) + \nu \quad i < j \tag{5.10}$$

The dimensional deviation x of the product is the dimensional deviation x_{m} of the product process after the completion of the last m station. Combined with Formulas (5.8) and (5.10), the following formula is obtained:

$$\begin{aligned}\Delta R &= F_{x \sim \Delta R}(x) + C + \nu \\ &= F_{x_m \sim \Delta R}(x_m) + C + \nu \\ &= F_{x_i \sim \Delta R}\left(x_i, e^s_{i+1}, \ldots, e^s_m, e^m_{i+1}, \ldots, e^m_m\right) + C + \nu\end{aligned} \tag{5.11}$$

The difference ΔR between the reliability of the manufactured product and the design requirements is caused by the dimensional deviation of the product processing process, the positioning deviation of the downstream manufacturing process, and the processing deviation and the manufacturing system itself can not meet the design requirements. Specifically, in Formula (5.11), x_i represents the process size deviation state of the product after the completion of processing at station i. $e^s_{i+1}, \ldots, e^s_m$ represents the positioning deviation from the downstream i + 1 station to the m station. $e^m_{i+1}, \ldots, e^m_m$ represents the corresponding downstream station processing deviation. C represents the loss of reliability due to the failure of the manufacturing system itself to meet the design requirements, material and process design defects.

According to the mathematical model [2, 3] established by Thornton on the impact of key characteristics of subsystems on system characteristics, when the size deviation of products is small, Taylor expansion is used and second-order terms are ignored. Formula (5.8) has a linear form. At the same time, the deviation flow model normalized by Eq. (2.19) is linear. Thus, Eq. (5.11) has a linear form when the process size deviation is small. As mentioned above, MTBF is used to describe the reliability of the product, which is expressed linearly by Eqs. (5.8) and (5.11) as follows:

$$\Delta MTBF = Lx + C + \nu \tag{5.12}$$

$$\begin{aligned}\Delta MTBF &= L_i x_i + \sum_{j=i+1}^{m} Q_j e^s_j + \sum_{j=i+1}^{m} P_j e^m_j + C + \nu \\ &= L_i x_i + B' u + C + \nu \\ &= L_i x_i + \sum_{j=i+1}^{m} b_j u_j + C + \nu\end{aligned} \tag{5.13}$$

where $u_k = [e^s, e^m]^T$.

Formula (5.12) is the correlation model between product reliability and product manufacturing process size based on deviation flow.

When building the correlation model between product reliability and product manufacturing process size, it is only necessary to determine the parameter vector L

in Eq. (5.12) and the parameter matrix $A(K)$ and $B(K)$ in Eq. (2.19), then L_i, B' in Eq. (5.13) can be obtained according to the recurrence formula. When determining the parameter vector L, it is necessary to analyze the specific product. Thornton pointed out that the product system can be decomposed, the impact of subsystems on the upper system can be investigated, the impact matrix can be established, and the impact of the characteristics of the lowest subsystem on the system characteristics can be calculated. Specifically in this book, the impact of product size state on product reliability can be calculated, that is, the value L can be given. Thornton believes that for products whose subsystems have a clear influence on the upper system, although the influence matrix dimension is relatively high, most of the influence matrices are sparse matrices, so the influence matrix can be filled. When the influence of the sub-system on its upper system is not clear, or the stratification of the system is complicated, the influence matrix can be established by slightly disturbing the sub-system and observing the change of the system characteristics. This book will use the experimental design method, which believes that each size of the product is a factor of the experiment, and obtain the average trouble-free working time of the product in different size deviation states through finite element simulation technology, and then carry out variance analysis and main effect analysis to obtain L value.

The responsibility of the manufacturing system is to follow the process specifications and produce the product with as little fluctuation as possible. For product reliability, according to Formula (5.13), the expectation of the average trouble-free working time of the manufactured product is as follows:

$$E(MTBF_e) = MTBF_d + C \tag{5.14}$$

Products are produced in the average trouble-free working time variance, which make the product of the average volatility of trouble-free working hours is as follows:

$$\nu(MTBF_e) = L_i^2 \nu(x_i) + \sum_{j+i+1}^{m} b_j^2 \nu(u_j) \tag{5.15}$$

On the other hand, when the processing of the ith station is completed, the product process size deviation x_i is measured. From Formula (5.13), it can be obtained that the deviation between the reliability of the product after processing and the expected average trouble-free working time of the product produced by the manufacturing system is as follows:

$$MTBF_\delta = L_i x_i + \sum_{j=i+1}^{m} b_j u_j + \nu \tag{5.16}$$

The presence of system white noise is noted and its expectation is:

$$\overline{MTBF_\delta} = L_i x_i + \sum_{j=i+1}^{m} b_j u_j \tag{5.17}$$

Equation (5.13) describes the fluctuation of product mean trouble-free working time caused by manufacturing process from a statistical point of view. Formula (5.17) shows the deviation of the mean trouble-free working time of a specific product. The fluctuation of product reliability can be described by the fluctuation of product mean trouble-free working time. Deviations in product reliability can be described using deviations in product mean trouble-free operating time. To control the fluctuation of product reliability, we should start from controlling the mean fault free working time deviation of each product, that is, reduce $\nu(MTBF_e)$ by reducing $\left|\overline{MTBF_\delta}\right|$, so as to reduce the reliability fluctuation $\nu(R_e)$.

5.2 Manufacturing Defect Detection Technology

5.2.1 Non-destructive Testing

Non-destructive testing (NDT) is a technique for detecting defects, chemical and physical parameters of materials, parts and equipment by means of X-ray, ultrasonic, infrared, electromagnetic and other technical instruments without damaging or affecting the performance of the tested object. The principle of non-destructive testing is to use the acoustic, optical, magnetic and electrical characteristics of the substance to detect whether there are defects or non-uniformity in the tested object without damaging or affecting the performance of the tested object, and give the size, location, nature and quantity of the defect information. Common nondestructive Testing methods are: Visual Testing (VT); Ultrasonic Testing (abbreviation UT); Radiographic Testing (RT); Magnetic particle Testing (MT); Penetrant Testing (PT); Penetrant testing Eddy Current Testing (ET); Acoustic emission (AE); Ultrasonic Diffraction Time Of Flight Diffraction (TOFD), etc.

Nondestructive testing has the following characteristics:

(1) Non-destructive, because it will not damage the performance of the tested object during detection;
(2) Comprehensive, because the detection is non-destructive, 100% comprehensive detection of the tested object can be carried out if necessary, which is impossible for destructive detection;
(3) With the whole process, destructive testing is generally only applicable to the testing of raw materials, such as tensile, compression, bending, etc., commonly used in mechanical engineering. Destructive testing is carried out for raw materials used in manufacturing. For finished products and supplies, unless they are not prepared to continue to serve, destructive testing cannot be carried out. Non-destructive testing does not damage the performance of the tested object.

Therefore, it can not only test the raw materials used in manufacturing, the intermediate process links, until the final product, but also test the equipment in service.

There are many non-destructive testing methods, according to NASA research and analysis, which can be divided into six categories of about 70 kinds. However, there are the following common ones in practical applications:

(1) Visual Inspection (VT)

Visual testing in the domestic implementation of relatively little, but in the international attention to the first stage of non-destructive testing of the primary method. According to international practice, visual testing should be done first to confirm that it will not affect the subsequent inspection, and then do the four routine tests. For example, BINDT's PCN certification has special VT1, 2, and 3 levels of assessment, and has special certification requirements. VT is often used for visual inspection of the weld, the weld itself has a process evaluation standard, can be done by visual inspection and direct measurement of the size of the preliminary inspection, found that the edge of the unqualified appearance defects, it is necessary to polish or repair, and then do other in-depth instrument testing. For example, the surface of the welded part and the surface of the casting more VT do more, and the forging is very little, and its inspection standards are basically consistent.

(2) Radiography (RT)

Radiography refers to the non-destructive testing method that uses X-ray or gamma ray to penetrate the specimen and uses film as the equipment for recording information. This method is the most basic and widely used non-destructive testing method. X-rays can penetrate substances that cannot be penetrated by the naked eye to make the film sensitive. When X-ray or R-ray irradiates the film, it can make the silver halide in the film emulsion layer produce latent shadows, just like ordinary light. Due to the different absorption coefficients of substances with different densities to the rays, the intensity of the rays irradiating to different parts of the film will also vary. Defects can be identified according to the blackness difference of the film after darkroom treatment. In general, the qualitative accuracy of RT is more accurate, there are intuitive images that can be preserved for a long time, the overall cost is relatively high, and the radiation is harmful to the human body, and the inspection speed will be slower.

(3) Ultrasonic Detection (UT)

Through the interaction between ultrasonic wave and specimen, the reflected, transmitted and scattered waves are studied, and the specimen is detected and characterized by macroscopic defect detection, geometric property measurement, organizational structure and mechanical property change, and then its specific application is evaluated. It is suitable for non-destructive testing of metal, non-metal and composite materials. It can detect the internal defects of specimens in a wide thickness range. For metal materials, thin wall pipes and plates with a thickness of 1 ~ 2 mm can be detected, and steel forgings several meters long can also be detected. Moreover, the defect location is more accurate, and the detection rate of area defects is higher. High sensitivity, can detect the internal size of the specimen is very small defects; And the detection cost is low, the speed is fast, the equipment is light, harmless to the human body and the environment, and the field use is more convenient. However, it is difficult to detect the specimen with complex shape or irregular shape. In addition, the location, orientation and shape of the defect, as well as the material and grain size of the defect all have a certain influence on the detection results, and the detection results have no direct witness records.

(4) Magnetic Particle Detection (MT)

After the ferromagnetic material and the workpiece are magnetized, due to the existence of discontinuity, the magnetic field on the surface and near the surface of the workpiece is locally distorted and the leakage field is generated, the magnetic powder applied on the surface of the workpiece is adsorbed, and the magnetic marks are visually visible under the appropriate light, thus showing the position, shape and size of the discontinuity. Magnetic particle detection is suitable for detecting discontinuities that are difficult to see visually on the surface and near the surface of ferromagnetic materials with very small size and very narrow gap (such as cracks 0.1 mm long and micron wide can be detected); It can also detect raw materials, semi-finished products, finished workpieces and parts in service, and can also detect plates, profiles, pipes, bars, welding parts, steel casting and forged steel parts, and find defects such as cracks, inclusions, hair lines, white spots, folding, cold insulation and porosity. However, magnetic particle detection cannot detect austenitic stainless steel materials and welds welded with austenitic stainless steel welding rods, nor can it detect non-magnetic materials such as copper, aluminum, magnesium, and titanium. It is difficult to find shallow scratches on the surface, deep holes buried and layering and folding with an Angle of less than 20° from the workpiece surface.

(5) Penetration Testing (PT)

After the surface of the part is coated with penetrant containing fluorescent dye or colored dye, the penetrant can penetrate into the surface opening defect after a period of time under the action of capillary; After removing the excess penetrant on the surface of the part, the developer is applied on the surface of the part. Similarly, under the action of capillary, the developer will attract the penetrant retained in the defect, and the penetrant will permeate back into the developer. Under a certain light source (ultraviolet light or white light), the penetrant trace at the defect will be realistic (yellow-green fluorescence or bright red). The morphology and distribution of defects are detected.

Penetration testing can detect a variety of materials, metal, non-metal materials; Magnetic and non-magnetic materials; Welding, forging, rolling and other processing methods; It has high sensitivity (0.1 μm wide defects can be detected), and has intuitive display, easy operation and low detection cost. However, it can only detect the defects of the surface opening, and is not suitable for inspecting the workpiece made of porous and loose materials and the workpiece with rough surface; Only the surface distribution of defects can be detected, and it is difficult to determine the actual depth of defects, so it is difficult to make quantitative evaluation of defects, and the detection results are greatly affected by the operator.

(6) Eddy Current Testing (ECT)

The coil with alternating current is placed on the metal plate to be measured or set outside the metal tube to be measured, then the alternating magnetic field will be generated in and near the coil, causing the induction alternating current in the specimen to be swirled, which is called eddy current. The distribution and size of eddy current, in addition to the shape and size of the coil, the size and frequency of the AC current, also depends on the conductivity, permeability, shape and size of the specimen, the distance from the coil and the surface of the crack defects. Therefore, under the condition of keeping other factors relatively unchanged, using a probe coil to measure the magnetic field changes caused by eddy currents can infer the changes in the size and phase of eddy currents in the specimen, and then obtain information about the changes in electrical conductivity, defects, material conditions and other physical quantities (such as shape, size, etc.) or the existence of defects. However, because the eddy current is alternating current and has skin effect, the detected information can only reflect the situation on or near the surface of the specimen.

According to the shape of the specimen and the purpose of detection, different forms of coils can be used, usually through the type, probe type and inserted coil 3 kinds. Through coil is used to detect pipe, bar and wire, its inner diameter is slightly larger than the object to be tested, when used to make the object to be tested at a certain speed through the coil, cracks, inclusions, pits and other defects can be found. The probe coil is suitable for local detection of the specimen. When the coil is placed on the metal plate, tube or other parts, it can check the fatigue cracks on the inner barrel of the aircraft take-off and landing strut and on the turbine engine blades. The insert coil, also known as the internal probe, is placed in the hole of the pipe or part for inner wall detection, which can be used to check the degree of corrosion of the inner wall of various pipes. In order to improve the detection sensitivity, probe and insert coils are mostly equipped with magnetic cores. The eddy current method is mainly used for the rapid detection of metal tubes, rods and lines on the production line, as well as the detection of large quantities of parts such as bearing steel balls and valves (then in addition to the eddy current instrument must be equipped with automatic loading and unloading and conveying mechanical devices), material sorting and hardness measurement, and can also be used to measure the thickness of the coating and coating film. In eddy current testing, the coil does not need to be in direct contact with the measured object, which can be detected at high speed and is easy to be automated, but it is not suitable for parts with complex shapes, and can only detect surface and near-surface defects of conductive materials, and the detection results are also easy to be interfered by the material itself and other factors.

(7) Acoustic Emission (AE)

A nondestructive testing method for assessing the properties or structural integrity of a material by receiving and analyzing its acoustic emission signals. The stress wave phenomenon caused by rapid release of strain energy due to crack expansion, plastic deformation or phase transition in materials is called acoustic emission. In 1950, J. Kaiser of the Federal Republic of Germany conducted a systematic study on the acoustic emission phenomenon in metals. In 1964, the United States first applied acoustic emission detection technology to the quality inspection of rocket engine housing and achieved success. Since then, acoustic emission detection methods have developed rapidly. This is a new non-destructive testing method, through the sound of crack expansion inside the material to detect. It is mainly used to detect the defects of equipment and devices in use, that is, the development of defects, to judge its good performance.

Acoustic emission technology has been widely used. Acoustic emission can be used to identify different types of deformation, study the fracture process and distinguish the fracture mode, detect crack growth less than 0.01 mm long, study stress corrosion fracture and hydrogen embrittlement, detect martensitic phase transition, evaluate the brittleness of the surface chemical heat treatment infiltration layer, and monitor the formation and propagation of cracks after welding, etc. In industrial production, acoustic emission technology has been used for hydraulic inspection of large components such as pressure vessels, boilers, pipelines and rocket engine shells to assess the risk level of defects and make real-time alarms. During production, PXWAE acoustic emission technology enables continuous monitoring of the integrity of components such as high-pressure vessels, nuclear reactor vessels and subsea oil recovery units. Acoustic emission technology is also used to measure the combustion rate of solid rocket engine powder and study the combustion process, detect leakage, study rock fracture, monitor mine collapse, and forecast mine safety.

(8) Ultrasonic Diffraction Time Difference Method (TOFD)

TOFD technology was first proposed in the 1970s by Dr. Silk of the National Nondestructive Testing Center in Harwell, UK, and its principle stems from Dr. Silk's research on the diffraction signal of the crack tip. In the same period, the Chinese Academy of Sciences also detected the diffraction signal of the crack tip, and developed a set of technology methods for crack height measurement, but did not develop the common TOFD detection technology. TOFD technology is first of all a detection method, but the instrument that can meet the requirements of this detection method has not been developed. The details are explained in the next section. TOFD requires the probe to achieve sufficient signal-to-noise ratio when receiving weak diffraction waves, the instrument can record the A-sweep waveform throughout the whole process, form the D-scan pattern, and can convert the A-sweep time value into the depth value by solving the triangle method. During the same period, the technical level of industrial inspection has not reached the level that can meet these technical requirements. Until the 1990s, when the development of computer technology made the digital ultrasonic flaw detector mature, it was possible to develop a portable and cost-effective TOFD detector. But even so, there are significant technical differences between TOFD instruments and ordinary A-super instruments. It is a method to detect defects by relying on the diffraction energy obtained from the "end Angle" and "end point" of the internal structure of the specimen to be tested (mainly referring to the defect), which is used for the detection, quantification and location of defects.

5.2.2 *Physicochemical Detection Technology*

Physical and chemical testing is the use of physical and chemical properties of substances, using a variety of test and analysis methods, the organization of substances, structure, components and their distribution, damage and failure forms to analyze, and study its detection methods and provide relevant information of the technical science. Such as: mechanical properties test, microstructure, chemical composition and component analysis, damage and failure analysis and detection technology.

(1) Mechanical Properties Test

Mechanical property testing technology is an experimental technology to measure various mechanical property criteria (indexes) of the tested object through different tests. Mechanical properties of materials refer to the mechanical characteristics of materials under different environments (temperature, medium, humidity), under various external loads (tensile, compression, bending, torsion, impact, alternating stress, etc.), generally speaking, the mechanical properties of metals include: Brittleness, strength, plasticity, hardness, toughness, fatigue strength, elasticity, ductility, rigidity, yield point or yield stress, etc.

(2) Physical Performance Test

Physical properties test includes density (body density, surface density, linear density), viscosity (viscosity coefficient), particle size, melting point, boiling point, freezing point, ignition point, flash point, thermal conductivity (specific heat, thermal conductivity, linear expansion coefficient), electrical conductivity (resistivity, conductivity, resistance temperature coefficient), magnetic properties (magnetic induction intensity, magnetic field strength, coercivity, iron loss) and other physical properties Performance testing.

(3) Chemical Analysis

Analysis based on chemical reactions of substances is called chemical analysis. Chemical analysis has a long history, is the basis of analytical chemistry, also known as classical analysis. Chemical analysis is absolutely quantitative, according to the amount of sample, the amount of reaction products or the amount of reagents consumed and the stoichiometric relationship of the reaction, by calculating the amount of components to be measured.

Chemical analysis can be divided into titration analysis (titrimetry) and gravimetry (gravimetry) according to the different methods of operation. In recent years, another analytical concept has been formed in China, which is called "microspectral analysis" technology.

(i) Titration analysis: According to the concentration and volume of the standard solution consumed by titration and the chemical reaction relationship between the measured substance and the standard solution, the content of the measured substance is found. This analysis is called titration analysis. Also known as volumetry. The use of solution 4 major equilibria: acid-base (ionization) balance, REDOX balance, complexation (coordination) balance, precipitation-dissolution balance.

(ii) Gravimetric analysis: According to the chemical properties of the substance, select the appropriate chemical reaction, the measured components into a fixed composition of precipitation or gas form, through passivation, drying, burning or absorbent absorption and a series of processing, accurate weighing, to find the content of the measured components, this analysis is called gravimetric analysis.

(4) Metallographic Microanalysis

Metallographic analysis is one of the important means of experimental research of metal materials. Based on the principle of quantitative metallography, the three-dimensional spatial morphology of the alloy structure is determined by measuring and calculating the microstructure of the two-dimensional metallographic sample surface or film, so as to establish the quantitative relationship between the alloy composition, microstructure and properties. The application of computer to image processing has the advantages of high precision and fast speed, which can greatly improve the work efficiency. Metallographic microanalysis is the use of metallographic microscope to observe the metallographic sample specially prepared for analysis, by magnifying tens of times to thousands of times to study the organization of materials. The main instruments for modern metallographic microanalysis are optical microscope and electron microscope.

When metallographic microscope is used to examine and analyze the microstructure of materials, it is necessary to prepare the analyzed materials into a sample of a certain size, and through grinding, polishing and corrosion processes, in order to carry out the observation and research of the microstructure of materials. The preparation process of metallographic samples generally includes the following steps: sampling, Mosaic, rough grinding, fine grinding, polishing and corrosion.

(5) Microstructure Analysis

Microstructure analysis mainly analyzes the microscopic crystal structure of the material, that is, which kinds of crystals the material is composed of, how the crystal cell size is, how much the relative content of various crystals and so on. The commonly used structural analysis methods are: XRD, TEM, TG, DTA, infrared and so on.

(6) Material Performance Test

Material performance test refers to the test of the above characteristics of the material under certain environmental conditions, also known as material performance test. The performance data measured by the material performance test not only depends on the material itself, but also depends on the conditions of the test.

The properties of materials can be divided into two categories, one is the characteristic properties, which belong to the inherent properties of the material itself. Including thermal properties (heat capacity, thermal conductivity, heat of melting, thermal expansion, melting boiling point, etc.), mechanical properties (elastic modulus, tensile strength, impact strength, yield strength, fatigue strength, etc.), electrical properties (conductivity, resistivity, dielectric properties, breakdown voltage, etc.), magnetic properties (paramagnetism, diamagnetism, ferromagnetism), optical properties (light reflection, refraction, absorption, etc.), transmission and luminescence, fluorescence and other properties, chemical properties (that is, the activity and ability of the material to participate in chemical reactions, such as corrosion resistance, catalytic performance, ion exchange performance, etc.). The other is functional property, which refers to the property that when a certain effect is applied to a material within a certain condition and a certain limit, the effect is transformed into another form of function through the material. Including thermoelectrical conversion performance (thermistor, infrared detection, etc.), light-thermal conversion performance (such as the conversion of sunlight into heat flat plate collectors), light-electrical conversion performance (solar cells), force-electrical conversion performance, magneto-optical conversion performance, electric-optical conversion performance, acousto-optical conversion performance, etc.

(7) Material Structure and Composition Analysis

Material structure and composition analysis refers to the analysis of the structure, chemical composition, surface and micro-area morphology, mechanical properties and physical and chemical properties of materials through precision testing instruments.

5.3 Key Process Characteristic Control Approach

5.3.1 Reliability Driven Deviation Monitoring Framework for Manufacturing Process

1. **Requirements and Characteristics of Reliability-Driven Manufacturing Process Monitoring**

The reliability of the product is determined by the manufacturing process of the product [4]. When the manufacturing process fluctuates, the size of the product will fluctuate, and further, the reliability of the product will also fluctuate. Therefore, it is very important to use process monitoring to find process fluctuations timely and effectively, so as to provide reference for further process control.

Statistical Process Control (SPC) is a commonly used tool for process monitoring with the help of mathematical statistics [5]. It analyzes and evaluates the manufacturing process, finds the symptoms of systemic factors in time according to the feedback information, and takes measures to eliminate their effects, so that the process is maintained in a controlled state only affected by random factors, so as to achieve the purpose of monitoring the process. It holds that when the process is only affected by random factors, the process is in a statistical control state. When there are systematic factors in the process, the process is in a state of statistical control. Because of the statistical regularity of process fluctuation, when the process is controlled, the characteristics of the process generally obey the stable random distribution. When a process is out of control, the process distribution changes. Statistical process control is to analyze and control the process by using the statistical regularity of process fluctuation. Therefore, it emphasizes the operation of processes in a controlled and capable state, so that products and services can steadily meet customer requirements.

Compared with traditional quality-focused statistical process control, reliability-driven process monitoring [6] has four main features:

(1) The manufacturing processes involved in reliability-driven process monitoring are mostly high-quality processes. Traditional quality control process monitoring usually focuses on the qualified rate of the product leaving the factory, while reliability-driven process monitoring needs to consider the ability of the entire use process to complete the function after the product is produced. Compared with traditional process statistical monitoring, reliability-driven process monitoring puts forward higher requirements for product quality reliability and manufacturing process stability. The manufacturing process with low process capability usually needs to be improved before applying reliability-driven process monitoring.
(2) If the control chart sends an alarm, diagnostic support can be provided to the process according to the monitoring results of the control chart.
(3) The number of monitored features is large, requiring the control chart false alarm rate to be as small as possible. For example, if there are five monitored dimensions, the average running chain length of the classic mean control chart is 370.3, that is, an average of 370.3 points will be monitored and a false alarm will occur. When five mean control charts were set up to monitor the process size separately, an average of 74.5 points were monitored, and one false alarm occurred in five control charts. False alarm will lead to machine shutdown maintenance, will make the site workers and quality management personnel lose confidence in the control chart, no longer trust the alarm issued by the control chart.
(4) The monitored characteristics do not need to adhere strictly to the normal distribution.

According to the above four characteristics, how to design and select the control chart and how to design the parameters of the control chart are the technical focus of reliability driven process monitoring.

2. **Reliability Driven Manufacturing Process Size Deviation Monitoring Process**

The reliability-driven manufacturing process dimension deviation monitoring process can be divided into four steps:

(1) Select the monitoring object;
(2) Design a new applicable control chart or select an existing applicable control chart;
(3) Design the selected control chart parameters;
(4) Implement control chart and process monitoring.

According to the correlation model between product reliability and product manufacturing process size, when the product blank is processed at station i, the size deviation x_i can be obtained by measuring the size X_i of the blank. Theoretically, the statistical process control chart can be used to monitor X_i and x_i, but the dimensions of X_i and x_i are usually large, and the use of multivariate statistical process control chart is a huge amount of calculation. And it is very difficult to diagnose the process when the process is unstable. Therefore, it is first necessary to screen and select process sizes that are sensitive to the reliability of the product for monitoring. By analyzing the deviation that may be introduced in the process of station processing, and using the correlation model of product reliability and process size, the influence of a certain size fluctuation on product reliability is evaluated and sorted, and then the process size that needs to be paid attention to in the manufacturing process is identified, providing input for monitoring and further control.

When designing a new applicable control chart design or selecting an existing applicable control chart, the focus should be placed on "applicable". Different from traditional process statistical monitoring, reliability driven process monitoring has its own characteristics. When reflected in the control chart, it is required to be able to monitor non-normal observed values, and to have a high controlled average operating chain length and a low false alarm rate.

After selecting the control chart, you need to set the parameters of the control chart. The traditional control chart parameter setting methods include economic method and statistical method, which respectively pursue the lowest average quality cost and the longest controlled average operating chain length. In the process monitoring of reliability driven, on the one hand, economy should be taken into account, on the other hand, the control chart should be designed to make its false alarm rate as low as possible, so as to adapt to its reliability driven process monitoring characteristics of high quality manufacturing.

When implementing control chart and process monitoring, it is necessary to carefully identify the true or false alarm. When true alarm occurs, the information of control chart and manufacturing site is used to provide support for process diagnosis.

3. **Select Monitoring Objects**

When selecting monitoring objects, observe the following principles:

(1) The monitored object is easy to lose control;
(2) After the monitored object is out of control, it will cause a large reliability deviation of the product;
(3) The monitored object can be monitored.

This allows the concept of risk to be used to assess each process size. If risk is marked as r, as follows:

$$r = \sum \text{Probability of accident}\ i \times \text{Impact on product reliability deviation}$$

The absolute value $\left|\overline{MTBF_\delta}\right|$ of the average trouble-free working time of the product is used to evaluate the effect of process size deviation on product reliability deviation. The purpose of process improvement is to control reliability deviation and reduce reliability fluctuation, so absolute value is used to adjust it.

5.3.2 *General Event Interval Control Chart Approach*

1. **The General Event Time Interval Control Chart Principle**

In reliability-oriented process monitoring, the monitoring object has four characteristics:

(1) The manufacturing process involved in reliability-driven process monitoring is mostly high-quality process;
(2) If the control chart sends an alarm, diagnostic support can be provided to the process according to the monitoring results of the control chart;
(3) The number of monitored features is large, requiring the control chart false alarm rate to be as small as possible;
(4) The monitored characteristics do not need to adhere strictly to the normal distribution.

In the actual manufacturing process, in addition to measurement monitoring such as size monitoring, there may also be counting monitoring such as defect monitoring. There are some defects in using traditional control charts for reliability oriented process monitoring. Quantitative control charts, such as mean control charts, EWMA control charts, CUSUM control charts and single value control charts in small-batch control charts, are only aimed at process loss of control caused by changes in mean value. To detect changes in process standard deviation, other control charts, such as range control charts and standard deviation control charts, are required. The simultaneous use of multiple control charts will improve the monitoring ability of process loss of control, but it will also lead to the increase of the first type of error probability and false alarm probability, and the parameter design of multiple control charts will be more complicated before implementation. In the face of high quality process monitoring, the numerical control chart is no longer suitable because of its own quality limitation. Another limitation is that almost every control chart first assumes that the process follows a normal distribution, which is contrary to the fourth characteristic of the monitored object.

This chapter extends the traditional event time control chart. The word "event" in the traditional event time control chart refers to the manufacture of nonconforming products, which is the main reason why the event time control chart is limited to counting data monitoring. A nonconforming product is a product where the measured value X of a quality characteristic exceeds the tolerance line, and this event can be denoted as $\{X : X \in (-\infty, LSL) \cap (USL, +\infty)\}$, where USL is the Upper Specification Limit and LSL is the Lower specification Limit. This event is extended beyond the nonconforming product, but the measured value falls into some pre-specified interval Ω, that is, the event is extended from $\{X : X \in (-\infty, LSL) \cap (USL, +\infty)\}$ to $\{X : x \in \Omega\}$. An event is considered to have occurred when the measured value X falls within a specified interval Ω. A control chart monitors the duration of this extended event; this type of chart is called a Generalized Time-Between-Events Charts (GTBE Charts).

Ω can be any set of possible observations. It can be continuous or it can be discrete. For example, in a traditional TBE control chart, the Ω range is $(-\infty, LSL) \cap (USL, +\infty)$ and is continuous. Ω can also be some possible observed enumeration value, for example Ω can be defined as $\{X : X = 2, 3, 4, 5, X \textit{ is the number of pinholes on the coated surface}\}$. Two forms of Ω events are studied here, $(-\infty, \mu - \varsigma\sigma) \cup (\mu + \varsigma\sigma, +\infty)$ and $(\mu + \varsigma\sigma, +\infty)$, where $\varsigma > 0$, assuming that the observed values during manufacturing follow a normal distribution with a mean of μ and a variance of σ_2. For the observed values that do not obey the normal distribution, or other forms of Ω events, it is only necessary to recalculate the corresponding Ω event probability and design the control limit to apply the GTBE control chart.

It turns out that the general event time interval control chart has been extended to be suitable for reliability-driven process monitoring, and its applicability is explained in detail later in this section.

The symbols used in this section are described:

μ	process mean value under controlled condition;
σ^2	process variance under control;
σ	standard deviation of a controlled process;
$\Delta\mu$	change in the process mean when out of control, usually $\Delta\mu$ can be expressed as $\Delta\mu = K\mu_\sigma$;
$\Delta\sigma$	change in the standard deviation of the process when it is out of control, usually $\Delta\sigma$ can be expressed as $\Delta\sigma = k\sigma_\sigma$;
GTBE charts	general event time control charts;
Ω	extended event;
ς	set Ω standard deviation multiplier;
αGTBE control chart	GTBE control chart whose event form follows $(\mu + \varsigma\sigma, +\infty)$, denoted as Ω_α;
βGTBE control chart	GTBE control chart whose event form follows $(-\infty, \mu - \varsigma\sigma) \cup (\mu + \varsigma\sigma, +\infty)$ is denoted as Ω_β;
γGTBE control chart	GTBE control chart whose event form follows $(-\infty, \mu - \varsigma\sigma)$, is denoted as Ω_γ;
DGTBE	GTBE control chart for binary monitoring;
Y	the number of observations between two consecutive Ω events;
Z	monitoring statistics of GTBE control chart;
r	the number of Ω events that occur in Z;

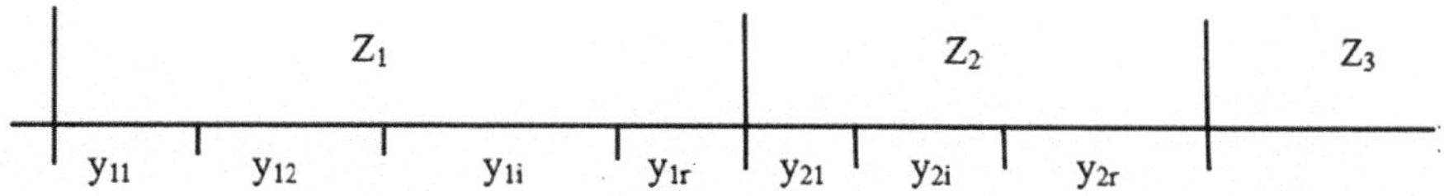

Fig. 5.3 The common event interval control chart monitors the process

LCL	lower limit of control limit, abbreviated as δ;
UCL	upper limit of control, abbreviated as η;
DTMC	discrete Markov chain model;
ARL	average running chain length.

For the control chart, there are three elements: monitoring variable, control limit and running chain length. As mentioned above, the "alarm" event is extended, an interval is set in advance according to the demand, after the measured value is obtained in the manufacturing process, it is considered that the measured value falls into a certain interval, which is the occurrence of Ω events, and the time interval of the occurrence of the event is monitored, so that the control chart is not limited to the monitoring of the alarm rate, but also extends to the monitoring of the occurrence interval of Ω events. Further, the interval for monitoring the occurrence of two consecutive Ω events is extended to the interval for monitoring the occurrence of r events. The actual control chart monitoring process is shown in Fig. 5.3.

The event Ω interval is Y. In the actual monitoring process, the time interval Z of two consecutive or r consecutive events can be monitored, and the formula is as follows:

$$z_i = \sum_{j=1}^{r} y_{ij} \tag{5.18}$$

The Ω event interval Y follows a geometric distribution, while the monitoring variable Z follows a negative binomial distribution.

$$Y \sim f(y) = (1-p)^{y-1}p \quad y = 1, 2, \ldots \tag{5.19}$$

where p is the probability of Ω event occurring at each observation.

Since Ω events obey different forms, the formula is as follows:

$$p = p(\Delta u, \sigma) = p(k_u, k_\sigma) = \begin{cases} \Phi\left(\frac{k_u-\xi}{1+k_\sigma}\right) & , \alpha GTBE \\ \Phi\left(\frac{k_u-\xi}{1+k_\sigma}\right) + \Phi\left(\frac{-k_u-\xi}{1+k_\sigma}\right) & , \beta GTBE \end{cases} \tag{5.20}$$

where $\Phi(X)$ is the cumulative distribution function of the standard normal distribution.

When the process is under control, $\Delta\mu = \Delta\sigma = 0$, Eq. (5.19) degenerates to:

$$p = \begin{cases} \Phi(-\varsigma) & , \alpha GTBE \\ 2\Phi(-\varsigma) & , \beta GTBE \end{cases} \tag{5.21}$$

Z follows negative binomial distribution, therefore:

$$Z \sim f(z) = \binom{z-1}{r-1}(1-p)^{z-r}p^r \quad z = r, r+1, r+2, \ldots \tag{5.22}$$

And

$$E(Z) = r/p \tag{5.23}$$

$$V(Z) = r(1-p)/p^2 \tag{5.24}$$

Determining control limits is an important part of control chart construction. When the control limit is far from the center line, the probability of the first type of error decreases, but the probability of the second type of error increases. When the control limit is close to the center line, the probability of the first type of error increases and the probability of the second type of error decreases. There are generally two methods for the formulation of control limits, one is similar to the traditional mean control chart of upper and lower $k\sigma$ limiting method, in order to balance the first and second types of errors, k is generally taken as 3. The other is to directly select the corresponding points corresponding to the first type of error probability. Here, because the monitoring object Z follows a negative binomial distribution, which is not symmetrical, the second method, the probability control limit, can be used to determine the upper control limit according to the first type of allowable error α. For example, if the allowable type I error α is 3%, then the upper control limit is the 98.5% probability line and the lower control limit is the 1.5% probability line. The formula is given as follows:

$$P[Z < LCL_\alpha] = \sum_{z=r}^{LCL_\alpha} \binom{z-1}{r-1}(1-p)^{z-r}p^r < \alpha/2 \tag{5.25}$$

$$P[Z > UCL_\alpha] = \sum_{z=UCL_\alpha}^{+\infty} \binom{z-1}{r-1}(1-p)^{z-r}p^r < \alpha/2 \tag{5.26}$$

It should be pointed out that the first type of error α mentioned here refers to the probability of Z exceeding the upper control limit or less than the lower control limit when the process is controlled, and is not the actual false alarm rate. For example, in the implementation of a GTBE control chart, the upper and lower control limits are 22, 75 respectively. Five sets of Z values were observed, respectively 43, 23, 55, 64, 75. In the continued observation, no Ω event occurred, that is, Z_5 exceeded the upper control limit 75, where $\alpha = 1/5 = 0.2$ can be calculated, and the actual false alarm probability is 1/(42 + 23 + 55 + 64 + 75) = 0.386%. Because when it comes to the alarm, 42 + 23 + 55 + 64 + 75 = 259 observations X have been monitored.

The GTBE control chart monitors the number of observations X that occur until the rth Ω event occurs. If the process is in a controlled state, then the probability p of Ω event occurrence is certain, and correspondingly, if the monitoring object Z is kept in a state that is not too large or too small, then the process is considered controlled. Otherwise, it is assumed that the process is out of control, and some attributable jump has occurred. In the GTBE control chart, control limits UCL and LCL are calculated by Eqs. (3.9) and (3.10), which can be used to determine whether Z is controlled. When the judgment is wrong, a false alarm occurs.

When the process is out of control, the probability p of the Ω event will also change, $p = p(k_\mu, k_\sigma)$, Y and Z will also change at the same time, you can draw a function diagram for $p = p(k_\mu, k_\sigma)$. Figure 5.4 shows the $p = p(k_\mu, k_\sigma)$ function diagram at 1.5. Figure 5.4 shows how the probability of an Ω event changes when the process is out of control. It can be considered that the detected object "event occurrence interval Z" is positively correlated with the reciprocal of the occurrence probability p of Ω event. In other words, as p gets bigger, Z gets smaller; As p gets smaller, Z gets bigger. In order to detect process variation as soon as possible, when process variation occurs, if probability p increases and Z decreases below LCL, process alarm occurs. If the probability p increases when the process is out of control, then from a statistical point of view, Z will increase. Only when Z exceeds the UCL will the process control chart alarm, which will make the runaway mean run chain longer than the UCL. Figure 5.4 shows the probability p in a controlled state. According to the above judgment, when setting the form of Ω event, we should increase the probability p when the process is out of control as much as possible, or relax the conditions appropriately, and increase the probability p when the process is out of control as much as possible when the process mean is shifted or the process standard deviation is larger. This is due to the fact that the standard deviation of the usual process is smaller, which means that the process capacity is increased, and usually this situation is not so urgent as to require the alarm of the control chart. Therefore, when designing the form of Ω events, $(\mu - \varsigma\sigma, \mu + \varsigma\sigma)$ is not chosen. If the contour plot of the occurrence probability p of Ω events is drawn, it will be found that the occurrence probability p of Ω events in the form of $(\mu - \varsigma\sigma, \mu + \varsigma\sigma)$ is mostly decreasing when the process is out of control. αGTBE control chart and βGTBE control chart also have defects in control, and the darker color region in the contour map represents the region with low probability, and the corresponding process loss in this part of the region is insensitive to αGTBE control chart and βGTBE control chart.

Consider the γGTBE control chart with Ω event form symmetric with the αGTBE control chart, set $\varsigma = 1.5$, and plot its $p = p(k_\mu, k_\sigma)$ function graph, as shown in Fig. 5.5. As can be seen from Fig. 5.5, the sensitive areas monitored by the γGTBE control chart and the αGTBE control chart complement each other. In the actual monitoring process, the process can be monitored by using the γGTBE control chart and the αGTBE control chart at the same time. When the monitoring results of the two control charts are combined, the process is considered to be out of control when an

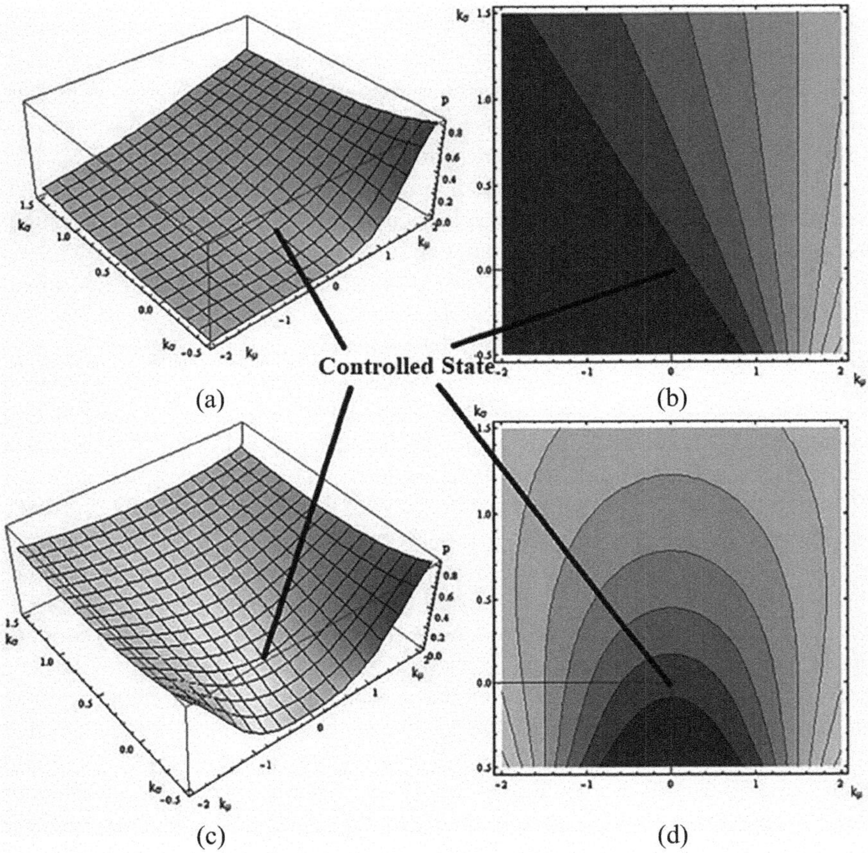

Fig. 5.4 $p = p(k_\mu, k_\sigma)$ function diagram. **a** αGTBE control diagram, 3D diagram; **b** αGTBE, contour line; **c** βGTBE, three-dimensional diagram; **d** βGTBE, contour line

alarm occurs in either control chart. This monitoring method is called Dual Universal Event Interval Control (DGTBE). This is similar to using both a mean control chart and a range control chart for process monitoring.

2. **The General Event Time Interval Control Chart Average Running Chain Length**

The run chain length is generally used to evaluate the control chart, and the average run chain length is an important parameter to characterize the run chain length. When the process is in a controlled state and an alarm occurs, the alarm is a false alarm, and the running chain length at this time is called the controlled running chain length. When the process enters the out-of-control state, the control chart sends an alarm, and the running chain length experienced at this time is the out-of-control running chain length. Usually when the control chart is designed, the controlled run chain length is increased as much as possible and the runaway run chain length is reduced.

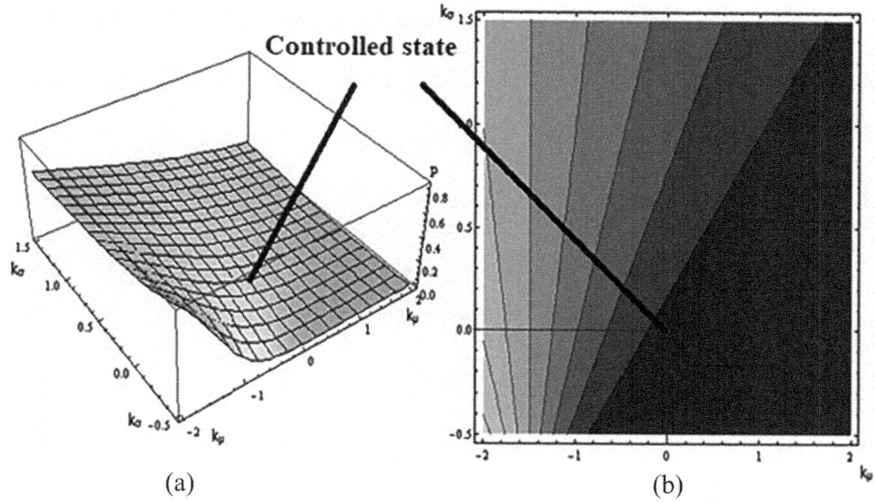

Fig. 5.5 Gamma GTBE control diagram $p = p(k_\mu, k_\sigma)$ function diagram: **a** three-dimensional diagram; **b** contour lines

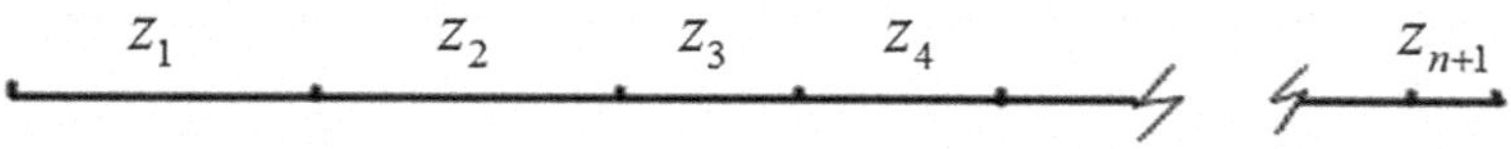

Fig. 5.6 Monitoring process

This section discusses the GTBE control chart and the average run chain length of the DGTBE.

Figure 5.6 shows the monitoring process of the general event interval control chart. During the operation of the process, each Z_i will be recorded to determine whether the Z_i is out of bounds. When the Z_i is out of bounds, an alarm will be issued, and the operation will end.

Detected $Z_1, Z_2, \ldots, Z_n, Z_{n+1}$, where $Z_1, Z_2, \ldots, Z_n$ are located in the [LCL, UCL] interval. No alarm occurs in the process. When Z_{n+1} is detected, an alarm is issued because Z_{n+1} exceeds the control limit, then the running chain length is:

$$RL = \sum_{i=1}^{n+1} z_i = \sum_{i=1}^{n} z_i + z_{n+1} \quad (5.27)$$

Both Z and N are random variables. It should be noted that when Z_{n+1} exceeds the upper control limit, the control chart will alarm, so Formula (5.27) can be revised as follows:

$$RL = \sum_{i=1}^{n+1} z_i = \sum_{i=1}^{n} z_i + \min(z_{n+1}, UCL + 1) \quad (5.28)$$

$\eta = UCL, \delta = LCL$. Z follows negative binomial distribution, and the cumulative probability distribution function with Z is:

$$F(\tau) = \sum_{z=r}^{\tau} f(z) = \sum_{z=r}^{\tau} \binom{z-1}{r-1} (1-p)^{z-r} p^r \quad \tau = r, r+1, r+2, \ldots \tag{5.29}$$

Therefore, average running chain length:

$$\begin{aligned} E(RL) &= E_{\{n,z\}}\left(\sum_{i=1}^{n} z_i + \min(z_{n+1}, \eta + 1) | z_i \in [\delta, \eta], \right. \\ &\qquad \left. z_{n+1} \in (-\infty, \delta) \cup (\eta, +\infty), i = 1, 2, \ldots, n \right) \\ &= E_{\{n,z\}}\left(\sum_{i=1}^{n} z_i | z_i \in [\delta, \eta], i = 1, 2, \ldots, n \right) \\ &\quad + E_{\{n,z\}}(\min(z_{n+1}, \eta + 1) | z_{n+1} \in (-\infty, \delta) \cup (\eta, +\infty)) \\ &= E_{\{n,z\}}\left(\sum_{i=1}^{n} z_i | z_i \in [\delta, \eta], i = 1, 2, \ldots, n \right) \\ &\quad + E_{\{z\}}(\min(z, \eta + 1) | z \in (-\infty, \delta) \cup (\eta, +\infty)) \end{aligned} \tag{5.30}$$

Formula (5.30), the first term is the expectation of $\sum_{i=1}^{n} z_i$, and $z_i \in [\delta, \eta]$, therefore:

$$\begin{aligned} & E_{\{n,z\}}\left(\sum_{i=1}^{n} z_i | z_i \in [\delta, \eta], i = 1, 2, \ldots, n \right) \\ &= \sum_{c=1}^{+\infty} p_c E_{\{z\}}\left(\sum_{i=1}^{c} z_i | z_i \in [\delta, \eta], i = 1, 2, \ldots, n \right) \\ &= \sum_{c=1}^{+\infty} p_c c E_{\{Z\}}(z | z \in [\delta, \eta]) \\ &= E_{\{z\}}(z | z \in [\delta, \eta]) \sum_{c=1}^{+\infty} p_c c \end{aligned} \tag{5.31}$$

where:

$$p_c = [F(\eta) - F(\delta - 1)]^c [1 - F(\eta) + F(\delta - 1)]$$

$$E_{\{z\}}(z | z \in [\delta, \eta]) = \sum_{z=\delta}^{\eta} z p\{z | z \in [\delta, \eta]\} = \sum_{z=\delta}^{n} z \frac{f(z)}{F(\eta) - F(\delta - 1)}$$

The last term of Eq. (5.30) is the expectation of $\min(z_{n+1}, UCL+1)$, and $z_{n+1} \in (-\infty, \delta) \cup (\eta, +\infty)$, therefore:

$$\begin{aligned} & E_{\{z\}}(\min(z, \eta+1)|z \in (-\infty, \delta) \cup (\eta, +\infty)) \\ & = \sum_{z=r}^{\delta-1} zp\{z|z \in (-\infty, \delta)\} \\ & + \sum_{z=\eta+1}^{+\infty} (\eta+1)p\{z|z \in (\eta, +\infty)\} \end{aligned} \tag{5.32}$$

where:

$$p\{z|z \in (\eta, +\infty)\} = 1 - F(\eta)$$

Therefore:

$$\begin{aligned} E(RL) = & \sum_{z=\delta}^{\eta} z \frac{f(z)}{F(\eta) - F(\delta-1)} \sum_{c=1}^{+\infty} [F(\eta) - F(\delta-1)]^c \\ & [1 - F(\eta) + F(\delta-1)]c + \sum_{z=r}^{\delta-1} Z \frac{f(z)}{F(\delta-1)} + (\eta+1)[1 - F(\eta)] \\ = & \frac{1}{1 - F(\eta) + F(\delta-1)} \sum_{z=\delta}^{\eta} zf(z) + (\eta+1)[1 - F(\eta)] \end{aligned} \tag{5.33}$$

According to Eq. (5.33), the average running chain length of the GTBE control chart can be calculated with the help of the mathematical tool Wolfram Mathematica.

Discrete Time Markov Chain (DTMC) Model is introduced to calculate the average running chain length of DGTBE monitoring. Use S_0 to mark the state in which the control chart emits an alarm, then S_0 is an absorbed state, marked $S_0 = \{v_0\}$. Mark S as the state in which no alarm is issued by the control chart, and the state space $\tilde{S}$ can be considered as the union of S and S_0. The transition probability matrix can be written as:

$$\tilde{P} = \begin{pmatrix} P & P_0 \\ 0 & 1 \end{pmatrix} \tag{5.34}$$

where P is the transition probability matrix of S, and $P_0 = (\mathrm{I} - \mathrm{P})e$, $e = (1, 1, \ldots, 1)^T$.

Define the initial state probability vector $\pi(0) = (\pi, \pi_0)$, where $\pi = (\pi_1, \pi_2, \ldots, \pi_m)$, $\pi_i \geq 0$, $\sum \pi_i = 1$. m is the number of elements in space S. Run the chain length RL as a random variable.

$$RL = \inf\{n: n \geq 0, X_n \in S_0\} \tag{5.35}$$

It is said that RL follows the phase-type distribution, so that the average value of the running chain length RL, the second order moment and the higher order moment can be obtained. The average running chain length is obtained:

$$E(RL) = \pi(\mathrm{I} - \mathrm{P})^{-1}e \tag{5.36}$$

To calculate the average running chain length E (RL), it is necessary to first determine the state in the S-space, the initial state probability vector, and the transition probability matrix. In DGTBE monitoring process using αGTBE control chart and γGTBE control chart simultaneously, a state is represented by a four-dimensional vector space, denoted as $\nu = (C_\alpha, Z_\alpha, C_\gamma, Z_\gamma)$. Where C_α represents the number of Ω events that have occurred in the current monitoring cycle of the αGTBE control chart; Z_α represents the number of observations that have already occurred in the current monitoring cycle of the αGTBE control chart; C_γ represents the number of Ω events that have occurred in the current monitoring cycle of the gamma GTBE control chart. Z_γ represents the number of observations that have already occurred during the current γGTBE control chart monitoring cycle. For distinction, subscripts α or γ are added after UCL, LCL, multipliers ς, Ω events, etc. By definition, C_α, Z_α, C_γ, and Z_γ are positive integers.

$$\begin{aligned} S = \{\nu = (C_\alpha, Z_\alpha, C_\gamma, Z_\gamma) &: 0 \le C_\alpha \le r_\alpha, C_\alpha \in \mathit{Integers}; \\ & C_\alpha \le Z_\alpha \le UCL_\alpha, Z_\alpha \in \mathit{Integers}; \\ & 0 \le C_\gamma < r_\gamma, C_\gamma \in \mathit{Integers}; \\ & C_\gamma \le Z_\gamma \le UCL_\alpha, Z_\gamma \in \mathit{Integers}; \} \end{aligned} \tag{5.37}$$

Z_α, Z_γ will not exceed their corresponding upper control limits, and C_α, C_γ will not exceed their corresponding Ω event occurrence times r_α, r_γ in a single cycle. When C Omega events are observed at the same time, at least C observations must have been made, so Z must be at least larger than C. There is $C_\alpha \le Z_\alpha$ and $C_\gamma \le Z_\gamma$ limit. In this way, the dimension m of matrix *P* can be calculated as:

$$m = \frac{1}{4}(r_\alpha + 1)(2UCL - r_\alpha + 2)(r_\gamma + 1)(2UCL_\gamma - r_\gamma + 2) \tag{5.38}$$

ς and m increase dramatically as r and ς increase, because big ς means big UCL, and as you can see from Eq. (5.37), UCL has an important influence on m. The increase of dimension will increase the difficulty of calculating the average running chain length. Fortunately, the probabilistic transition matrix *P* is a sparse matrix with at most three non-zero elements per row. This is because for each state transition, there are only three possible outcomes: an ωα event occurs, an ωγ event occurs, and no Ω event occurs.

If the probability of an Ωα event is pα and the probability of an Ωγ event is pγ, then the probability of no ωα event is $q = 1 - p_\alpha - p_\gamma$. This creates the transition matrix. Figure 5.7 shows the one-step transition probability diagram. According to Fig. 5.7, the transition probability matrix can be filled.

Because starting from the state ν = (0, 0, 0, 0), assign the elements corresponding to $\pi(0)$ to 1 and the rest to 0, the initial state probability vector is obtained. The probability transition matrix and initial state probability vector are obtained, that is, the average running chain length E (*RL*) can be calculated.

Note that the average running chain length can also be calculated using methods such as Monte Carlo simulation. By studying $ARL = ARL(k_\mu, \Gamma)$, the monitoring ability of control chart in the face of process mean changes is discussed. Γ is the design parameter vector. For GTBE control charts and DGTBE monitoring, Γ is (r, ς).

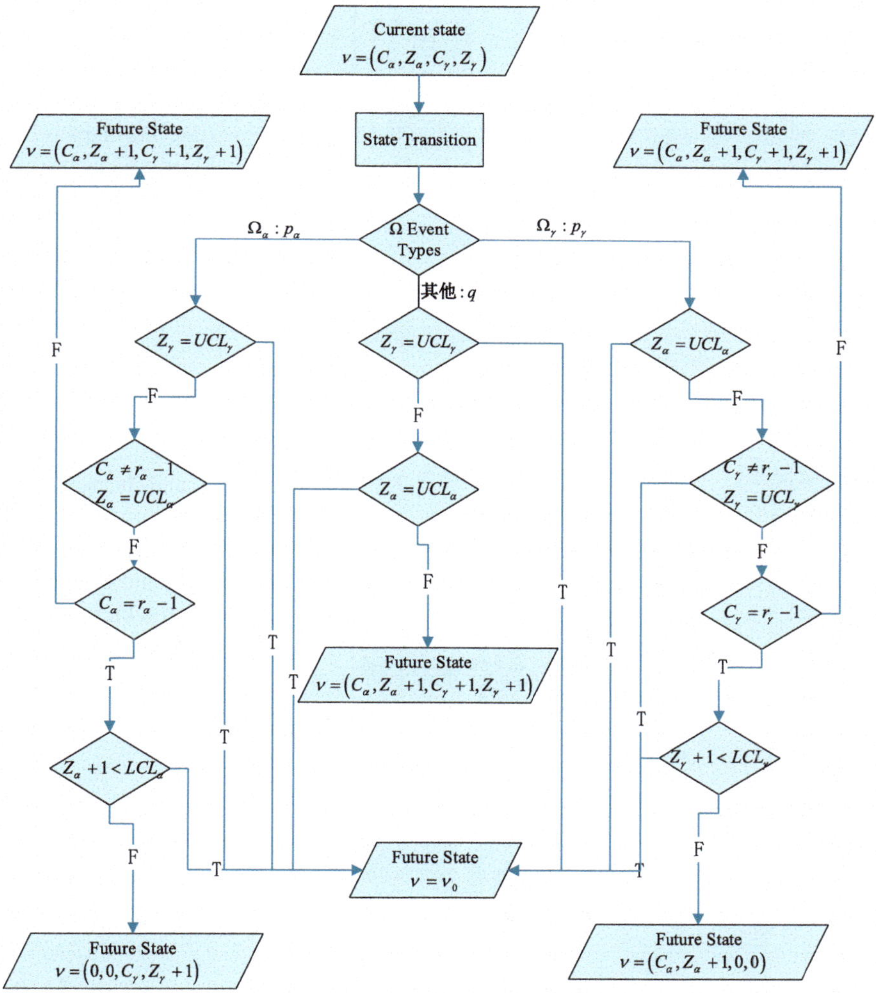

Fig. 5.7 One-step transition probability diagram

According to Formula (5.37) and Monte Carlo simulation, the average running chain length is given. It is concluded that the parameters used in the αGTBE control chart and γGTBE control chart involved in DGTBE monitoring are consistent. In general, the average running chain length of these two monitors is larger, so logarithmic graph paper is used to plot. Figure 5.8 shows the function diagram.

According to the graph above, the following conclusions can be drawn:

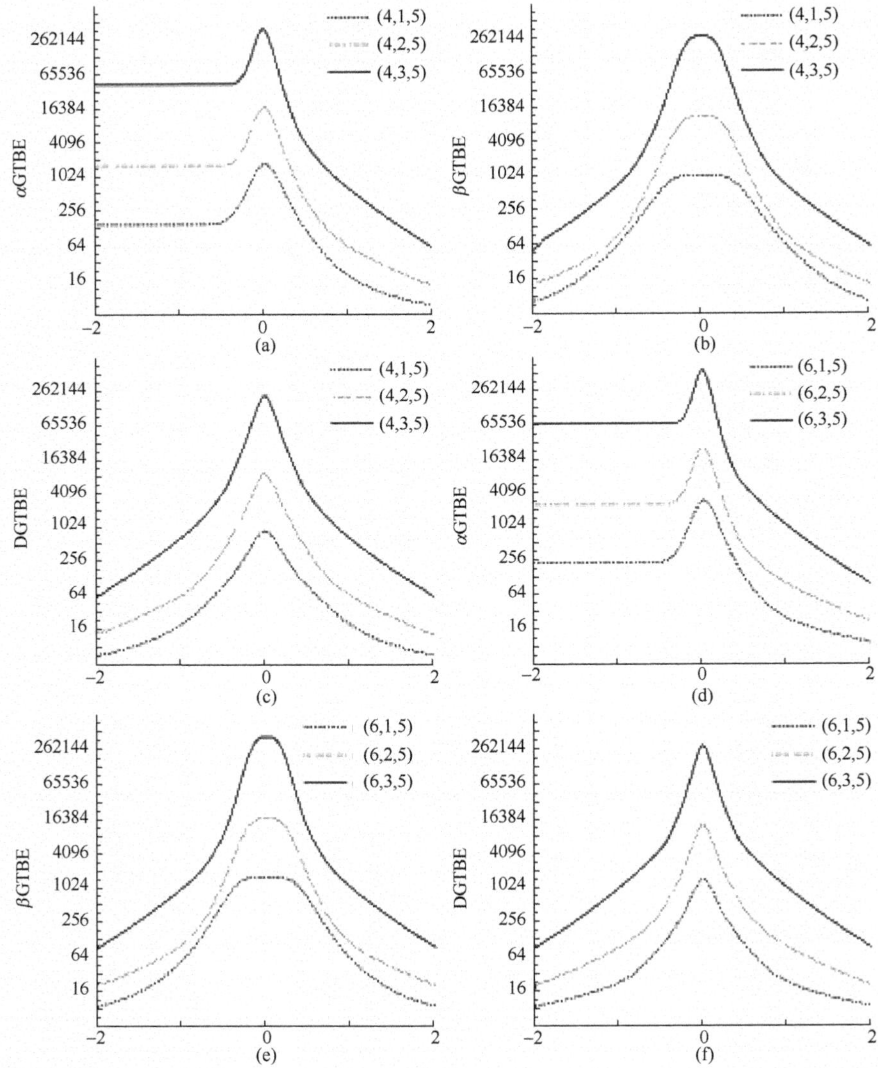

Fig. 5.8 Average run chain length monitored by the GTBE control chart and DGTBE when the process mean changes

(1) The ability of βGTBE control chart and DGTBE to monitor process mean changes is symmetric. αGTBE is not sensitive to process mean reduction.
(2) Larger *r* or ς will result in a larger controlled average running chain length. Since the $ARL = ARL(k_{\mu}, \Gamma)$ function is continuous, a large controlled mean operating chain length makes the control chart insensitive to small mean changes. Therefore, the control chart of GTBE type is more suitable for monitoring the change of $k_{\mu} > 0.5$. A large controlled average operating chain length will greatly reduce the false alarm rate. This increases management's confidence in the control chart and reduces the cost of maintenance outages due to false alarms.
(3) If there is no prior knowledge of process changes, it is recommended to use DGTBE monitoring, whose ability to monitor process mean changes is more balanced. At the same time, DGTBE monitoring is more sensitive than βGTBE control chart when the mean change is small, and DGTBE monitoring performance is better than αGTBE control chart when the mean decrease during monitoring process.
(4) A large *r* value is recommended for parameter setting. When ς is fixed, a larger r value increases the controlled mean running chain length and has less effect on the runaway mean running chain length, that is, less reduces the ability of the control chart to detect the runaway process. When a larger average running chain length is required, adjusting r is preferable to adjusting ς.

Using similar steps to plot $ARL = ARL(k_{\mu}, \Gamma)$, we can examine the ability of GTBE control charts and DGTBE monitors to detect changes in standard deviations for processes. As shown in Fig. 5.9.

According to the graph, the following conclusions can be drawn:

(1) All three kinds of monitoring can monitor the decrease or increase of process standard deviation. When the process standard deviation decreases, the runaway mean running chain length is slightly larger than the upper control limit UCL.
(2) All three monitors are sensitive to large process standard deviation increases. When the process standard deviation variation coefficient reaches 0.5, the runaway mean running chain length soon drops below 100.
(3) It is recommended to use the βGTBE control chart to monitor the change of process standard deviation. When $k_{\sigma} \in (0.5, 1.2)$, it is more sensitive than DGTBE monitoring.

3. **Implementation Steps and Case Study**

Figure 5.10 shows the GTBE control diagram.

Univariate measurement data is generally considered to obey normal distribution, while multivariate measurement control chart will be transformed into univariate data types such as chi-square distribution through Hotelling monitoring variables and other methods in monitoring. In this way, according to the specific distribution differences, the general event time interval control chart can be used for the univariate and multivariate measurement data monitoring of high-quality processes.

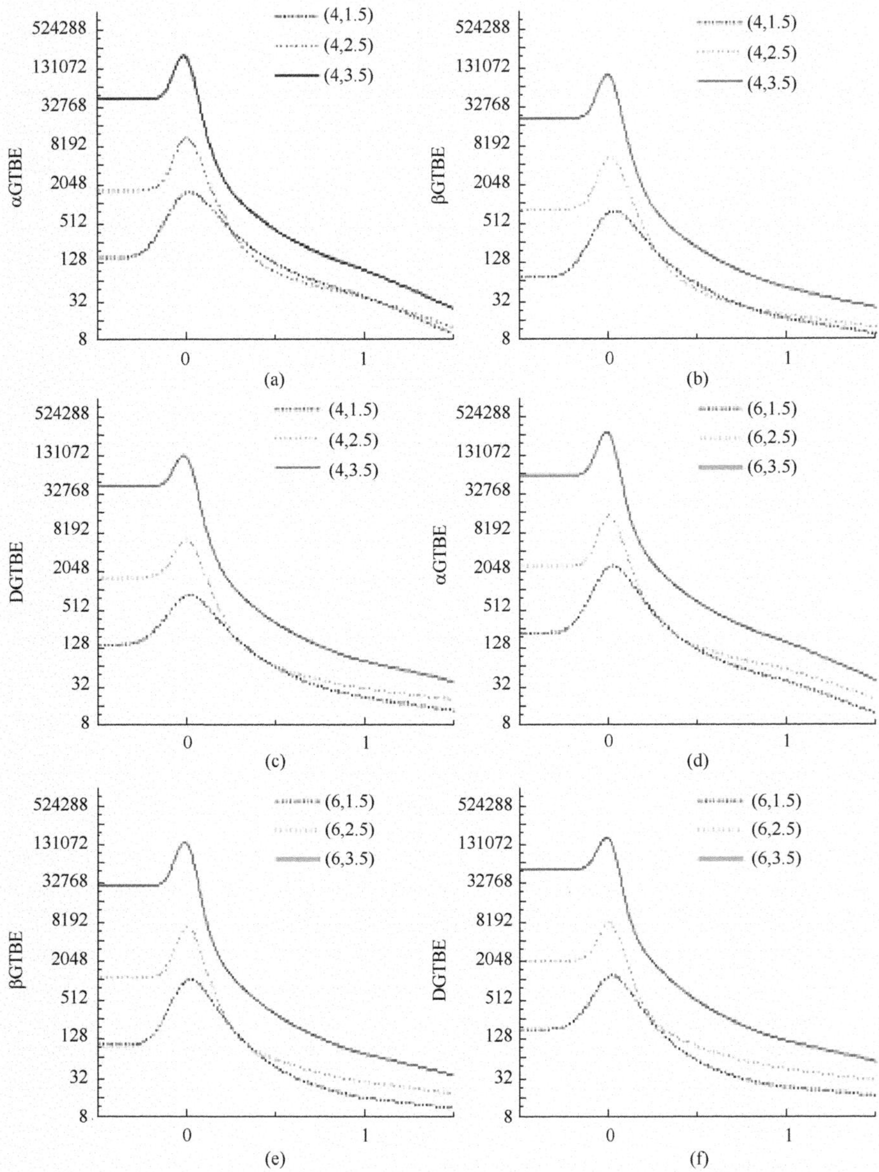

Fig. 5.9 Average run chain length monitored by the GTBE control chart and DGTBE when the process standard deviation changes

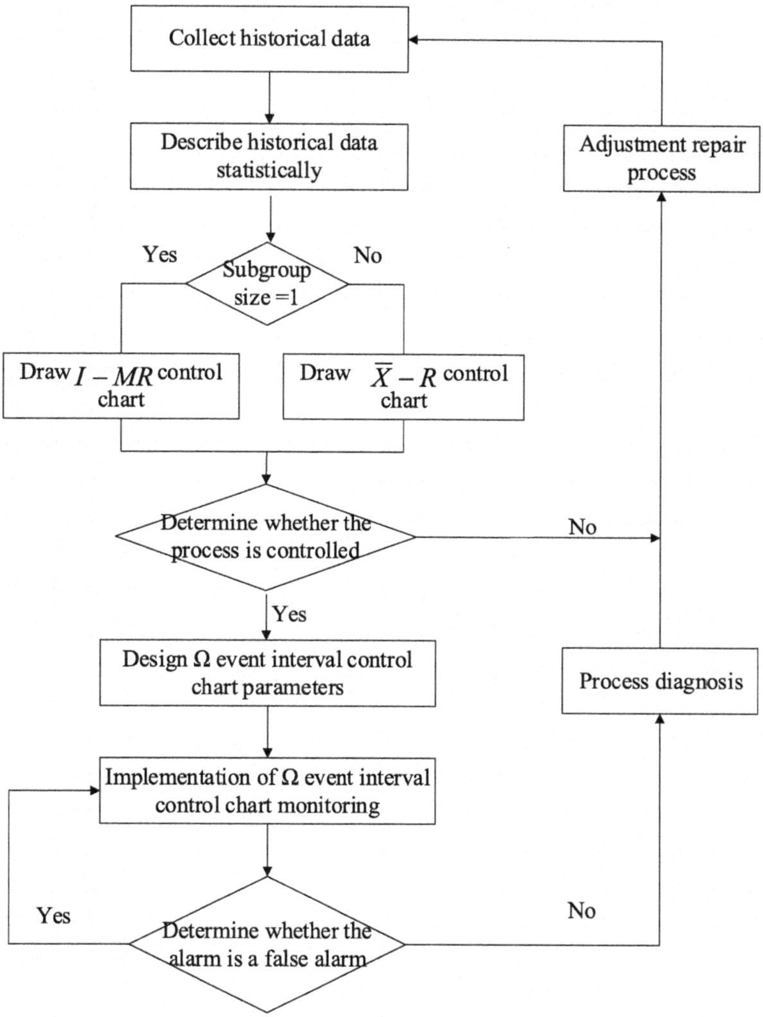

Fig. 5.10 Procedure for implementing a general event interval control chart

Combined with the traditional TBE control chart for counting data monitoring, it achieves the purpose of all-round monitoring of high quality process measurement data and counting data. More importantly, the common event time interval control chart fundamentally reduces the occurrence of false alarms, which is very beneficial for reliability-driven process monitoring. The following details the GTBE control chart implementation steps.

Step 1: Collect historical data. Usually, you can query past quality records here. The amount of data collected usually ranges from 20 to 70. For example,

60 historical time series data of a certain quality feature are collected, as shown in Table 5.3.

Step 2: The results verify that the data collected above are independent and subject to normal distribution. It can be calculated that the mean is 19.919 and the standard deviation is 0.972. Since the size of the subgroup is 1, the I-MR control chart is drawn, as shown in Fig. 5.11. It can be seen from the figure that all data points are within the bounds and do not exceed the *UCL* or *LCL* control limits. It can be concluded that the current process is controlled and the general event time interval control chart is applied for long-term monitoring.

Step 3: Design common event interval control chart parameters. Generally speaking, the selection of control chart parameters follows such a rule, on the one hand, the control chart can have a better ability to detect process anomalies, on the other hand, it is hoped that the false alarm rate of the control

Table 5.3 Instance data

18.938	18.761	19.998	20.501	20.155	20.439	19.910	19.946	20.331	18.285
20.662	20.636	21.573	20.223	18.622	20.097	21.654	18.786	19.592	19.515
18.366	20.433	20.839	19.918	19.396	20.401	20.712	19.024	18.448	20.663
20.138	19.554	19.832	22.126	19.979	20.910	19.221	19.774	21.212	18.454
18.858	18.974	19.104	19.198	19.199	20.272	18.657	20.354	21.764	20.794
21.166	18.356	18.952	18.717	20.223	20.548	21.337	20.354	21.124	18.986

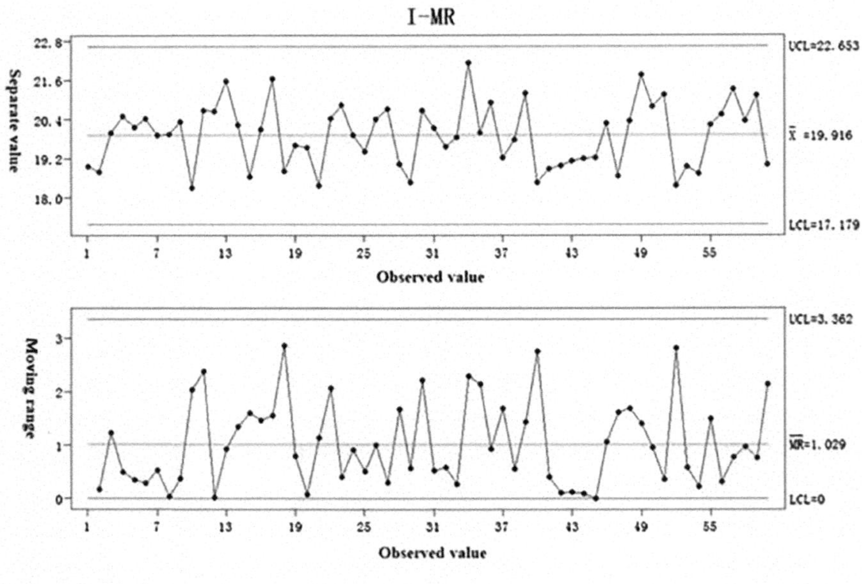

Fig. 5.11 I-MR control chart view historical data

chart is low, which is reflected in the average running chain length, that is, the longer the controlled average running chain of the control chart, the shorter the uncontrolled average running chain length of the control chart, the better. The procedure for calculating a general event interval control chart is as follows.

Combined with the above formula, the control limits of the calculated general event time interval control chart are given in Table 5.4. After the control limit, r, ς is given, the βGTBE control chart is selected according to the result, using the parameter combination of $r = 5$, $\varsigma = 2.5$, $UCL = 883$, $LCL = 116$.

After the monitoring variables and control limits are given, the control chart can be used in the actual manufacturing process. The average run chain length is often used to evaluate the efficiency of the control chart run.

Step 4: Monitor. Figure 5.12 shows the implementation diagram of the general event interval control chart.

Assuming that the quality characteristic to be monitored is the height of the machined part, set the laser transmitter ULG and LLG, corresponding to the height

Table 5.4 Generic event interval control chart limits

Control chart type		Ω-I				Ω-II			
	α	0.03		0.05		0.03		0.05	
		UCL	LCL	UCL	LCL	UCL	LCL	UCL	LCL
ς = 1.5	r = 3	56	5	51	6	115	9	105	10
	r = 5	78	13	73	14	161	23	150	26
	r = 7	100	21	94	23	204	40	191	44
	r = 9	120	31	113	33	245	59	232	64
ς = 2.5	r = 3	632	42	579	51	1267	83	1160	101
	r = 5	883	116	821	132	1769	230	1646	263
	r = 7	1116	206	1048	229	2236	409	2099	455
	r = 9	1339	305	1265	334	2683	608	2534	665

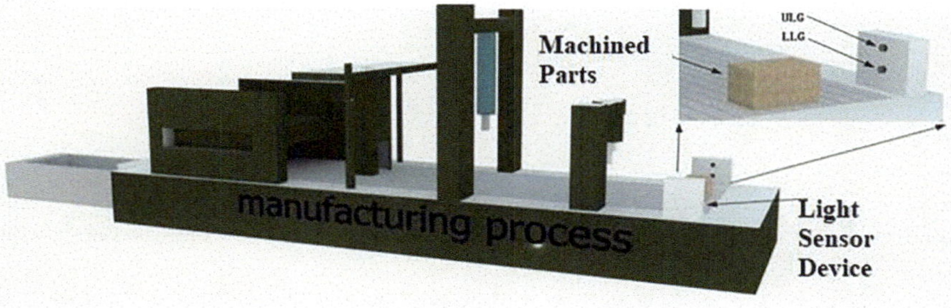

Fig. 5.12 Implementation diagram of the general event interval control chart

of $\mu + \varsigma\sigma$ and $\mu - \varsigma\sigma$, and arrange the optical sensor device on the other side of the station. If the quality characteristic exceeds $\mu + \varsigma\sigma$, the machined part blocks the ULG and LLG, the optical sensor device can not receive the laser signal. If the quality characteristic is less than LLG, the optical sensor can receive both ULG and LLG laser signals at the same time. When the counter reaches 5, it means that five Ω events have occurred, and the number of parts manufactured during the occurrence of these five Ω events is recorded, which is Z. When Z exceeds the upper control limit UCL or is less than the lower control limit LCL, an alarm is issued. In this way, the monitoring data obtained are as follows:

$$439,\ 305,\ 423,\ 160,\ 307,\ 309,\ 428,\ 360,\ 337,\ 265,$$
$$252,\ 355,\ 647,\ 730,\ 429,\ 392,\ 601,\ 260,\ 423,\ 217,$$
$$748,\ 497,\ 343,\ 379,\ 588,\ 557,\ 216,\ 266,\ 262,\ 101.$$

Step 5: Draw a general event interval control diagram, as shown in Fig. 5.13. Note that an alarm is issued when Z exceeds the upper control limit UCL or is less than the lower control limit LCL. The last detected event interval is 101, less than the lower limit of 116, at which time an alarm is issued. Check whether the process is really out of control. If the alarm is false, go back to Step 4. Otherwise, go to Step 6. Background knowledge of the engineering staff is required here.

Step 6: Analyze the cause of process loss and adjust the repair process to bring it back into a controlled state. Background knowledge of the engineering staff is required here.

The applicability of the general event time interval control chart.

Compared with the traditional control chart, the general event time interval control chart is more practical in reliability driven process monitoring. The characteristics of reliability driven process monitoring are described here.

(1) The Common Event Time Interval Control Chart Has a Low False Alarm Rate

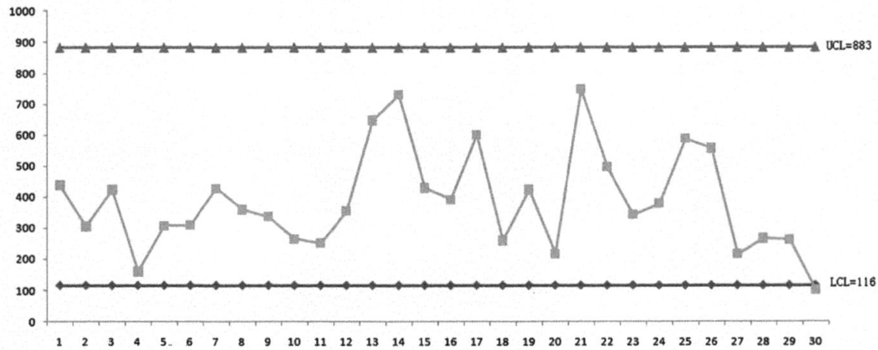

Fig. 5.13 General event interval control chart

It can be considered that the observation point falling outside the 3σ line is also an Ω event, so the control chart will issue an alarm only when the probability of the occurrence of the Ω event is abnormal. In other words, the general event time interval control chart can monitor the false alarm rate of the traditional control chart, and it is considered that the false alarm is a statistically bound event, and only when the probability of the event changes, the process will lose control. This greatly increases the controlled average operating chain length of the control chart and reduces the false alarm rate.

(2) General Event Time Interval Control Chart AIDS Process Diagnosis

In the process of control chart monitoring, when an alarm occurs, the control chart should have the ability to assist the staff in process diagnosis.

For αGTBE and βGTBE control charts, assuming that the last observation before the alarm is Z, it can be assumed that the probability of Ω event occurrence is out of control at this time.

$$\frac{r}{Z} = p = p(\Delta\mu, \Delta\sigma) = p\left(k_\mu, k_\sigma\right) = \begin{cases} \Phi\left(\frac{k_\mu - \varsigma}{1+k_\sigma}\right) & \alpha GTBE \\ \Phi\left(\frac{k_\mu - \varsigma}{1+k_\sigma}\right) + \Phi\left(\frac{-\varsigma - k_\mu}{1+k_\sigma}\right) & \beta GTBE \end{cases} \tag{5.39}$$

The possible out-of-control states of the process (Δμ, Δσ) can be calculated.

For DGTBE control charts, it is assumed that the last observed quantities of αGTBE control charts and γGTBE control charts before alarm are Z_α and Z_γ.

$$\Phi\left(\frac{k_\mu - \varsigma}{1 + k_\sigma}\right) = \frac{r'}{Z_\alpha}\Phi\left(\frac{-\varsigma - k_\mu}{1 + k_\sigma}\right) = \frac{r''}{Z_\gamma} \tag{5.40}$$

(3) The General Event Time Interval Control Chart Is Universal

Because of the great freedom in formulating Ω events, the general event time interval control chart is also suitable for monitoring processes that do not follow the normal distribution. Even Ω events can be discrete, such as the number of ash dots during a painting job. The general event time interval control chart monitors the probability of Ω event occurrence. As long as the probability of occurrence is determined, the general event time interval control chart can be implemented.

5.3.3 *Reliability Driven Truncated Control Chart Control Technology*

The inherent reliability of complex products comes from their multi-station manufacturing process. Fluctuations between multiple stations in the manufacturing stage lead to fluctuations in key quality characteristics that characterize product quality, and further, the reliability indicators established and required at each design stage will also fluctuate. It can be seen that the stability of the manufacturing process is the primary condition for the formation of standard product quality characteristics, and it is also an important indicator to reflect the level of batch production quality uniformity. It is very important to use process monitoring to find process fluctuation in time and effectively, so as to provide reference for further process control. Traditional counting or metering control chart technology is often used to detect process fluctuations, and carry out targeted analysis, control or prediction of uncertainty factors through the statistical characteristics of sample observation information. Due to the influence of accidental factors, the manufacturing quality of actual products is bound to deteriorate compared with the design requirements. However, if the variation is outside the scope of this stable mode, the cause can be detected, corrected and controlled.

1. **The Impact of Process Truncated Quality Characteristics on Product Reliability Analysis**

As an evolutionary feature of quality characteristics, fluctuation generally exists in every stage of manufacturing and has a negative impact on the reliability of manufactured products. The monitoring of manufacturing process reliability often requires the detection means of manufacturing process to obtain state data. Limited to truncated factors such as time and cost, the key quality characteristic data collected during testing can be divided into key non-truncated quality characteristics and truncated quality characteristics according to the assignment characteristics. The non-truncated data corresponds to the real value of the quality characteristic, which is the completely normal data generally processed by the conventional control chart. Truncated data inevitably exists and corresponds to incomplete values of quality characteristics. There are two types of truncated data, timing truncated data and fixed truncated data, according to the different truncated methods. Timing truncated data, namely right truncated data, is widely used in reliability detection data. In this way, as incomplete quality data, the right truncated key quality feature contains less quality information and has great inaccuracy.

Reliability, as a quality in time dimension, has entered the comprehensive integration stage for the whole life, whole process and full characteristics, which is specifically manifested in the transmission evolution process from reliability in design stage, inherent reliability in manufacturing process, reliability in sales stage to reliability in use process as shown in Fig. 5.14 [6]. What drives the evolution of product reliability over the life cycle is the fluctuation of key quality characteristics, which have different characteristics at different stages. The design stage is mainly to identify and prevent the fluctuation of key quality characteristics, the manufacturing stage is mainly to form fluctuations and control fluctuations, the sales stage is mainly to achieve the lossless transmission of fluctuations, and the maintenance of fluctuations through maintenance support in the use stage, so as to avoid triggering failures. The chart below clearly illustrates the fluctuating evolution of key quality characteristics in relation to the evolution of product reliability.

In Fig. 5.14, the inherent reliability index of the manufactured product is formed through the operation of processing and assembly in accordance with the reliability

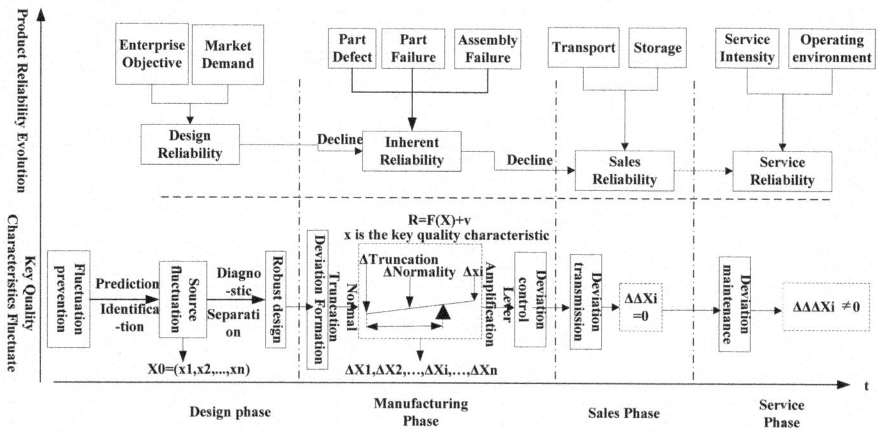

Fig. 5.14 Correlation analysis of key quality characteristic fluctuation and product reliability

index and related parameters defined in the design stage. The inherent reliability of the output of the manufacturing end is often difficult to meet the requirements of the design reliability, which is manifested as a decline in reliability, because the prevention design of manufacturing fluctuation factors is not in place in the design stage and the real-time control of the fluctuation factors is not in place in the manufacturing stage.

In particular, due to the limitation of quality inspection methods, there is a truncated phenomenon in the inspection data of components in the processing stage of the manufacturing process, and the corresponding key quality characteristics show the characteristics of truncated fluctuations. Considering that the uncertainty of fluctuations affects the level of reliability, it can be determined that the higher the uncertainty, the more difficult to ensure reliability. Because truncated characteristics contain less information, the uncertainty degree of truncated characteristics is greater than that of normal fluctuations. As reflected in the wave effect lever in Fig. 5.14, the fluctuation of truncated key quality characteristics has a more prominent effect on the decline of actual manufacturing reliability. How to deeply analyze the truncated key quality characteristics to restore the real state of the data as much as possible and establish the analysis and control technology of the truncated key quality characteristics fluctuation data is the key to control and prevent the decline of product reliability in the manufacturing stage and the resulting high early failure rate.

2. **Analysis of Truncated Characteristics by Asymptotic Reduction Method**

Truncated quality characteristics are different from traditional measurement and counting statistics because they do not simply record each quality characteristic value, nor do they simply count the number of qualified or unqualified products. Under normal circumstances, truncated quality characteristic data is a composite data mixed with measurement and counting. In view of this, the book fully confronts the problem of truncated data distortion, uses the mathematical principle of successive asymptotic in the approximation problem as a guide, and deeply analyzes the truncated property to restore the true state of the data as much as possible.

(1) The Principle of Asymptotic Reduction of Convergent CEV

When the truncated feature is truncated at the level C, that is, the observation value adopts the timing truncation mechanism to obtain the right truncated data, and the 0–1 variable γ is introduced. When the mass characteristic value meets $X_i \leq C$, $\gamma_i = 1$, and X_i is the actual observed value. When the mass characteristic value meets $X_i > C$, $\gamma_i = 1$, and X_i is the truncated observation value. At the same time, its value is recorded as the truncated horizontal value C. Therefore, the key quality feature data of right-truncated type can be characterized by (X_i, γ_i), while the total number of truncated quality features is $m = \sum_{i=1}^{n} \gamma_i$ and the total number of non-truncated quality features is $r = \sum_{i=1}^{n} (1 - \gamma_i)$. Then we have the sample observation $X = (x_1, x_2, \ldots, x_r, C_{n-r}, \ldots, C_n)$.

The mathematical principle of successive progressive reduction is to first determine the approximate range of the solution according to the conditions of the problem, and then gradually reduce the solution space by constantly improving the method or eliminating impossible situations, and finally obtain the optimal solution of the problem [7]. The quasi-true reduction method of converging CEV takes the asymptotic reduction of data as the basic idea, and carries on the function limit processing in the asymptotic analysis for the conditional expected value of the traditional CEV that is greater than the value of the truncated level C, and then determines the truncated expected value $v_{cc} = \lim\limits_{\theta \to \tilde{\theta}} E_{cc}\{(X, \theta)|(X, \theta) \geq C\}$ for the quasi-true characterization of the truncated data to ensure the effectiveness of the truncated data after processing, thus improving the efficiency of the control chart.

Therefore, the value of the initial truncated data is no longer referred to the truncated level, but is replaced by the estimated truncated expected value of the actual data that is higher than the truncated level. The use of truncated expected value instead of truncated data can satisfy that the mean value of sample data is equal to the mean value of actual process, which greatly reduces the estimation bias of skewed distribution caused by truncated data in traditional process parameter estimation.

(2) Convergent CEV Treatment of Truncated Quality Characteristics

The convergent CEV quasi-true reduction technique for truncated characteristics is based on the combination of maximum likelihood estimation method and truncated expected value method, and guided by the function limit, according to the following steps:

Step 1: Establish the density distribution function of right-censored mass characteristics $f(x_i)^{1-\gamma_i} \Pr(X_i > C)^{\gamma_i}$, which can also be written as. Then, according to the meaning of maximum likelihood function, the log-maximum likelihood function considering censored factors is:

$$\log L(\theta) = (N - r)S(C;\theta) + \sum i \in Df(x_i;\theta) \tag{5.41}$$

where, θ represents one or more quality characteristic distribution parameters, x_i is the sample quality characteristic value obtained by detection, C corresponds to the truncated level value of the quality characteristic, f represents the failure distribution of x_i, S represents the survival distribution of x_i, r represents the number of non-truncated samples, and D represents the set of non-truncated samples.

Step 2: Specify the correlation between the truncated expected value of the truncated data and the potential distribution form of the mass characteristics, as a function of the potential distribution parameters:

$$v_c = E\{(X,\theta)|(X,\theta) \geq C\} = \frac{\int_C^\infty xf(x;\theta)dx}{\int_C^\infty f(x;\theta)dx} = h(\theta) \tag{5.42}$$

Step 3: The value of the quality characteristic in the quasi-true reduction method of convergent CEV is defined as follows: the truncated data is replaced by the truncated expected value, and the non-truncated data keeps the original value unchanged, so that the sample information can be restored and updated.

$$v_i = \begin{cases} x & if\ x \leq C \\ v_c & if\ x > C \end{cases} \tag{5.43}$$

Step 4: According to the restored sample information and the maximum likelihood equation, the estimated value of the distribution parameter is determined, which is represented as a function of the truncated expected value as shown in the following equation.

$$\hat{\theta} = g(v_i) \tag{5.44}$$

In the specific calculation, it is necessary to assign an appropriate initial value to parameter θ, and then calculate $\hat{\theta}$ according to the formula sequence of Steps 2–4, and re-replace the obtained new parameter values into the formula of Steps 2–4, so many iterations of calculation until the parameter values converge to the required precision. Furthermore, it can be determined that the truncated expected value under convergent CEV is:

$$v_{cc} = \lim_{\theta \to \hat{\theta}} E_{cc}\{(X,\theta)|(X,\theta) \geq C\} \tag{5.45}$$

Through the above method, an initial set of truncated quality characteristic data $X = (x_1, x_2, \ldots, x_r, C_{n-r}, \ldots, C_n)$ is restored and replaced by $(x_1, x_2, \ldots, x_r, v_{cc}, v_{cc}, \ldots, v_{cc})$ to complete the quasi-true restoration of truncated quality characteristic data.

(3) Construction of Truncated Characteristic Control Chart and Its Performance Indicators

Due to the influence of truncation mechanism, the test statistics of truncated control chart are different from the simple arithmetic mean of quality characteristics of traditional control chart, and the test statistics of truncated control chart may have more complex forms and contain truncated data and its deformation. Then, it is difficult to determine the derivation of upper and lower control limits according to the principle of traditional $\mu \pm 3\sigma$. In addition, for truncated quality characteristics, the size of the truncated proportion in the truncated sample has a significant effect on whether the control chart type is $\overline{X}$, S or $\overline{X} - s$. Steiner and MacKay [8] have demonstrated that when the proportion of truncated samples is high (greater than 0.5), $\overline{X}$ truncated control chart is also suitable for detecting the change of process standard deviation. Assuming that the process variance increases, the truncated feature on the left side of the process mean will move to the side with a lower value, while the truncated feature on the right side of the process mean still has a uniform truncated threshold value v_{cc}, so it can be inferred that the overall sample value tends to decrease, that is, the corresponding process mean decreases. In this way, it can be seen that for the quality characteristics of high right truncated, the increase of process variance has similar behavior and effect as the decrease of process mean, so only the $\overline{X}$-truncated control chart can be analyzed and discussed.

Therefore, Monte Carlo simulation is used here to obtain control limits for different truncated proportions and sample sizes. The monitoring parameters created, the implementation steps of the control chart and the performance analysis indicators are not limited by the specific type of the control chart to control the truncated characteristics, and the applicability is wide.

Step 1: Take the estimated value $\hat{\theta} = g(v_i)$ of the distribution parameter determined above as input, generate N random numbers conforming to the distribution of $f(\cdot)$B. Set the truncated value C according to the actual situation of the project, and replace all the values greater than C with C.

Step 2: Set n unsorted truncated random data generated in Step 1 into a group, calculate the test statistic $T_j = \sum_{i \in D} X_i + m\frac{C}{n}$, $(j = 1, 2, \ldots, \frac{N}{n})$ of each group, which is the monitoring parameter, and obtain the statistic matrix $TS = \left[T_1, T_2, \ldots, T_{\frac{N}{n}}\right]^T$.

Step 3: Sort the elements in C from small to large and give the sequence statistic matrix $TS = \left[T_1, T_2, \ldots, T'_{\frac{N}{n}}\right]^T$ to meet the size requirements of $T'_1 \leq T'_2 \leq, \ldots, \leq T'_{\frac{N}{n}}$. When the position scalar $p = [p_1, p_2] = \left[\frac{N}{n} * R_f, \frac{N}{n} * (1 - R_f)\right]$ is defined and the expected false alarm rate is R_f, the estimated control limits can be determined as follows:

$$\begin{aligned} &Lower\ Control\ Limit\colon LCL = \frac{1}{2}\left(T'_{p_1} + T'_{p_1+1}\right) \\ &Upper\ Control\ Limit\colon LCL = \frac{1}{2}\left(T'_{p_2} + T'_{p_2+1}\right) \end{aligned} \tag{5.46}$$

Here, the determination of the upper and lower limits of the control is mainly based on the most basic Type I error, that is, the expected false alarm rate, and the boundary division of the sequential sample values on the upper and lower boundaries. In particular, the control lower limit defines the error boundary for the smaller observations in the sample and the control upper limit defines the error boundary for the larger observations in the sample.

Step 4: Repeat step Step 1–Step 3 for many times to calculate the arithmetic mean value of the control limits obtained each time. As the final control limit, the updated sample point information can be input into the control chart to complete the construction of the control chart.

According to the number of control limits contained in the studied control chart, there are unilateral control charts such as mean control chart and bilateral control charts such as mean-range or mean standard difference. The truncated control chart presented in this paper is a unilateral control chart, which is caused by the characteristics of the truncated data itself. When the truncation mechanism is right truncated, the quality characteristics show the characteristics of high expectation. On the one hand, the upward shift of the process mean will increase the truncation proportion, resulting in less information contained in the observed value of the sample itself, and it is difficult to judge whether the mean value of the manufacturing process in which it is located has been shifted. On the other hand, in engineering applications, the characteristics of large-scale products often focus on the attenuation trend of the characteristics without worrying that the product characteristics are better than the design specification value, so it is reasonable and practical to establish A unilateral control limit $LCL = \frac{1}{2}\left(T'_{p_1} + T'_{p_1+1}\right)$ for the quality characteristics of right-truncated type.

In engineering applications, Average run length (ARL) is often regarded as the primary index to measure the effectiveness of control charts. Based on the results, the effect index only measures the observed values that are extracted in a statistical sense. When the process state is close to the anomaly but not obvious, the ARL comparison is no longer suitable. Therefore, this book will use a new index different from ARL, The Optimal Reaction Distance (R_dB), to evaluate the performance of truncated control chart from the perspective of preventive control. Where $R_d = Min|Test\ Statistic - Control\ Limit|$ refers to the point closest to the control limit among all sample points in the initial controlled state, that is, the most dangerous alarm point, corresponding to the optimal response distance. In this way, even when the process is preliminarily determined as steady state, the manufacturing process can be diagnosed and analyzed according to the most dangerous alarm points, and preventive improvements can be made, thus providing a strong guarantee for the manufacturing quality.

3. **Analysis of Numerical Examples**

The computer board is selected to generate the test data of the truncated characteristics of its reliability life collected on site. With reference to the method proposed in this book, the control chart based on the idea of successive progressive reduction is constructed to analyze the test data of the truncated characteristics, so as to effectively guarantee the output quality of the computer board components.

Table 5.5 (only the top 20 groups are listed) shows the initial sample data of 100 groups with a capacity of 5 whose potential distribution conforms to the Weibull distribution with shape parameter and size parameter both being 1, totaling 500 data. Among them, the sample data truncation ratio p_c reaches 75%, and the truncation level value is $C = 0.279$.

Using the convergent CEV asymptotic reduction step shown in the above formula, Matlab programming iterative operation is carried out to estimate the parameter values of Weibull distribution. For 500 initial sample data in Table 5.5, after initial value $\alpha = 0.5,\ \beta = 0.5$ is assigned to participate in iterative operation, the parameter value of Weibull distribution obtained is $\hat{\alpha} = 0.9916,\ \hat{\beta} = 0.7315$, and the truncated expected value $\hat{v}_{cc} = 3.1483$ determined by CEV quasi-true reduction method is converged. Then, the final truncated quality characteristic data is restored and updated according to the above formula.

Thus, according to the process parameter $\hat{\alpha} = 0.9916$ of the iteration output, set the number of iterations $t = 10^5$, and expand the programming according to the determination method of the control lower limit in the previous section, the control lower limit summary matrix $T\_LCL = [LCL_1, LCL_2, \ldots, LCL_t]$ can be obtained. From the summary matrix T_LCL, the control lower limit $LCL = 0.6345$ of convergent CEV can be obtained.

Further, the following effect analysis can be carried out.

The traditional Weibull CEV method is a conditional expected value method based on Weibull coordinate scale. When constructing the control chart, the corresponding process parameter is determined as $\hat{\alpha}' = 0.9989,\ \hat{\beta}' = 0.9935$. Because the standard form $f(w; 1, \beta)$ of Weibull distribution still depends on the shape parameter β, and the relationship with β is very complicated, it is impossible to separate the parameters. Therefore, for different shape parameters, different control limits of CEV $\overline{X}$ control chart will be simulated, and there is no unified general chart as a reference. Since

Table 5.5 Initial sample data (top 20 groups)

No.	Sample value					No.	Sample value				
1	0.2790	0.2448	0.2790	0.2790	0.2790	11	0.2790	0.2790	0.2790	0.2790	0.2790
2	0.2790	0.2790	0.2790	0.2790	0.2790	12	0.2790	0.2790	0.2790	0.2790	0.2790
3	0.0030	0.2790	0.1312	0.0686	0.1992	13	0.2790	0.2790	0.2790	0.2790	0.2790
4	0.2790	0.1705	0.2790	0.2095	0.2790	14	0.1725	0.2790	0.2790	0.2428	0.2790
5	0.2790	0.0809	0.2790	0.2790	0.2790	15	0.2790	0.2790	0.2093	0.1246	0.2790
6	0.2790	0.2601	0.0372	0.2787	0.2790	16	0.2790	0.2790	0.2790	0.0902	0.2790
7	0.2790	0.2790	0.2790	0.2790	0.2790	17	0.2790	0.2790	0.2790	0.2790	0.2790
8	0.2790	0.2790	0.0274	0.0286	0.0345	18	0.0436	0.1868	0.0255	0.2790	0.2790
9	0.2790	0.2790	0.2790	0.0121	0.2790	19	0.2790	0.2654	0.2790	0.2790	0.1788
10	0.2790	0.2790	0.2790	0.1460	0.2790	20	0.0784	0.0678	0.2790	0.1057	0.2790

the standard form of the exponential distribution model is independent of the scale parameter α and shape parameter β of the original Weibull distribution, it is possible to draw the upper and lower limit curves of the standard control under the standard distribution model. Then through a simple formula transformation, we can get that the control limit of CEV $\overline{X}$ control chart under any parameter is $LCL = \left(\alpha lcl^{X}\right)^{\frac{1}{\beta}}$. so the control lower limit of control chart determined by traditional CEV method is $LCL' = (0.9989 * 0.289)^\wedge\left(\frac{1}{0.9935}\right) = 0.2866$. For actual process control, the control limits and CEV control limits of convergent CEV asymptotic reduction are reflected on the control diagram, as shown in Fig. 5.15.

Figure 5.15 shows that all the points are above the control limit, and the process is preliminarily judged to be controlled. The truncated characteristic control chart based on the asymptotic reduction idea of convergent CEV proposed in this book can be used to monitor the process. Then, the optimal reaction distance of the two is compared. The control limit (dashed line) under the asymptotic reduction of convergent CEV in Fig. 5.15 corresponds to $R_d = Test\ Statistic_{(3)} - LCL$, that is,$R_d = 0.71006 - 0.6345 = 0.07556$. The initial control limit (solid line) corresponds to $R_d = 0.3771$, which shows that the control limit under convergent CEV asymptotic reduction can locate the most dangerous alarm point in time and show better sensitivity. At the same time, Monte Carlo simulation was used to compare the average run length O-ARL indicator size of traditional CEV control chart and convergent CEV control chart under the same truncated data, as shown in Fig. 5.16 (left) and (right).

It can be seen that under the same truncated ratio, the O-ARL level of the convergent CEV control chart is lower than that of the traditional CEV control chart, indicating that the convergent CEV control chart can detect the abnormal points and unstable states of the process as early as possible.

To sum up, in view of the need for more accurate analysis and monitoring of truncated characteristics in the manufacturing process of washing machine computer

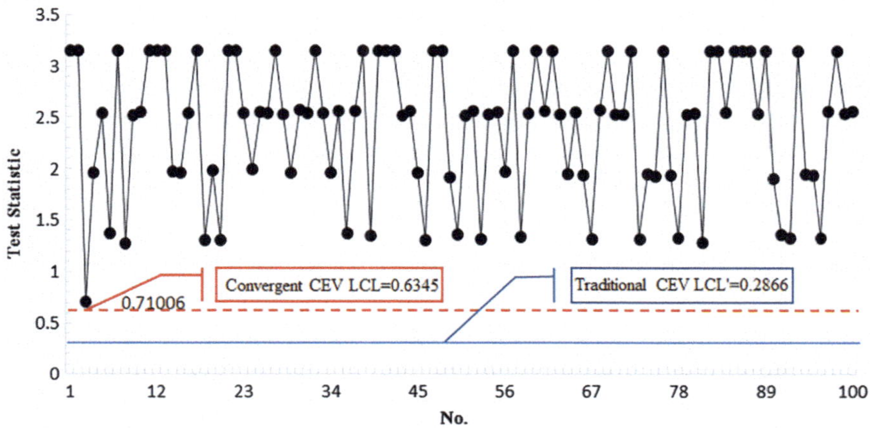

Fig. 5.15 Comparison of truncated characteristic control charts that affect the life of a computer board in a washing machine

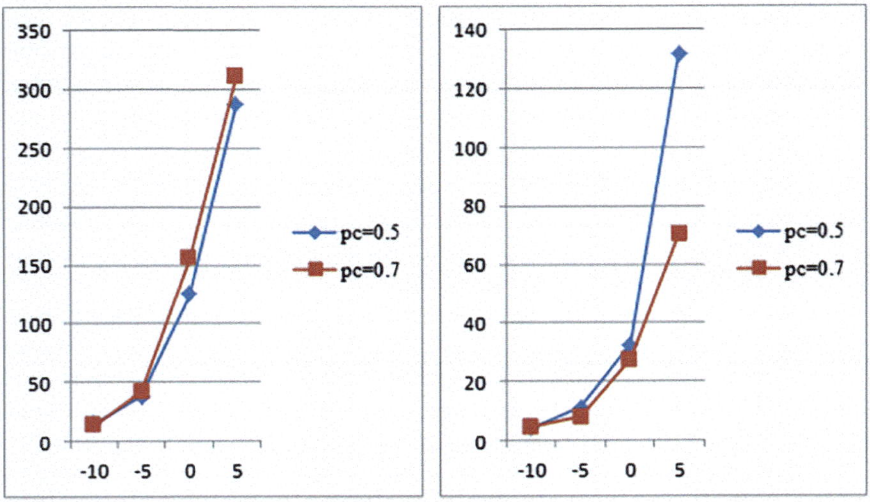

Fig. 5.16 Traditional CEV (left) and convergent CEV (right) control diagram O-ARL

board, a quasi-true reduction method of truncated characteristics based on convergent CEV is proposed based on the idea of data asymptotic. On this basis, the parameter estimation method of truncated Weibull distribution model and the detailed steps of design control chart are given. The effectiveness of the research method is verified by controlling the truncated characteristics which affect the reliability life of the computer board. The results show that the convergent CEV asymptotic reduction method is a universal truncated data processing method, which breaks through the limitations of the traditional CEV method in the type of control chart, and can give the general monitoring parameters, construction steps and performance analysis indicators, and obtain the applicability and operability of the truncated control chart. For monitoring truncated quality characteristics, compared with the traditional CEV control chart, the process control limit is more accurate, and the implementation steps are more clear. In the real-time monitoring of the quality of the manufacturing process of the washing machine computer board, the dynamic change trend of the studied truncated characteristics can be keenly found, which creates conditions for the first time to discover process anomalies and determine the design prevention point to capture the focus of process monitoring.

5.3.4 Manufacturing Process Product Reliability Decline Risk Control Technology Based on Profile Monitoring

1. **Reliability Slide Risk Control Framework Based on Profile Monitoring**

The content of manufacturing process product reliability decline risk control is based on the identification of manufacturing process reliability decline risk factors, and evaluation of reliability decline risk factors on the actual impact of manufacturing process product reliability decline risk, and corresponding to the content of Chaps. 2 and 3 of this book. Generally, the methods to identify the factors that lead to the risk of reliability decline in the manufacturing process include process FMEA, potential defect detection in the manufacturing process, aging technology, etc. Firstly, the weak process links are identified by analyzing the risk priority number of the relevant fault modes in the production process through process FMEA, and then the potential defect detection and aging technology in the manufacturing process are used. It can detect and identify manufacturing defects during the manufacturing process and before the product is delivered to the user, and provide data support for the establishment of the reliability decline mechanism of the manufacturing process. At the same time, in order to evaluate the actual impact of factors affecting reliability decline risk on the reliability decline risk of manufacturing process products, this book puts forward a reliability decline risk assessment method based on quality deviation from the perspective of manufacturing deviation transmission to quantitatively assess the reliability decline risk of manufacturing process products.

The risk of product reliability decline in manufacturing process mainly comes from the quality deviation of people, machines, materials, methods, loops and measurements in manufacturing process. In order to make the reliability of the products produced by the manufacturing process meet the design requirements, it is necessary to monitor the KQCs (including the abnormal components of the manufacturing system and the deviation of the quality characteristics of the manufacturing process) that cause the risk of product reliability decline in the manufacturing process. Based on the influence relationship of KQCs deviation in the manufacturing process in the RQR chain on the reliability of the manufacturing process products, the reliability of the products produced in the manufacturing process can be ensured by monitoring this influence relationship to meet the reliability requirements in the design stage. Therefore, how to monitor this influence relationship will be the key to control the reliability decline risk of the manufacturing process products.

In this book, the profile monitoring technology [9], which aims to monitor the functional relationship between variables, is intended to describe and monitor this influence relationship. As shown in the upper part of Fig. 5.17, the main reason for the decline risk of product reliability in manufacturing process is the decline in manufacturing process quality caused by abnormal manufacturing system, and then the accumulation of quality deviation in manufacturing process leads to the decline risk of product reliability, which is an RQR chain deviation transmission relationship. Therefore, in order to control the decline risk of manufacturing process product reliability, it is necessary to use profile monitoring technology to monitor the relationship between manufacturing system reliability R and manufacturing process quality Q in the RQR chain (RQ effect), and the relationship between manufacturing process quality Q and manufacturing process product reliability R (QR effect).

As shown in the lower part of Fig. 5.17, the profile monitoring technology is used to analyze and monitor the influence relationships in phase I and phase II.

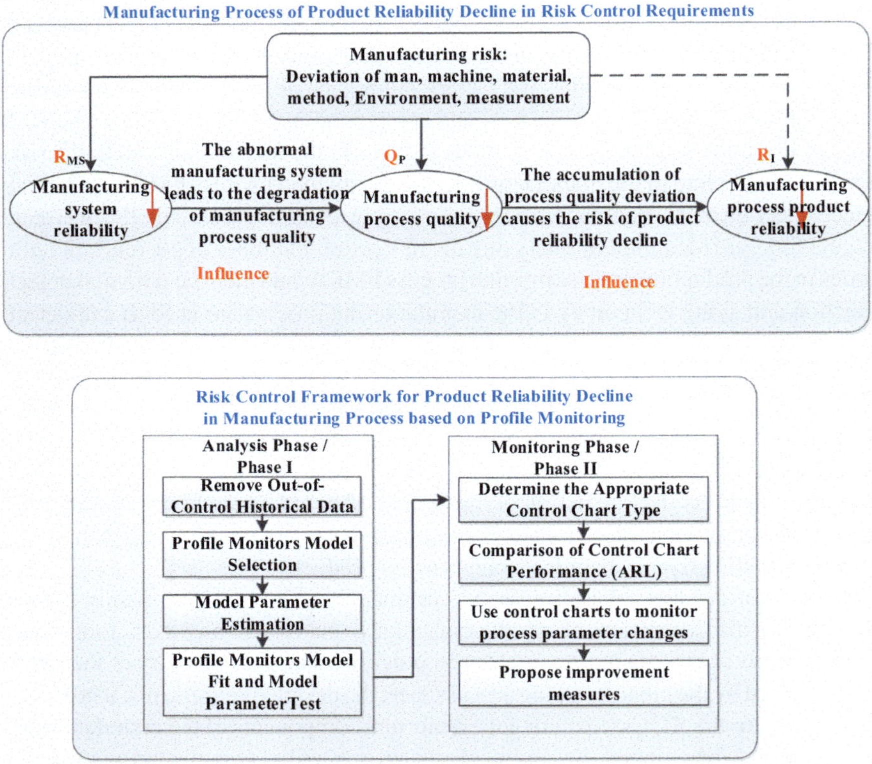

Fig. 5.17 Profile monitoring based manufacturing process product reliability decline risk control framework

(1) Analysis stage. The analysis stage of profile monitoring is mainly to complete the analysis, establishment and testing of the profile monitoring model: After distinguishing the controlled and uncontrolled KQCs data of manufacturing process, the uncontrolled KQCs data are eliminated, the corresponding profile monitoring model is established by using the controlled KQCs data, the main parameters of the model are estimated, and the established model and the estimation are tested. Here, profile monitors the category of the model. According to the function types formed between the description variables and the response variables in profile, it can be divided into linear and nonlinear categories. And further can be subdivided into Simple linear (Simple linear), Multiple linear (Multiple linear), Binary (Binary), general nonlinear (Non-linear), Polynomial (Polynomial) and other types.

(2) Monitoring stage. The monitoring stage of profile monitoring is mainly based on the parameters of the profile monitoring model estimated in the analysis stage to determine the appropriate monitoring control chart so as to monitor the anomalies of the KQCs sample data in the manufacturing process as soon as possible. Generally, Average run length (ARL) is used as an evaluation index

to evaluate the performance of control charts. Then, the control chart is used to monitor the KQCs data of manufacturing process, and the corresponding improvement measures are proposed.

2. **The Downside Risk Monitoring Model for Binary Profile**

According to the profile monitoring model, the response variables are mostly assumed to be continuous, and the case of discrete variables is less considered, but in actual production, the response variables are likely to be discrete, such as Binary, that is, the data only qualified and unqualified two cases. This type of data is becoming more and more common in actual production, but at the same time, it is often easy to be ignored, resulting in a series of problems. Therefore, in the case of binary response variables, this book uses profile monitoring technology to monitor the relationship between manufacturing system reliability R in the RQR chain and manufacturing process quality Q (RQ effect), and the relationship between manufacturing process quality Q and manufacturing process product reliability R (QR effect). It is expected that a more complete and general monitoring and control method of slide risk for binary profile analysis stage and monitoring stage can be proposed.

To this end, first of all, in the analysis stage, this book establishes a slide risk monitoring model for binary profile through logistic regression model, and estimates and tests the relevant parameters of the model according to maximum likelihood estimation (MLE) and Chi-square goodness of fit test, respectively. Finally, the effectiveness of the slide risk monitoring model is verified by using an example of aviation alloy fasteners.

(1) Model Establishment

(i) Construction of Binary Profile Monitoring Model

Suppose n relatively independent experiments are carried out, and the descriptive variable X_i of each experiment has p dimensions,$X_i = \left(x_{i1}, x_{i2}, \ldots, x_{ip}\right)^T$, $i = 1, 2, \ldots, n$. The corresponding response variable here is defined as Z_i, and Z_i follows the Bernoulli distribution with parameter π_i, whose expectation and variance can be expressed as $E(z_i) = \pi_i$, $\text{var}(z_i) = \pi_i(1 - \pi_i)$. The relationship between π_i and x_i can be expressed by the logistic regression model function $g(\pi_i)$, as follows:

$$g(\pi_i) = \log \frac{\pi_i}{1 - \pi_i} = \beta_1 x_{i1} + \beta_2 x_{i2} + \cdots + \beta_p x_{ip} \tag{5.47}$$

where $\beta = \left(\beta_1, \beta_2, \ldots, \beta_p\right)^T$ is the final parameter to be estimated by the model, the log operation represents the natural logarithm ln, and π_i represents the success probability of the Bernoulli distribution, while $x_{i1} = 1$ is set for Formula (5.47) and β_1 becomes the intercept of the model.

At the same time, the exponent with base e on both sides of Eq. (5.47) is calculated, then (where $\eta_i = X_i\beta = \sum_{j=1}^{p} \beta_j x_{ij}$):

$$\pi_i = \frac{\exp(x_i^T \beta)}{1 + \exp(x_i^T \beta)} = \frac{\exp(\eta_i)}{1 + \exp(\eta_i)} \tag{5.48}$$

At the same time, suppose that there are m_i ($i = 1, 2, \ldots, n$) observations corresponding to each i experiment, and mark $M = \sum_{i=1}^{n} m_i$ as the total number of observations in the experiment, then $y_i = \sum_{j=1}^{m_i} z_{ij}$ can be obtained, where z_{ij} is the *ith* observation in the *jth* experiment. Thus each experiment follows the Bernoulli distribution of (m_i, π_i) with an expectation equal to $E(y_i) = m_i\pi_i$.

Assuming that m_i observations in each experiment are relatively independent, the joint probability density function of $y_1, y_2, \ldots, y_n$ is:

$$L(\pi; y) = \prod_{i=1}^{n} \binom{m_i}{\pi_i} \pi_i^{y_i} (1 - \pi_i)^{m_i - y_i} \tag{5.49}$$

where $\pi = (\pi_1, \pi_2, \ldots, \pi_n)^T$, $Y = (y_1, y_2, \ldots, y_n)$. Taking logarithms of both sides of Eq. (5.49) at the same time, and using $\eta_i = X_i\beta = \sum_{j=1}^{p} \beta_j x_{ij} = \log\left(\frac{\pi_i}{(1-\pi_i)}\right)$ to substitute for simplification, Eq. (5.49) can be rewritten as:

$$l(\beta; y) = \sum_{i=1}^{n} \log\binom{m_i}{\pi_i} + \sum_{i=1}^{n}\sum_{j=1}^{p} y_i\beta_j x_{ij} - \sum_{i=1}^{n} m_i \log\left[1 + \exp\left(\sum_{j=1}^{p} \beta_j x_{ij}\right)\right] \tag{5.50}$$

Then take the partial derivative of Eq. (5.50) with respect to β:

$$\frac{\partial l(\beta; y)}{\partial \beta} = X^T y - \sum_{i=1}^{n} m_i\pi_i X_i = X^T(y - \mu) \tag{5.51}$$

For Formula (5.51), where $\mu = (\mu_1, \mu_2, \ldots, \mu_n)^T = E(y) = (m_1\pi_1, m_2\pi_2, \ldots, m_n\pi_n)^T$ and $X = (X_1, X_2, \ldots, X_n)^T$ is a $n \times p$ C-dimensional matrix.

It is asked to solve the maximum likelihood estimator (MLE) of the final parameter β, which can be solved from the formula of $X^T(y - \mu) = 0$, where $0 = (0, 0, \ldots 0)$ is the p-dimensional zero vector. Let $\hat{\beta} = (\hat{\beta}_1, \hat{\beta}_2, \ldots, \hat{\beta}_p)^T$ be the estimate of β. Therefore:$(\hat{\eta}_1, \hat{\eta}_2, \ldots, \hat{\eta}_n)^T = X\hat{\beta}$, $\hat{\pi}_i = \frac{\exp(\hat{\eta}_i)}{1+\exp(\hat{\eta}_i)}$, $i = 1, 2, \ldots, n$.

Let's also assume $\hat{W} = diag\{m_1\hat{\pi}_1(1 - \hat{\pi}_1), m_2\hat{\pi}_2(1 - \hat{\pi}_2), \ldots, m_n\hat{\pi}_n(1 - \hat{\pi}_n)\}$, where $\hat{W}$ is a diagonal matrix of $n \times n$ dimensions. Also assume $q = (q_1, q_2, \ldots, q_n)^T$, where $q_i = \hat{\eta}_i + \frac{(y_i - m_i\hat{\pi}_i)}{[m_i\hat{\pi}_i(1-\hat{\pi}_i)]}$, $i = 1, 2, \ldots, n$. Rewrite the formula of q_i:

$$q = \hat{\eta}_i + (\hat{W})^{-1}(y - \hat{\mu}) = X\hat{\beta} + (\hat{W})^{-1}(y - \hat{\mu}) \tag{5.52}$$

where $\hat{\mu} = (m_1\hat{\pi}_1, m_2\hat{\pi}_2, \ldots, m_n\hat{\pi}_n)^T$. Multiply both sides of Eq. (5.52) by $X^T\hat{W}$:

$$X^T\hat{W}q = X^T\hat{W}X\hat{\beta} + X^T\hat{W}(\hat{W})^{-1}(y-\hat{\mu}) = X^T\hat{W}X\hat{\beta} + X^T(y-\hat{\mu}) \qquad (5.53)$$

Assuming $X^T(y-\mu) = 0$, the estimated value of the final parameter β satisfies:

$$\hat{\beta} = (X^T\hat{W}X)^{-1}X^T\hat{W}q \qquad (5.54)$$

(ii) Iterative Solution of Monitoring Model Parameters

In the final solution and establishment of qualified parameter β, the process satisfies the following algorithm:

(a) When solving parameter β, it needs to be initialized. The initial label is $\hat{\beta}^0$, and $\hat{\beta}^0$ can be obtained by ordinary least squares estimation (OLS) [10], and the initial setting is i = 0;
(b) Based on $\hat{\beta}^i$, calculate $\hat{\eta}^i$, $\hat{\pi}^i$, $\hat{\mu}^i$ and $\hat{W}^i$;
(c) Calculate $\hat{q}^i = \hat{\eta}^i + \left(\hat{W}^i\right)^{-1}\left(y - \hat{\mu}^i\right)$;
(d) The number of items is increased by $\beta^{(\hat{i}+1)} = \left(X^T\hat{W}^iX\right)^{-1}X^T\hat{W}^i\hat{q}^i$. The parameter is updated by β, and the iterative calculation is set by $i = i + 1$;
(e) Repeat steps (b)–(d), l times, until $\frac{\left\|\hat{\beta}^l - \beta^{\hat{l}-1}\right\|}{\left\|\beta^{\hat{l}-1}\right\|} \le \alpha$, where $\|v\|$ is the Euclidean distance of the parameter v, and α will be a small quantity by default, such as $\alpha = 10^{-5}$.

The last $\hat{\beta} = \hat{\beta}^l$ is the β estimate that we need to find.

Based on the estimated value of β, estimates of $\hat{\pi}_i = \frac{\exp(\hat{\eta}_i)}{1+\exp(\hat{\eta}_i)}$, $i = 1, 2, \ldots, n$, $\hat{W}$ and $\hat{W} = diag\{m_1\hat{\pi}_1(1-\hat{\pi}_1), m_2\hat{\pi}_2(1-\hat{\pi}_2), \ldots, \hat{\pi}_n(1-\hat{\pi}_n)\}$ can be obtained, as well as estimates of other process parameters.

(2) Numerical Example Verification

(i) Aviation Alloy Fasteners

Alloy fasteners are a general term for a class of mechanical parts used when two or more parts (or components) are tightly connected into a whole. In modern aircraft, whether it is military or civilian, a large number of mechanical connection methods of fasteners are used to connect the parts, components and components of the aircraft as a whole. The use of fasteners is as little as hundreds of thousands of pieces, and more than one million pieces. Its type structure is diverse, and it must be selected according to the different use parts, use environment, use load, life requirements, and installation methods.

At present, the world's aviation fasteners are mainly monopolized by the world's largest fastener companies such as Alcoa, the United States Precision Casting (SPS) company, the French Lisi Hi-shear, the United States Boeing, and the United States Monogram Company and other professional companies. The Russian and Chinese aerospace fasteners have their own systems for domestic use. Alcoa's specific products are as shown in Table 5.6.

The number of aviation alloy fasteners is large, the degree of universality is high, and the reliability of its product quality is of great importance to social security and national economic development.

The practical examples in this book are from Montgomery et al.'s Introduction to Linear Regression Analysis (2006, p. 466) [11], which selected qualified data for products under controlled conditions. The method is to apply a certain pressure to the aviation alloy fasteners, determine the limit value of the pressure, the result is binary type, only "pressure" and "fracture" these two results, according to its performance to determine the stability of its quality.

Where the describing variable x is the number of Pounds per square inch (PSI) of pressure loaded, the applied pressure x is subject to 2500, 2700, ..., 4300. The response variable y is expressed as the number of product fractures in n independent repeated measurements of the *ith* experiment, which follows the Bernoulli distribution. This experiment is carried out 10 times cumulatively. The pressure, number of independent repeated measurements and number of fractures of each experiment are shown in Table 5.7.

(ii) Initialization of Model Parameters

According to the logistic regression model and the actual calculation examples and Eq. (5.47) of aviation fasteners in Table 5.7, the model can be obtained:

$$\log(\pi/[1-\pi]) = \beta_0 + \beta_1 \log(x) \tag{5.55}$$

where is A single measurement of the probability of alloy fastener load fracture, and set the final need to estimate the parameter $\beta = (\beta_0, \beta_1)^T$, the description variable x in the calculation process is expressed as a matrix, and at the same time according

Table 5.6 Typical types of aviation fasteners

Number	Fastener name	Product category
1	Bolt/screw	Countersunk bolts with various drive slots and hexagonal, twelve-square, D-head and splined first-class bolts
2	Nut	Self-locking nut, pallet nut, K-FastTM nut, rivet free plate nut
3	Single side fastener	Core-pulling rivets and threaded screws for metal structures and composite structures
4	Quick release fastener	Remove fastener
5	Ring groove fastening system	GP, LGP, XPL^{TM} ring slot nail fastening system
6	Bolting system	Edi-bot2 fastening system, lightweight high lock bolt fastening system
7	Special rivet	Plate single side rivets, titanium niobium rivets, bimetal rivets

Table 5.7 Loading test data of aviation alloy fasteners

Pressure, x(psi)	Sample size, n	Number of breaks, y
2500	50	10
2700	70	17
2900	100	30
3100	60	21
3300	40	18
3500	85	43
3700	90	54
3900	50	33
4100	80	60
4300	65	51

to Eq. (5.47) habitual setting $x_{i1} = 1$, the first column of the matrix is equal to 1.

$$X = \begin{pmatrix} 1 & 1 & \cdots & 1 & 1 \\ \log(2500) & \log(2700) & \cdots & \log(4100) & \log(4300) \end{pmatrix}^T \tag{5.56}$$

When solving parameter β, it is necessary to use ordinary least square method for initialization. Here, we use the data in Table 5.7 and Matlab software to solve, and the code of least square initialization is as shown in Table 5.8.

After the operation, we get the value $\hat{\beta}_0 = (41.3328, -5.0835)^T$ of the initial β.

At the same time, with $a = \log(x)$ as the horizontal coordinate and $b = \log\left(\frac{\pi}{[1-\pi]}\right)$ as the vertical coordinate, the effect diagram of least square method is obtained, as shown in Fig. 5.18. According to the effect diagram, the fitting effect is good.

(3) Calculation of Model Parameters

After initializing parameter β, its initial least squares estimate $\hat{\beta}^0$ is obtained, which is substituted into the iterative calculation of steps (b)–(e) in the iterative solution of

Table 5.8 Common least squares Matlab initialization code

```
y = [10, 17, 30, 21, 18, 43, 54, 33, 60, 51]
n = [50, 70, 100, 60, 40, 85, 90, 50, 80, 65]
p = y./n
p1 = log(p./(1 − p))
x1 = [log(2500), log(2700), log(2900), log(3100), log(3300), log(3500), log(3700), log(3900),
log(4100), log(4300)]
poo = polyfit(x1, p1, 1)
pa1 = polyval(poo, 0)
pa = polyval(poo, x1)
plot(x1, p1, 'o', x1, pa)
```

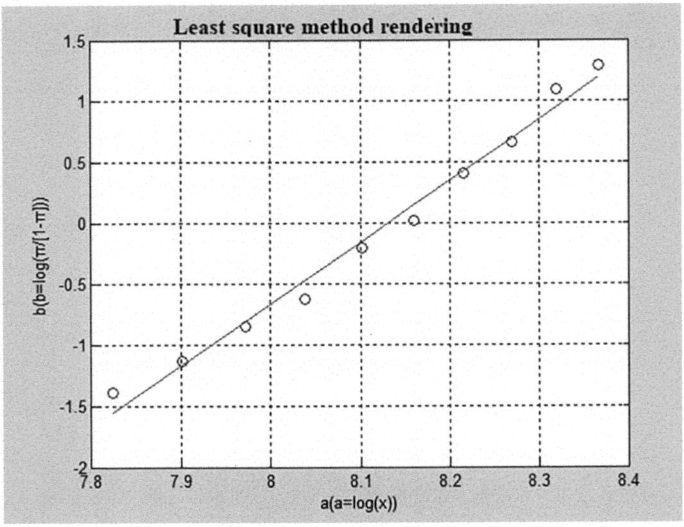

Fig. 5.18 Effect of the common least square method

monitoring model parameters. Also using Matlab to achieve logical calculation, the cyclic function is compiled as shown in Table 5.9.

The function is named "yc.m", where it is based on the final expression $\hat{\beta} = \left(X^T\hat{W}X\right)^{-1}X^T\hat{W}q$ of and the loop cutoff condition $\frac{\left\|\hat{\beta}^{l}-\beta^{\hat{l}-1}\right\|}{\left\|\beta^{\hat{l}-1}\right\|} \le \alpha$, where $\alpha = 10^{-5}$.

In the specific calculation process, first enter the initial estimate yc([41.3328; –5.0835]) in the main interface of Matlab, call the loop function, carry out the first step of operation, and obtain the first step result:

Table 5.9 Matlab calculation code of model parameters

```
function betai = f(beta0)
x = [1, log(2500); 1, log(2700); 1, log(2900); 1, log(3100); 1, log(3300); 1, log(3500); 1,
log(3700); 1, log(3900); 1, log(4100); 1, log(4300)]
m = [50; 70; 100; 60; 40; 85; 90; 50; 80; 65]
y = [10; 17; 30; 21; 18; 43; 54; 33; 60; 51]
eata = x * beta0
a = exp(eata)
pie = a./(1 + a)
zw = 1 – pie
w1 = m. * pie. * (1 – pie)
u = m. * pie
w = diag(w1)
q = x * beta0 + inv(w) * (y – u)
betai = inv((x' * w * x)) * x' * w * q
k = abs(sum((betai – beta0).^2)/(sum((beta0).^2)))
```

$$betai = (-60.7879, 7.4685)^T k = 6.1042$$

where k is not taken root, the final result must be less than 10^{-10}, while shown in Table 5.10.

After performing the operation 4 times, the required value of β is obtained, and the result is as follows:

$$\hat{\beta} = \hat{\beta}^l = (-42.1106,\ 5.1772)^T k = 7.5526 \times 10^{-13}$$

Then the value of the final demand parameter β of the model for the aviation alloy fastener example is: $\beta = (-42.1106,\ 5.1772)^T$.

(4) Testing of Models and Parameters

(i) Model Goodness of Fit Test

This book uses logistic regression model to fit the model of aviation alloy fasteners. In order to test The validity of model fitting, the chi-square goodness of fit test [11] was used to test the model.

Goodness of Fit testing is used to test the agreement between the actual number of observations and the theoretical number of observations calculated according to a certain hypothesis or model, through the hypothesis testing process. Based on the example of aviation alloy fasteners, the null hypothesis H_0 and the alternative hypothesis H_1 are proposed:

H_0 There is basically no difference between the sample and the theoretical model distribution.

H_1 The sample distribution is different from that of the theoretical model.

Chi-square goodness of fit test needs to judge whether the original hypothesis H_0 is rejected, and Pearson χ^2 statistic is used to measure the degree of fit, so as to judge. The formula is as follows:

$$\chi_p^2 = \sum_{i=1}^{k} \frac{(actual\,frequency - theoretical\,frequency)^2}{theoretical\,frequency} = \frac{(O_1 - E_1)^2}{E_1} + \frac{(O_2 - E_2)^2}{E_2} + \cdots + \frac{(O_k - E_k)^2}{E_k}$$

At the same time, a certain significance level a is usually given, and the critical value χ_a^2 can be obtained by checking the distribution table of χ^2 to make $P = \left(\chi^2 > \chi_a^2\right) = a$, so that the critical value χ_a^2 is compared with Pearson's

Table 5.10 Cyclic calculation of model parameter 1

```
i = 1; while i < 10
yc(ans)
i = i + 1
```

statistic χ_p^2 to obtain the rejection domain of the Chi-square goodness of fit test. If the Pearson statistic is greater than the critical value, the null hypothesis is rejected: $\chi_p^2 > \chi_a^2(\nu) = \chi_a^2(k - r - 1)$. Where ν represents the number of degrees of freedom, k represents the number of experimental groups in the sample, and r represents the number of parameters calculated when calculating the theoretical distribution.

The observed number y is tested, and the calculation process of Chi-square goodness of fit test is shown in Table 5.11.

Where n and y correspond to the number of groups and observations in Table 5.11 respectively, π can be obtained by $\pi_i = \frac{\exp(x_i^T\beta)}{1+\exp(x_i^T\beta)} = \frac{\exp(\eta_i)}{1+\exp(\eta_i)}$ in the previous section, theoretical frequency $yl = n \times \pi$, chi-square component equal to $(O_i - E_i)^2$, chi-square statistic equal to $\frac{(O_i-E_i)^2}{E_i}$, where $i = 1, 2, \ldots, 10$.

Finally, by calculating the Pearson χ^2 statistic, χ_p^2 is equal to 0.753187322, which is much smaller than the critical value of 14.067 for χ_a^2 (where the degree of freedom ν is equal to $10 - 1 - 2 = 7$, and the significance level a is 0.05). Therefore, the original hypothesis H_0 cannot be rejected, that is, the logistic regression model adopted is basically no different from the sample distribution, and the fitting degree is good.

(ii) Model Binomial Coefficient Test

For logistic regression models, we initially assume $\log\left[\frac{\pi}{(1-\pi)}\right] = \beta_0 + \beta_1 \log(x)$, but do not verify whether the model has a binomial coefficient $\log\left[\frac{\pi}{(1-\pi)}\right] = \beta_0 + \beta_1 \log(x) + \beta_2(\log(x))^2$. Next, the Chi-square test is also used for verification, and the principle is similar to the Chi-square goodness of fit test:

$$H_0\colon \beta_2 = 0 \quad versus \quad H_1\colon \beta_2 \neq 0$$

Table 5.11 Chi square goodness of fit test table

Group number (n)	Observed number (y)	Theoretical probability (π)	Theoretical frequency (yl)	χ^2 component number (a)	χ^2 statistic (b)
50	10	0.1674	8.37	2.6569	0.317431302
70	17	0.2304	16.128	0.760384	0.047146825
100	30	0.3024	30.24	0.0576	0.001904762
60	21	0.3797	22.782	3.175524	0.139387411
40	18	0.4583	18.332	0.110224	0.006012655
85	43	0.5344	45.424	5.875776	0.129353998
90	54	0.6048	54.432	0.186624	0.003428571
50	33	0.6677	33.385	0.148225	0.004439868
80	60	0.7225	57.8	4.84	0.083737024
65	51	0.7691	49.9915	1.01707225	0.020344904

Theoretical frequency must meet $yl > 5$. If there is one or more $yl < 5$, it should be combined with neighboring, until it is greater than or equal to 5.

By calculation, the chi-square statistic is 2.595×10^{-7}, much less than $\chi^2_{1,0.05} = 3.841$ (Degree of freedom is 1, and significance level is 0.05). The null hypothesis can not be rejected, so it is judged that there is no quadratic term in the logistic regression model for the sample of aerospace alloy fasteners.

3. **Downside Risk Control for Binary Profile**

This section mainly calculates the average running chain length (ARL) of the two control charts by selecting the Shewhart control chart and EWMA control chart based on the likelihood ratio test suitable for monitoring binary profiles, and based on the parameters of the logistic regression model obtained in the previous section and the Monte-Carlo simulation to be used next. The performance of the control chart is evaluated to provide effective method support for the downside risk control of binary profile.

(1) Two Control Charts Based on Likelihood Ratio Test

(i) Likelihood Ratio Test for Binary Profile

The likelihood ratio test (LRT) is usually used to evaluate the fit between the two models to be compared and the collected data, and the model is judged mainly through the test of parameters. In a binary profile, you can first assume that the response variable Y follows the Bernoulli distribution of the parameter, and at the same time test the applicability of the model parameters by assuming:

$$A = \{0.05, 0.07, 0.09, 0.09, 0.2, 0.1, 0.15, 0.1, 0.15\}$$
$$A_1 = \{0.3, 0.3, 0.4\}$$

Similar to $\pi_i = \frac{\exp(x_i^T\beta)}{1+\exp(x_i^T\beta)} = \frac{\exp(\eta_i)}{1+\exp(\eta_i)}$ in "4.2.1 Model Building", there is $p(x) = \frac{\exp\left(\beta_0+\sum_{j=1}^{p}\beta_i x_i\right)}{1+\exp\left(\beta_0+\sum_{j=1}^{p}\beta_i x_i\right)}$, $p^* = \frac{\exp\left(\beta_0^*+\sum_{j=1}^{p}\beta_i^* x_i\right)}{1+\exp\left(\beta_0^*+\sum_{j=1}^{p}\beta_i^* x_i\right)}$. Distinguish H_0 and H_1 at the same time, and assume $(\beta_0, \beta_1, \ldots, \beta_p) \neq \left(\beta_1^*, \beta_2^*, \ldots, \beta_p^*\right)$. Or at least one of the parameters between the two is different, and $(\beta_0, \beta_1, \ldots, \beta_p)$ is known and $\left(\beta_1^*, \beta_2^*, \ldots, \beta_p^*\right)$ is estimated during the monitoring phase.

For H_0 and H_1, the likelihood function satisfies:

$$H_0{:}\, L_{H_0} = p(x)^{\sum_{i=1}^{n} Y_i}(1-p(x))^{n-\sum_{i=1}^{n} Y_l} H_i{:}\, L_{H_1} = \hat{p^*}(x)^{\sum Y_i}\left(1-\hat{p^*}(x)\right)^{n-\sum_{i=1}^{n} Y_l}$$

where the estimates satisfy: $\hat{p^*} = \frac{\exp\left(\hat{\beta_0}^*+\sum_{j=1}^{p}\hat{\beta_i}^* x_i\right)}{1+\exp\left(\hat{\beta_0}^*+\sum_{j=1}^{p}\hat{\beta_i}^* x_i\right)}$, where $\left(\hat{\beta_0}^*, \hat{\beta_2}^*, \ldots, \hat{\beta_p}^*\right)$ is the maximum likelihood estimate of $\left(\beta_1^*, \beta_2^*, \ldots, \beta_p^*\right)$.

Taking the logarithm of both sides of the equation:

$$H_0: L_{H_0} = \sum_{i=1}^{n} Y_i \log(p(x)) + \left(n - \sum_{i=1}^{n} Y_i\right) \log(1 - p(x))$$

$$H_1: L_{H_1} = \sum_{i=1}^{n} Y_i \log\left(\hat{p^*}(x)\right) + \left(n - \sum_{i=1}^{n} Y_i\right) \log\left(1 - \hat{p^*}(x)\right)$$

At the same time, the likelihood ratio test statistic is defined as λ and:

$$\lambda = L_{H_1} - L_{H_0}$$

If λ is greater than the parameter c determined according to the significance level α, H_0 will be eliminated. In the analysis phase, at least one parameter of $(\beta_0, \beta_1, \ldots, \beta_p)$ is out of bounds, so it needs to be re-calculated.

(ii) Shewhart Control Chart Based on Likelihood Ratio Test

The main difference between the Shewhart control chart based on likelihood ratio test and the traditional Shewhart control chart is the choice of control limits. In general, its lower control limit is set to 0, and its upper control limit is related to the asymptotic distribution of the likelihood ratio statistics λ and 2λ. At the same time, when the sample size n is relatively large, the default error rate of 2λ is a, and the chi-square distribution of freedom k is $\chi^2_{k,a}$.

For the solution of the UCL of the Shewhart control chart based on the likelihood ratio test, the UCL can be approximately equal to χ^2_α when the sample size n is relatively large. In the case of relatively small sample size, UCL can be obtained by Bartlett correction factor method and Parametric bootstrap method. Only Bartlett correction factor method is introduced here.

The Bartlett correction factor method is very direct and simple to solve UCL, and the estimated value of UCL can be expressed as: $\hat{UCL} = \left(\frac{1+b}{n}\right)\chi^2_{k,a}$.

Where n is the sample size, b can be simulated by Monte Carlo method, and $\chi^2_{k,a}$ is the corresponding Chi-square distribution that 2λ follows when n is relatively large.

(iii) EWMA Control Chart Based on Likelihood Ratio Test

An EWMA (Exponentially Weighted Moving Average) control chart is an exponentially weighted moving Average (Exponentially weighted Moving Average) chart with good results for large migrations, but insensitive to small migrations. EWMA charts are exponentially more suitable for monitoring small migrations.

The statistics of the EWMA control chart can be constructed as follows:

$$EWMA_1 = \theta * 2\lambda_i + (1 - \theta)EWMA_{i-1}$$

where $i \geq 1$, and θ is the smoothing coefficient, satisfying $0 \leq \theta \leq 1$. The initial $EWMA_0$ is k, which is equal to the asymptotic average of $EWMA_1$.

Similarly, for the control limits of EWMA control charts based on likelihood ratio tests, the lower control limit LCL is set to 0, and the upper control limit UCL is usually solved by Monte Carlo method. At the same time, UCL can also be solved by bi-sectional search method or approximate estimation method, which will not be introduced in detail here.

(iv) Advantages of Likelihood Ratio Test Control Chart

Traditional control charts are suitable for a relatively narrow range of data types, but control charts based on likelihood ratio tests are suitable for both linear and nonlinear profile types as long as the likelihood function of the profile model can be established. And the range of distribution of the response variable Y is not necessarily ordinary, such as linear or normal distribution, in this case the response variable can be binary.

Meanwhile, in multivariate linear profiles, T^2 plots for multivariate variables and Shewhart-type.LRT control plots (i.e., Shewhart-type.LRT control plots) (i.e., EWMA-type.LRT control plots i.e., EWMA control plots based on likelihood ratio tests) generally perform equally, while EWMA-type.LRT control plots (i.e., EWMA control plots based on likelihood ratio tests) perform better than multivariate T2 plots. As the number of data samples N to be analyzed increases, the control chart based on likelihood ratio test is significantly better than the multivariate T2 control chart. Further, when dealing with binary profiles, control charts based on likelihood ratio tests can be very sensitive and effective in monitoring data variation or drift, which will be further analyzed and discussed in the following section.

(2) Performance Evaluation of the Likelihood Ratio Test Control Chart for Aviation Alloy Fasteners

For the loading experimental data of the aviation alloy fasteners shown above, its state is controlled and qualified, following the Bernoulli distribution of the parameter p(x), and then $\beta = (-42.1106, 5.1772)^T$, it can be known that $p(x) = \frac{\exp\left(\beta_0+\sum_{j=1}^{p}\beta_i x_i\right)}{1+\exp\left(\beta_0+\sum_{j=1}^{p}\beta_i x_i\right)} = \frac{\exp(-42.1106+5.1772\log(x))}{1+\exp(-42.1106+5.1772\log(x))}$, where x is 2500, 2700, …, 4100 arithmetic data.

At the same time, the change of parameter $\beta = (\beta_0, \beta_1)^T$ is simulated as follows:

$$\beta_0 \to \beta_0 + \Delta\beta_0, \Delta\beta_0 = 0.02\ ,\ 0.04\ ,\ \cdots\ ,\ 1.98\ ,\ 2.00\ ;$$

$$\beta_0 \to \beta_0 + \Delta\beta_0^{'}, \Delta\beta_0^{'}=\text{-}2.00\ ,\ -1.98\ ,\ \cdots\ ,\ -0.04\ ,\ -0.02\ ;$$

$$\beta_1 \to \beta_1 + \Delta\beta_1, \Delta\beta_1 = 0.02\ ,\ 0.04\ ,\ \cdots\ ,\ 1.98\ ,\ 2.00\ ;$$

$$\beta_1 \to \beta_1 + \Delta\beta_1^{'}, \Delta\beta_1^{'}=\text{-}2.00\ ,\ -1.98\ ,\ \cdots\ ,\ -0.04\ ,\ -0.02\ ;$$

Here, through the parameters of the logistic regression model, and in the case of the number of samples n = 100, the simulation of the specific value of the aviation alloy fasteners conforming to the Bernoulli distribution is expressed as shown in Table 5.12.

Table 5.12 Simulation data of aerospace alloy fasteners

Pressure, x(psi)	Sample size, n	Breaking number, y
2500	100	17
2700	100	24
2900	100	30
3100	100	38
3300	100	46
3500	100	53
3700	100	61
3900	100	67
4100	100	72
4300	100	77

At the same time, when the number of samples n = 100, we define the smoothness coefficient of EWMA-type.LRT control chart to be equal to 0.2, and the first type error rate α of the two control charts is defined as 0.0027 with the accuracy of 3σ, that is, ARL should be equal to $ARL_0 = \frac{1}{\alpha}$, that is, 370, when the control chart is controlled. Based on Monte Carlo method, we use Matlab to calculate the average running chain length ARL of EWMA-type.LRT and Shewhart-type.LRT control charts, and then compare the ARL_1 when the process is out of control. Correspondingly, the smaller the ARL_1 is, the ARL_0 of the two control charts is fixed. Indicates the more effectively the control chart detects this Shift.

The two control charts were simulated for 10,000 times respectively. The following are the specific changes of the ARL of EWMA-type.LRT and Shewhart-type.LRT control charts drawn by Excel when parameters β_0 and β_1 were offset, as shown in Figs. 5.19 and 5.20.

In Figs. 5.19 and 5.20, β_0 and β_1 are used as positive and negative transformations with deviation greater than zero and less than zero respectively. When parameter deviation occurs, it is hoped that the control chart can detect it as soon as possible. When β_0 is used for positive and negative transformation, the two figures in Fig. 5.19 are generally symmetric, indicating that positive and negative transformation of β_0 has no obvious difference in the changes of ARL. Meanwhile, with the gradual increase of the deviation of β_0, the average sample number in the out-of-control state of the test, namely ARL1, becomes smaller. In addition, for the two control charts, the ARL1 of EWMA-type.LRT control chart for β_0 offset is smaller on the whole than that of Shewhart-type.LRT control chart, that is, EWMA-type.LRT control chart will be more sensitive and effective in detecting β_0 offset here.

When parameter β_1 is offset by positive and negative transformations, the change of Fig. 5.20 is more gentle than that in Fig. 5.19, that is, the change of ARL is more sensitive to the change of parameter β_0. Similarly, the ARL1 of EWMA-type.LRT control chart when parameter β_1 is out of control is smaller than that of Shewhart-type.LRT control chart. Both Figs. 5.19 and 5.20 indicate that EWMA-type.LRT control chart is more effective for parameter detection here.

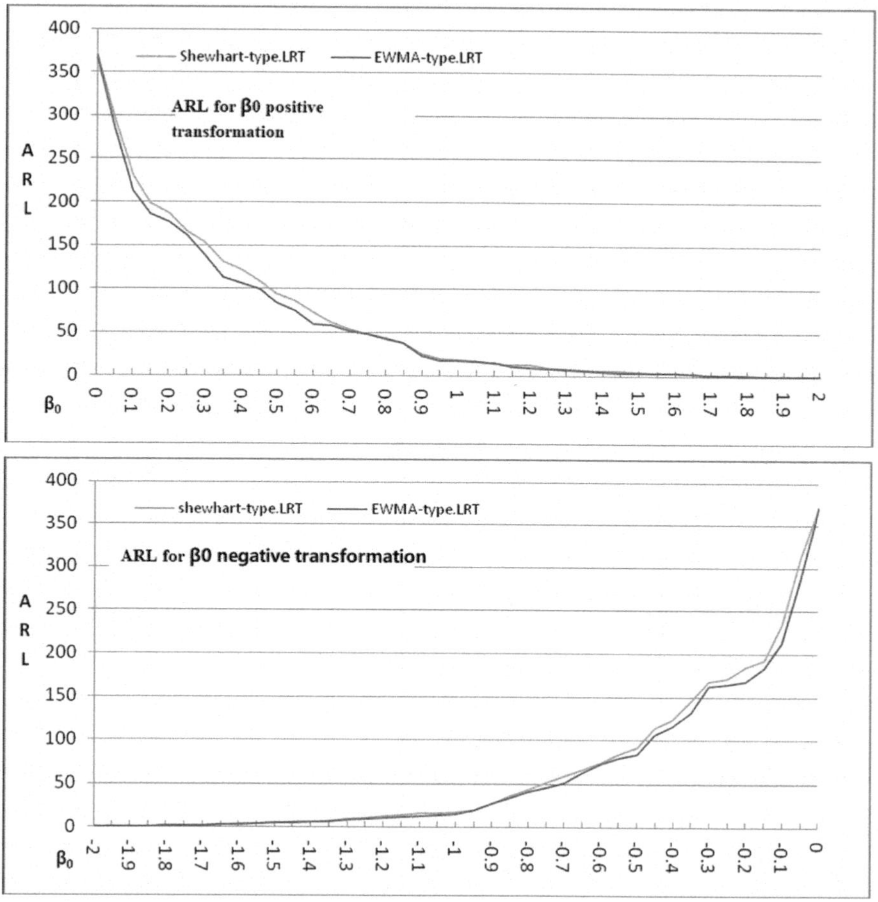

Fig. 5.19 Control chart ARL changes during β_0 positive/negative transformation

The following is part of the data from the Monte Carlo simulation calculation of two kinds of control chart ARL. The same conclusion can be reached by analyzing the data, the specific data are as shown in Table 5.13.

Figure 5.21 shows the scatterplot of 10,000 Monte Carlo simulation ARL calculations when the Shewhart-type.LRT control chart is controlled, that is, when β_0 and β_1 are offset. The horizontal axis is the specific value of 10,000–10,000 times, and the vertical axis is the specific ARL value of each simulation, whose ARL result is 372.67 in Table 5.13. From the scatter plot, we can find that the data is most closely distributed when it is close to 372.67, and only a few of the simulation calculations need to be done above 1000. In this experiment, the maximum number of samples is 3763. The average sample size is 372.67, which is the value of ARL we need, as shown in follows.

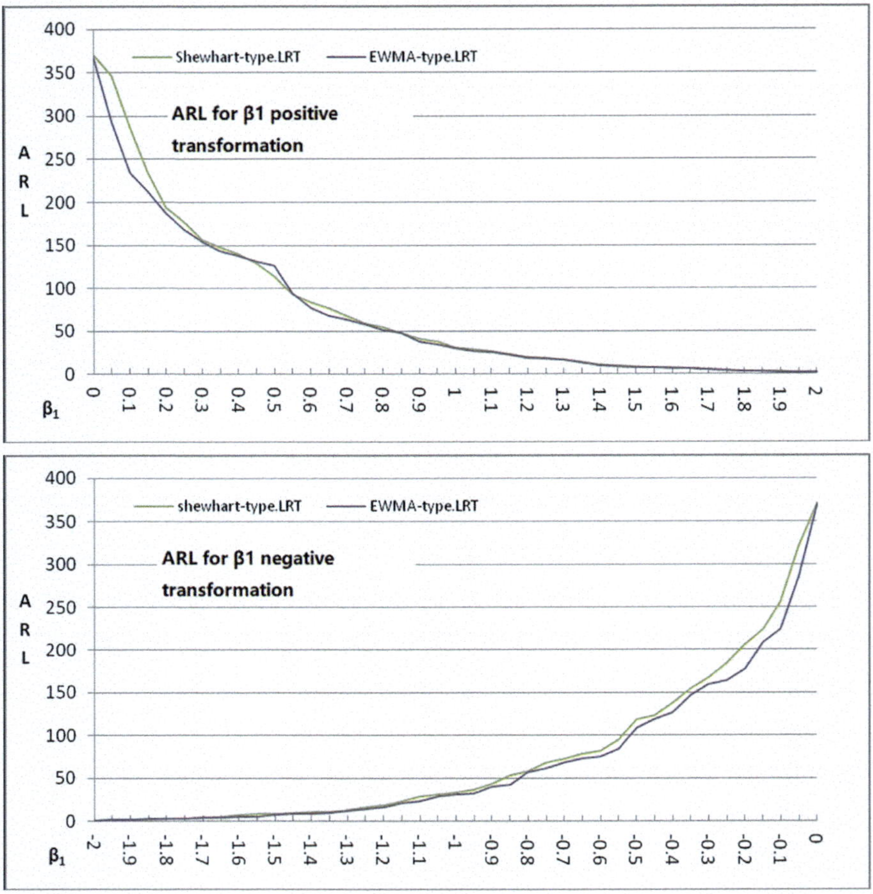

Fig. 5.20 Control chart ARL changes during β_1 positive/negative transformation

Table 5.13 Partial data of the control chart ARL calculated by Monte-Carlo method

$\Delta\beta_0$	$\Delta\beta_1$	$ARL_{Shewhart.LRT}$	$ARL_{EWMA.LRT}$	$\Delta\beta_0$	$\Delta\beta_1$	$ARL_{Shewhart.LRT}$	$ARL_{EWMA.LRT}$
− 2	0	0.85	0.88	0	− 2	1.53	1.47
− 1.8	0	1.74	1.69	0	− 1.8	3.24	3.13
− 1.6	0	4.17	3.08	0	− 1.6	7.38	5.2
− 1.4	0	6.59	6.24	0	− 1.4	10.89	9.36
− 1.2	0	13.13	13.11	0	− 1.2	18.62	16.48
− 1	0	17.86	15.84	0	− 1	33.47	31.26
− 0.8	0	43.63	41.25	0	− 0.8	58.25	57.04
− 0.6	0	74.37	73.3	0	− 0.6	82.01	75.42
− 0.4	0	123.9	116.52	0	− 0.4	137.9	126.54
− 0.2	0	184.52	168.4	0	− 0.2	206.43	176.87
0	0	372.67	369.82	0	0	371.24	368.5
0.2	0	187.43	177.54	0	0.2	193.43	187.52
0.4	0	121.72	106.76	0	0.4	140.35	137.4
0.6	0	73.21	59.43	0	0.6	83.78	77.56
0.8	0	43.8	42.2	0	0.8	54.36	51.24
1	0	18.54	17.82	0	1	31.26	30.5
1.2	0	12.91	9.45	0	1.2	20.3	18.77
1.4	0	6.75	5.24	0	1.4	11.39	9.94
1.6	0	4.24	4.1	0	1.6	7.32	7.13
1.8	0	1.65	1.57	0	1.8	3.18	3.06
2	0	0.87	0.9	0	2	2.6	2.4

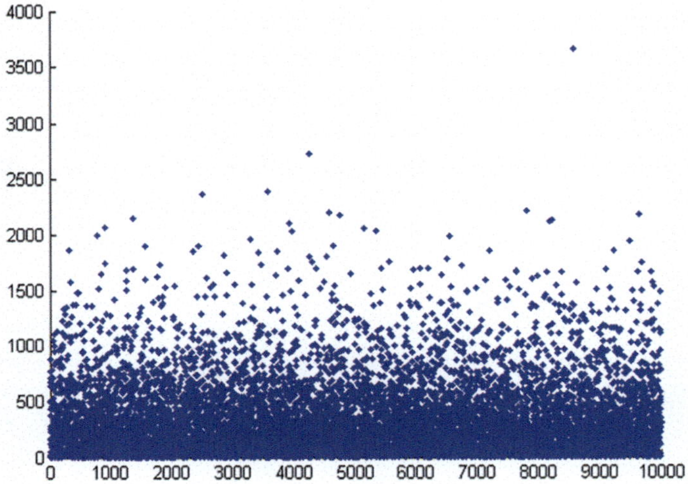

Fig. 5.21 Shewhart-type.LRT control chart scatterplot of ARL calculation when controlled

References

1. Kang Rui, He Yihai. Basis of Quality Engineering Technology. Beijing University of Aeronautics and Astronautics Press, 2012.
2. Thornton A C. A Mathematical Framework for the Key Characteristic Process. Research in Engineering Design, 1999, 11 (3): 145–157.
3. He Yihai, Tang Xiaoqing, Chang Wenbing, 2010. Technical decomposition approach of critical to quality characteristics for product design for six sigma. Quality and Reliability Engineering International. 26(4):325–339.
4. Yihai He, Changchao Gu, Zhenzhen He, Jiaming Cui. Reliability-oriented quality control approach for production process based on RQR chain. Total Quality Management & Business Excellence. 2018, 29(5–6):652–672.
5. Kourti T, Macgregor J F. Multivariate SPC methods for process and product monitoring. Journal of Quality Technology, 1996, 28 (4): 409–428.
6. Murthy, D. N. P. New research in reliability, warranty and maintenance. The 4th Asia-Pacific International Symposium on Advanced Reliability and Maintenance Modeling (APARM 2010), McGraw-Hill International Enterprises, 2010: 504–515.
7. Steiner, S. H., R. J. MacKay. Monitoring processes with data censored owing to competing risks by using exponentially weighted moving average control charts. Journal of the Royal Statistical Society. Series C: Applied Statistics, 2001, 50(3): 293–302.
8. Steiner, S. H., R. J. MacKay. Monitoring Processes with Highly Censored Data. Quality Control and Applied Statistics, 2001, 46: 139–140.
9. Woodall W H, Spitzner D J, Montgomery D C. Using control charts to monitor process and product quality profiles. Journal of Quality Technology, 2004, 36(3):309–320.
10. Wang Fuchang, Cao Huirong, Zhu Hongxia. Classical least squares and full least squares and their parameter estimation. Statistics and Decision, 2009, 2009(1):16–17.
11. Yang Zhenhai. Goodness of Fit Test. Science Press, 2011.

Chapter 6
Development and Application prospect

6.1 Development Trend of Reliability Technology in Manufacturing Process

On March 25, 2015, Premier Li Keqiang pointed out at the executive meeting of the State Council that China is in the process of accelerating industrialization, and manufacturing is an important pillar and foundation of the national economy. The core of the "Made in China 2025" [1] strategy that is currently being vigorously promoted is high-reliability smart manufacturing. It is of great significance to promote China's manufacturing from big to strong, make China's manufacturing include more factors created in China, rely more on Chinese equipment and Chinese brands, and promote economic growth at a medium-to-high speed and move towards a medium-to-high-end level.

With the increasing maturity of advanced technology systems such as big data technology [2], cloud computing technology, cyber-physical system, and Internet of Things system, advanced manufacturing systems will have the ability to collect and process big data in the manufacturing process in real time and dynamically, driving a new round of transformation in the manufacturing industry—the rapid arrival of Industry 4.0 [3], at the same time, information and network technology are also driving tremendous changes in the manufacturing industry.

On the basis of the manufacturing process reliability technology framework proposed in this book, and according to the RQR chain model [4, 5] proposed by the author of this book, a manufacturing process reliability technology development framework as shown in Fig. 6.1 is proposed.

As shown in Fig. 6.1, in terms of the manufacturing environment, the core of smart manufacturing based on Industry 4.0 and big data is high-quality manufacturing, so ensuring high-quality manufacturing needs will be the core of future manufacturing technology [6] development. It provides fertile ground for manufacturing reliability theory and technological development. Especially for China, a big manufacturing country, in the process of becoming a manufacturing power, the strategic significance

Y. He et al., *Reliability Theory and Technology in Manufacturing Process*,
https://doi.org/10.1007/978-981-96-2735-6_6

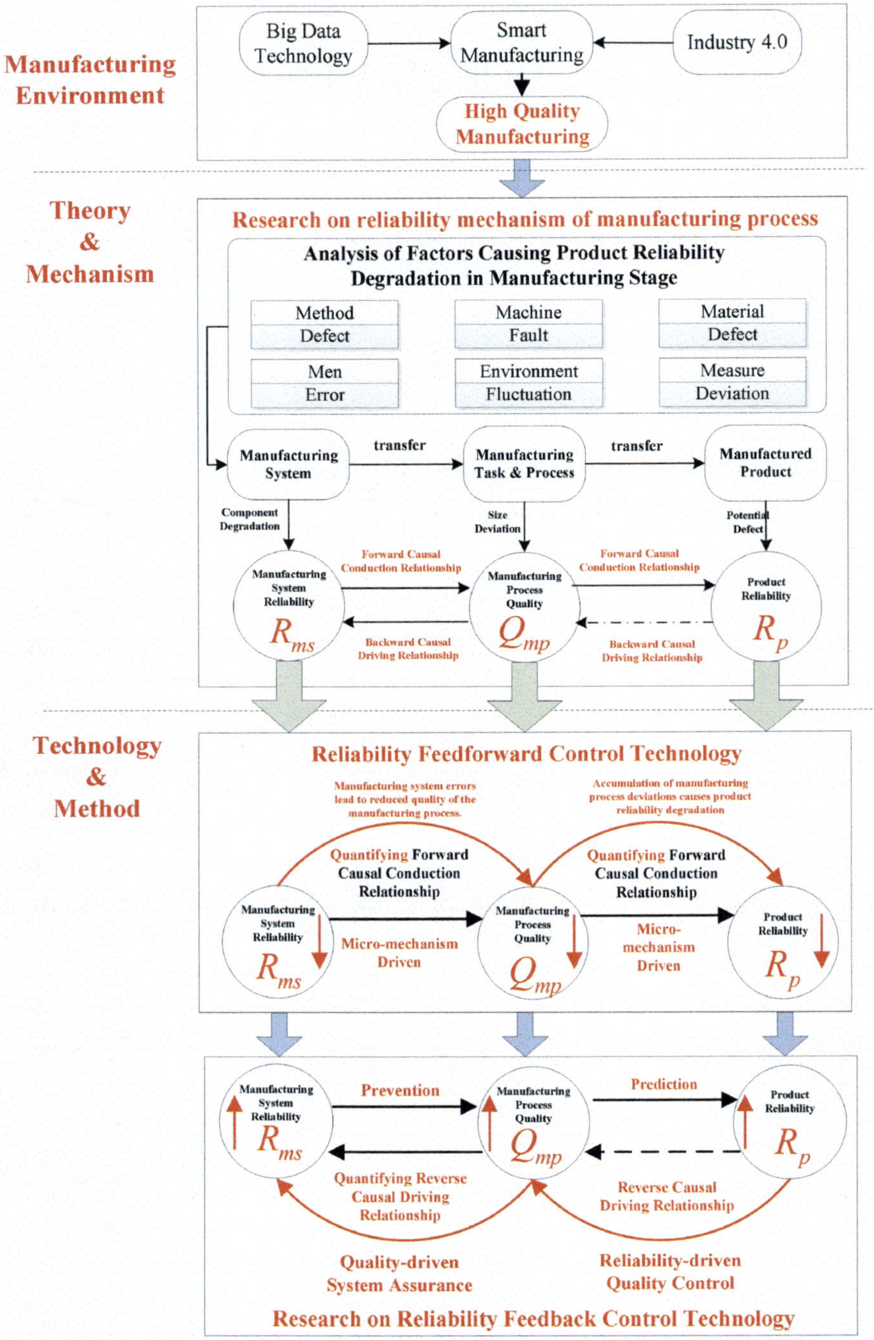

Fig. 6.1 The development route of reliability technology in manufacturing process

of giving priority to the development of high-quality manufacturing technology is of great significance; in terms of mechanism and theory, the development of big data technology and nanoscale and small-scale measurement and detection technology provide an important technical guarantee for us to know the manufacturing system, manufacturing task and process, the reliability degradation mechanism of the manufactured products and the influence relationship among the three, and establish the mathematical relationship of quantitative analysis; in terms of methods and technologies, the clarity of manufacturing systems, manufacturing tasks and processes, and the clarity of the reliability degradation mechanism of manufactured products, on the one hand, provide valuable prior information for us to continuously improve the reliability of manufacturing systems, manufacturing tasks and processes, and manufactured products; From the point of view of system engineering, we use the internal relationship of the three parties to provide us with convenient conditions for establishing the feedforward and feedback control technology for product reliability. Combined with the development trend of reliability system engineering technology, the key prospects for the development of reliability technology in manufacturing process in the future are as follows:

(1) Research on the Analysis Method of Product Reliability Mechanism in the Manufacturing Process Based on the Fusion of Big Data and Fault Physics

Among the three main objects of manufacturing process reliability research, manufacturing system, manufacturing task and process, and manufactured product, the formation mechanism of manufactured product reliability in the manufacturing stage is the foundation and focus. Existing research assumes that the high early failure rate caused by manufacturing factors is reasonable, and only passively uses burn-in tests and run-in tests to passively remove it. The main bottleneck is that the formation mechanism of the reliability of manufactured products has not been clearly studied. It also seriously restricts the development of manufacturing reliability theory and technology. Fortunately, the emergence of big data technology and new industrial 4.0 technologies based on cyber-physical systems has made it possible for us to study and solve the difficult problem of product reliability mechanism in the manufacturing process. Therefore, the reliability mechanism analysis method of manufacturing process products based on the fusion of big data and fault physics is the primary problem to be solved in the future development of manufacturing process reliability technology.

(2) High-Quality Control Technology for Digital Manufacturing Process

In the case that the manufacturing process factors affect the reliability mechanism of the manufactured product, in order to ensure the reliability of mass-produced products, the high-quality control technology of the digital manufacturing process will be the key. Because the product is the result of the process, and reliable products depend on high quality manufacturing process. Therefore, in the future advanced manufacturing process based on the cyber-physical system, how to develop the corresponding high-quality production process capability analysis and process characteristic deviation control technology driven by product reliability in the environment of massive manufacturing quality data, will also be a key issue that must be solved in the future development of manufacturing process reliability technology.

(3) Research on Intelligent Analysis and Optimization Technology of Manufacturing System Reliability

The manufacturing system is the matrix of manufactured products and the material basis of all advanced manufacturing models and activities. Most of the traditional manufacturing system reliability research and maintenance technology research are based on the basic reliability of equipment, and there are few researches on task reliability with polymorphic and dynamic characteristics, which directly restricts the real-time and predictability of reliability analysis of manufacturing systems and greatly restricts the real-time preventive and predictive maintenance of the manufacturing system for the quality and reliability of manufactured products to ensure that manufacturing tasks are completed with quality. Considering that intelligent manufacturing is the mainstream of the future manufacturing model, dynamic prediction and adjustment capabilities are the core. Therefore, the intelligent analysis and optimization technology of manufacturing system reliability will be the focus problem that must be solved in the future development of manufacturing process reliability technology.

(4) Research on Integrated Manufacturing Reliability Analysis and Control Technology

Most of the reliability research in the field of advanced manufacturing only focuses on the analysis and optimization of reliability, and pays little attention to the main manufacturing indicators such as cost, time and output. However, in the highly informatized and automated manufacturing sites in the future, requirements such as reliability, quality, output, environment, and cost all need to be considered and optimized at the same time. Reliability technology must be integrated with manufacturing activities. For this reason, how to carry out integrated analysis and optimization of the reliability of the manufacturing process together with other production technical indicators will also be a practical problem that must be solved in the future development of manufacturing process reliability technology, which will directly restrict the application and promotion of manufacturing reliability technology.

6.2 Application Prospects of Reliability Technology in Manufacturing Process

From the perspective of system engineering, the manufacturing system consisting of manufacturing equipment, the production process to realize the production task, and the produced product composed of work in process (WIP) are the three basic starting points for various analysis and optimization in the manufacturing stage. And from the RQR chain theory in Sect. 2.3 of this book, it can be concluded that the root cause of the reliability problem of the produced product lies in the production process, and the root cause of the reliability problem of the production process lies in the manufacturing system. Therefore, in the application and promotion of manufacturing process reliability technology in the future, we can coordinate the application and promotion of manufacturing system reliability technology, production process reliability technology and produced product reliability technology according to the application framework shown in Fig. 6.2. Strive to add the following work items to the General Requirements for Equipment Reliability (GJB450A-2004) as soon as possible, and provide technical guarantee for fundamentally ensuring the stable mass production of with high-reliability and long-life products.

As shown in Fig. 6.2, if an enterprise encounters the problem of reliability degradation of mass-produced products (generally manifested as a high early failure rate), it is necessary to systematically carry out the following manufacturing reliability work items in sequence during the manufacturing process:

(1) Early Product Failure Analysis and Improvement

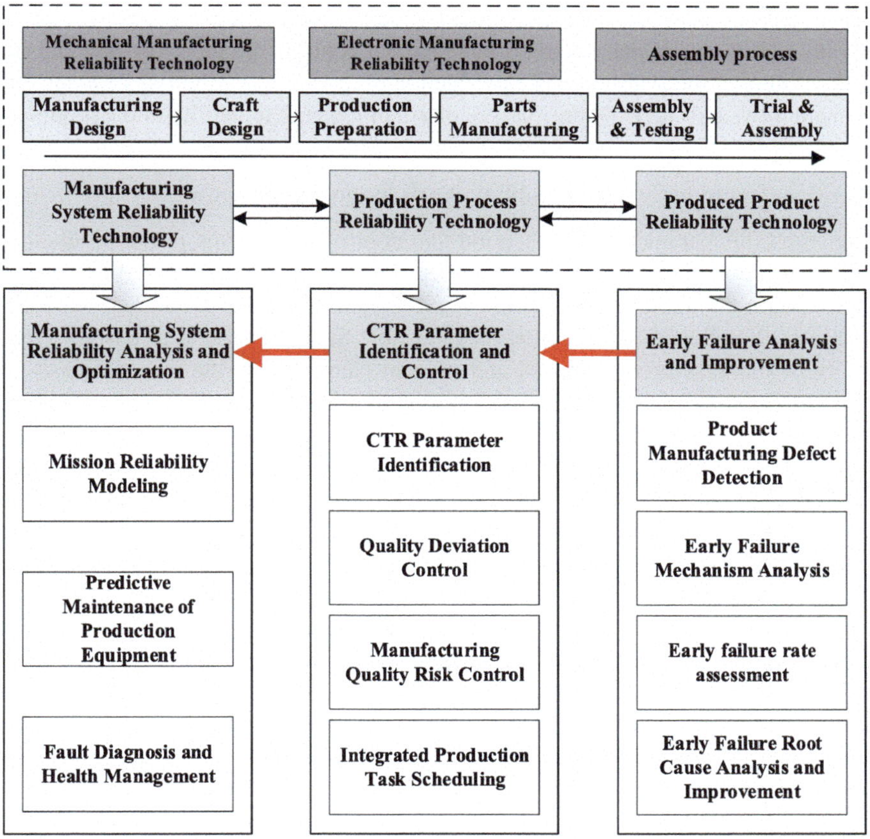

Fig. 6.2. Application framework of manufacturing process reliability technology

The uncontrolled reliability of the manufacturing process is mainly manifested in the high early failure rate of the produced products. Therefore, when implementing and applying the reliability technology of the manufacturing process, it is first necessary to carry out product manufacturing defect detection, early failure mechanism analysis, and early failure rate evaluation. and the identification and improvement of early failure root causes. It is a solid first step to solve both the symptoms and root causes of manufacturing reliability problems, and at the same time, it also provides a clear goal for the optimization and control of the targeted production process and manufacturing system.

(2) Identification and Control of CTR Parameters in the Production Process

When the early failure mechanism and root cause are clarified, it is necessary to go back to the assembly and processing process to carry out identification and control of key manufacturing parameters (CTR, Critical to Reliability) that affect product reliability, and reduce product manufacturing defects and early failure rate, specifically including identification of CTR parameters, quality deviation control, manufacturing quality risk control and integrated production task scheduling, etc. The main purpose of this part of the process-oriented manufacturing reliability work item is to further improve the stability of the production process, reduce the risk of manufacturing quality, and carry out preventive control of the high reliability of the produced products.

(3) Manufacturing System Reliability Analysis and Optimization

After the CTR parameter identification and control work items in the production process are completed, a series of product reliability problems caused by quality deviations in the manufacturing process will be greatly reduced, but there will still be a small number of uncontrolled inexhaustible factors that lead to unstable reliability of manufactured products, and its root cause lies in the material basis of the manufacturing process: the reliability degradation of the manufacturing system. Because the degradation and failure of manufacturing equipment in the manufacturing system will cause a large number of quality substandard and manufacturing defects. Therefore, in order to fundamentally solve the problem of product manufacturing reliability, we also need to conduct scientific analysis and optimization of the task reliability of the manufacturing system, including manufacturing system task reliability modeling, predictive maintenance of production equipment, fault diagnosis and health management of manufacturing system, etc. On the surface, this part of the work items has little to do with the reliability of the product being produced. In fact, a large number of deep-seated product reliability problems are caused by the reliability of the manufacturing system. Therefore, in order to fundamentally prevent and combine product manufacturing Reliability issues, we need to actively promote the analysis and optimization of manufacturing system reliability.

References

1. National Manufacturing Power Construction Strategy Advisory Committee, High Quality Manufacturing (Made in China 2025 Series), Electronic Industry Press.
2. Jay Lee, Industrial big data. China Machine Press, 2015.
3. Jianlin Yan, Dejing Kong. Analysis of "Industrial Internet" and "Industry 4.0" and Their Enlightenment for the Development of China's Manufacturing Industry. Engineering Sciences. 2015, 17(07):141–144.
4. Yihai He, Changchao Gu, Zhenzhen He, Jiaming Cui. Reliability-oriented quality control approach for production process based on RQR chain. Total Quality Management & Business Excellence. 2018, 29(5–6):652–672.
5. He Yihai, He Zhenzhen, Wang Linbo, Gu Changchao, 2015. Reliability modeling and optimization strategy for manufacturing system based on RQR chain. Mathematical Problems in Engineering. vol. 2015, Article ID 379098, 13 pages.
6. Behzad Esmaeilian, Sara Behdad, Ben Wang. 2016. The evolution and future of manufacturing: A review. Journal of Manufacturing Systems. 39:79–100.

The manufacturer's authorised representative in the EU is Springer
Nature Customer Service Centre GmbH, Europaplatz 3, 69115 Heidelberg,
Germany. If you have any concerns regarding our products, please
contact ProductSafety@springernature.com

Printed and bound by CPI Group (UK) Ltd, Croydon, CR0 4YY
30/09/2026
02225302-0010